Economics *of* Carbon Sequestration *in* Forestry

Edited by
Roger A. Sedjo
R. Neil Sampson
Joe Wisniewski

Critical Reviews
in
Environmental
Science and Technology

Terry J. Logan, Editor
School of Natural Resources
The Ohio State University
Columbus, Ohio

CRC Press
Taylor & Francis Group
Boca Raton London New York

CRC Press is an imprint of the
Taylor & Francis Group, an **informa** business

Critical Reviews in Environmental Science and Technology

Aims and Scope

Understanding and assessing the myriad environmental problems that face society today and devising rational strategies and methods for their control are two of the major international challenges of our times. There is a global demand for pollution abatement, but adequate and appropriate measures for pollution control, pollution prevention, and remediation of past environmental degradation will require sound technical knowledge of the factors and processes involved. Environmental science is the complex and dynamic interaction of diverse scientific disciplines, including earth and agricultural sciences, chemistry, biology, medicine, and engineering, and the development of new disciplines like environmental toxicology and risk assessment. This journal seeks to serve as an international forum for the critical review of current knowledge on the broad range of topics in environmental science. It addresses current problems of the day and the scientific basis for new pollution control technologies. *Critical Reviews in Environmental Science and Technology* was previously published by CRC as *Critical Reviews in Environmental Control*. While the name has been changed to reflect the broader scope of the journal, the purpose of *Critical Reviews* remains the same: comprehensive and timely reviews by the leading authorities in each subject area.

First published 1997 by Lewis Publishers

Published 2019 by CRC Press
Taylor & Francis Group
6000 Broken Sound Parkway NW, Suite 300
Boca Raton, FL 33487-2742

© 1997 by Taylor & Francis Group, LLC
CRC Press is an imprint of Taylor & Francis Group, an Informa business

First issued in paperback 2019

No claim to original U.S. Government works

ISBN 13: 978-0-367-44798-4 (pbk)
ISBN 13: 978-0-8493-1158-1 (hbk)
ISSN 1064-3389

Visit the Taylor & Francis Web site at
http://www.taylorandfrancis.com

and the CRC Press Web site at
http://www.crcpress.com

Critical Reviews in Environmental Science and Technology is covered in *Current Contents: Agricultural, Biology + Environmental Sciences; Science Citation Index; Scisearch; Research Alert;* abstracted and indexed in the BIOSIS database, in *Current Awareness in Biological Sciences,* and in *The Engineering Index and COMPENDEX;* and abstracted in *Cambridge Scientific Abstracts* and in *Chemical Abstracts Services;* and abstracted in *Wilson Applied Science and Technology Abstracts* and indexed in *Applied Science and Technology Index.*

CRITICAL REVIEWS in ENVIRONMENTAL SCIENCE and TECHNOLOGY

Volume 27 / Special Issue

SPECIAL ISSUE:
Economics of Carbon Seqestration in Forestry

TABLE OF CONTENTS

Part I: Workshop Overview

Economics of Carbon Sequestration in Forestry: An Overview S1
R. Neil Sampson and Roger A. Sedjo

Part II: Working Group Papers

An Economic Approach to Planting Trees for Carbon Storage S9
Peter J. Parks, David O. Hall, Bengt Kriström, Omar R. Masera,
Robert J. Moulton, Andrew J. Plantinga, Joel N. Swisher, and
Jack K. Winjum

Sequestering Carbon in Natural Forests S23
C. S. Binkley, M. J. Apps, R. K. Dixon, P. E. Kauppi, and
L.-O. Nilsson

Consideration of Country and Forestry/Land-Use Characteristics
in Choosing Forestry Instruments to Achieve Climate
Mitigation Goals S47
Kenneth R. Richards, Ralph Alig, John D. Kinsman, Matti Palo,
and Brent Sohngen

Conceptual Issues Related to Carbon Sequestration:
Uncertainty and Time S65
G. C. van Kooten, A. Grainger, E. Ley, G. Marland, and B. Solberg

Carbon Sequestration and Sustainable Forest Management:
Common Aspects and Assessment Procedures S83
Catrinus J. Jepma, Sten Nilsson, Masahiro Amano,
Yamil Bonduki, Lars Lönnstedt, Jayant Sathaye, and Tom Wilson

Part III: Individual Papers

Assessing Effects of Mitigation Strategies for Global Climate
Change with an Intertemporal Model of the U.S. Forest and
Agriculture Sectors S97
Ralph Alig, Darius Adams, Bruce McCarl, J. M. Callaway, and
Steven Winnett

TABLE OF CONTENTS
(continued)

Incorporating Climate Considerations into the National
Forest Basic Plan in Japan S113
Masahiro Amano

Economic Impacts of Climatic Change on the Global
Forest Sector: An Integrated Ecological/Economic Assessment S123
John Perez-Garcia, Linda A. Joyce, C. S. Binkley, and
A. D. McGuire

Silvicultural Options to Conserve and Sequester Carbon in
Forest Systems: Preliminary Economic Assessment S139
Robert K. Dixon

CO_2-Taxing, Timber Rotations, and Market Implications S151
Hans Fredrik Hoen and Birger Solberg

Compensating for Opportunity Costs in Forest-Based Global
Climate Change Mitigation S163
Alan Grainger

Optimal Subsidies for Carbon: Cost-Effectiveness and
Distributional Considerations S177
Eduardo Ley and Roger A. Sedjo

Carbon Sequestration and Tree Plantations:
A Case Study in Argentina S185
Eduardo Ley and Roger A. Sedjo

The Economic and Environmental Impact of Paper Recycling S193
Stig Byström and Lars Lönnstedt

Forest/Biomass Based Mitigation Strategies:
Does the Timing of Carbon Reductions Matter? S213
Greg Marland, Bernhard Schlamadinger, and Paul Leiby

Forestry Options for Sequestering Carbon in Mexico:
Comparative Economic Analysis of Three Case Studies S227
Omar Masera, Mauricio R. Bellon, and Gerardo Segura

The Physical Risks of Reforestation as a Strategy to
Offset Global Climate Change S245
Robert J. Moulton and John F. Kelly

Can Recycling of Waste Help us to Sequester Carbon
in Forestry? Experimental Results and Economic Visions S259
Lars Owe Nilsson

The Cost of Carbon Sequestration in Forests: A Positive Analysis S269
Andrew J. Plantinga

TABLE OF CONTENTS
(continued)

The Time Value of Carbon in Bottom-Up Studies S279
Kenneth R. Richards

Coercion and Enterprise in the Provision of Environmental Public Goods: The Case of Carbon Sequestration in the United States S293
Kenneth Richards

A Dynamic Model of Forest Carbon Storage in the United States during Climatic Change S309
Brent Sohngen and Robert Mendelsohn

Forest Biomass as Carbon Sink — Economic Value and Forest Management/Policy Implications S323
Birger Solberg

Incremental Costs of Carbon Storage in Forestry, Bioenergy and Land-Use S335
Joel N. Swisher

Assessing Timber and Non-Timber Values in Forestry using a General Equilibrium Framework S351
William A. Thompson, G. Cornelis van Kooten, and Ilan Vertinsky

ECONOMICS OF CARBON SEQUESTRATION IN FORESTRY: AN OVERVIEW

R. NEIL SAMPSON and ROGER A. SEDJO

American Forests and Resources for the Future, Washington, DC 20036 USA

ABSTRACT: This overview features the main issues discussed in the workshop, "Economics of Carbon Sequestration in Forestry," held in Bergendal, Sweden, May 15-19, 1995. The most important conclusion to emerge from the workshop was the need for a more widely-accepted approach to the methodology used in economic analysis. Conceptual issues were discussed, and approaches proposed. Details on those issues and proposals can be found in the working group papers that are contained in this volume.

KEY WORDS: carbon sequestration, economic analysis, tree planting projects, forestry.

1. INTRODUCTION

International concern over the accumulation of "greenhouse gases" in the Earth's atmosphere has led to consideration of ways that national and international policies can slow the rate of increase. One of the major greenhouse gasses — carbon dioxide (CO_2) — has increased significantly in recent decades, with the major contributing factor being the combustion of fossil fuels in the production of energy, and a second source being land use practices (e.g., deforestation) that reduce the stock of C held in forests and soils.

The CO_2 level in the atmosphere is directly linked to terrestrial ecosystems through the global C cycle. Green plants take up atmospheric CO_2 and convert it to cellulose and other organic compounds. Many of these compounds, such as wood and some soil organic compounds, are relatively stable, and maintain the C in inert form for many years, until the C is once again released in gaseous form due to biological respiration or combustion. In many forms, such as peat, coal, and petroleum, the C may be captured permanently in geologic formations or at least captured for very lengthy periods until extraction or other major change occurs.

Forests hold two-thirds of terrestrial C and, as the forest biomass increases over time, so does the stock of C sequestered in the standing forest and its soils. Changes (either additions or removals) in the C stock are called flows. Old growth forests, which may have large stocks of C, may only have small or negligible flows, since net biomass growth is modest or negligible. Alternatively, a young forest may have a relatively modest stock of C due to its small total biomass, but at the same time generate substantial flows into that stock due to the rapid growth of juvenile trees. Transitions from one ecological condition to another will generally produce substantial positive or negative flows.

1064-3389/97/$.50

The direct connection between atmospheric CO_2 levels and terrestrial ecosystems has encouraged governments and private entities to explore possibilities to increase the area or rate of growth in forests as one way of converting existing atmospheric CO_2 into sequestered terrestrial C stocks (Sedjo et al., 1995). In addition, the same effect can be achieved by slowing the rate of C release back into the atmosphere through the conservation of existing forests and the conversion of forest products into more stable forms. The use of sustainably-produced biomass as a fuel substitute for fossil energy keeps fossil C locked up while net C emissions from the biomass fuel arc offset by new replacement growth.

As public policy and private investment decisions about C management programs proceed, economic analyses are needed to assist in comparing the available options. These analyses are complicated by the fact that the primary benefit being considered — the sequestration and storage of C in terrestrial systems — is essentially an abstract, global, public good, for which the international community has yet to establish an agreed-upon value or market structure.

Additional uncertainties exist about the future rate of growth in forests under changing climatic conditions, the probability of natural disturbances such as insect or disease epidemics that kill growing trees, or the risk of wildfires which can convert much of the biomass back to CO_2. Forest management practices and decisions greatly affect forest C dynamics, through the selection of practices during the harvest and reforestation period, or the manner in which the biomass is utilized after harvest. Added to the biological and environmental uncertainties are the economic, political, legal and cultural uncertainties that can rival or even surpass those of the natural world.

2. FORESTRY MITIGATION OPPORTUNITIES

Mitigation opportunities in the forest sector may be categorized as follows:

- Increasing the standing inventory of forest and soil biomass, either through expanding the area of forests, increasing the growth rate of existing forests, or reducing the rate of forest loss and/or conversion to other land uses;
- Increasing the storage of C in long-lived forest products;
- Substituting wood products for other materials whose manufacture and usage causes more C to be emitted to the atmosphere; and
- Utilizing biomass energy as a replacement for fossil fuels.

Not considered at this workshop were forestry-related activities such as using trees to protect buildings and reduce heating and air conditioning demands, or agro-forestry.

Programs to encourage forest sector activities can range in scale and scope from local to national and international, and consist of efforts aimed at a single type of activity (e.g. an incentive program to encourage additional tree planting) to efforts that provide incentives for any forestry activity that can be shown to affect CO_2 balances in a positive way. In addition to public sector activities, some private companies and associations are supporting forestry projects as a potential means of offsetting their CO_2 emissions.

3. CONCEPTUAL ISSUES

Sound economic analysis of forest sector projects designed to sequester or store C raises several conceptual issues that need resolution. Scientific evidence points to the increase in atmospheric CO_2 as one contributing factor to potentially-damaging global climate changes. The question that arises is, how much damage? If data can be developed that provide a generally-accepted economic value on the amount of damages likely to occur, then we can use the marginal damage created by each additional tonne of CO_2 emitted to the atmosphere as a measure of the value of keeping that tonne in terrestrial storage, or sequestering an additional tonne from the atmosphere (van Kooten et al., 1995).

The major areas of scientific uncertainty surrounding future climate change and its effect on forests, as well as the continued uncertainty about the cause-effect relationships between C cycling and climate change, create significant challenges to economic analysis. This is compounded by the fact that forestry projects may extend many decades, if not centuries, into the future, with C benefits following a logarithmic growth curve that places the bulk of the benefit flow in the future. For those projects where high costs are incurred at the beginning, and the benefits depend on several decades of growth and survival, the risks of project failure are critical to an accurate economic appraisal. Thus, along with risks and uncertainties, arise questions about time and discounting. Should C benefits be discounted? If so, at what rates? The answers chosen to these questions will create critical differences in the outcome of the analysis.

4. ECONOMIC ANALYSIS IN POLITICAL DECISION MAKING

Policy makers have adopted broad goals for greenhouse gas reductions that reflect commitments to the international Framework Convention on Climate Change that was signed by over 150 Nations following the June 1992 Rio de Janeiro United Nations Conference on Environment and Development (UNCED). In this context, economic analyses may be useful in devising policy and program mixes to achieve the selected goals with the least net social cost.

In these assessments, the choice of policy instruments needs to be matched to the characteristics of the country involved, the nature of the forestry actions desired, and an understanding of the behavioral response likely to be elicited by each type of instrument (Richards et al., 1995). Since C sequestration is almost certain to be a secondary, rather than primary, purpose of national forestry policies, economic analysis aimed at helping policy makers will need to illustrate not just the economic and C impacts of the proposed policies, but their likely interactions and with effects upon other agriculture, forestry, environmental, energy, and trade policies, as well.

5. TWO APPROACHES TO ECONOMIC ANALYSIS

There is a considerable difference between analyzing a single mitigation project and analyzing a proposed national program or policy. For single or small projects, a "micro" or "bottom-up" methodology involves the identification of all of the various C flows or stocks that will be impacted by the project (e.g., change in forest or soil biomass), and calculating the amount of C stocks with and without the project. The net difference in each stock is calculated, and the net sum of changes in all stocks is the C impact of the project (Parks et al., 1995). A simple mathematical model can be constructed to show the various C flows or stocks to be considered, such as:

Let C_B = Biomass C, expressed as the total vegetative and soil organic C

C_P = Product C

C_S = Substitution Effect (the C emission savings associated with, for example, using a wooden structural member in place of steel or concrete, whose manufacture involves significant CO_2 emissions)

C_F = Fossil fuel combustion displaced

Then C_T (total C) = C_B + C_P + C_S + C_F

The net C impact of the project (C_N) can then be calculated as:

$C_N = (C_{B2} + C_{P2} + C_{S2} + C_{F2}) - (C_{B1} + C_{P1} + C_{S1} + C_{F1})$, where x_1 represents the condition without the project (the reference case) and x_2 is the condition expected with the project.

In quantifying each of the C factors, care must be used to prevent either double-counting or omissions. In some studies, biomass C in the forest will be quantified in compartments, such as standing timber, understory growth, dead and rotten material, forest floor, and soil C. Different parameters may be selected in an analysis based upon the available data, or the need to illustrate the different types of effect. An example might be a project where a significant increase is planned in terms of soil organic C content. Here, because of the importance of the soil to the analysis, it might be preferable to show soil C as a separate component of the C equation. If care is taken to calculate C_{B1} with the same components and methods as C_{B2}, the net result should be acceptable.

An additional factor to consider, however, is the effect of time on both the reference case and the project. Commonly, projects are evaluated over fairly long time periods. Since most forests take at least 20 years or longer to reach harvestable size, and many forests may be planted for long-term growth and protection, it is common for a project analysis to cover several decades, during which the reference case would undergo change as well. The analysis, then, becomes a time-step analysis over the analytic period chosen, with the net difference between the "with" and "without" situation being the project impact. As noted above, treatment of the discount rate is critical for long-term economic analysis. Costs are always discounted. There are questions, however, as to whether or how to discount C benefits.

The analysis gets somewhat more complicated as one considers a "macro" or "top-down" evaluation of national or international programs. Here, the cumulative effects of the program may be large enough to affect markets, prices, or producer behavior. In theory, if a national tree planting program results in a large new supply of timber products on the market, the resulting price effect may discourage other producers from continuing to plant trees or manage forest land. Thus, in addition to the project resulting in the net gain of new forest biomass that would be indicated by the "bottom up" analysis, some "offsets" or "leakage" of the intended effects may need to be quantified and factored into the final result.

Other difficulties in analyzing large national or international incentive programs occur in predicting the amount of incentive that is large enough to elicit the desired behavior, but not so large as to attract excess participation — or participation by producers who are less efficient or effective in meeting the program's goals. There are also social impacts — winners and losers — in large programs. If forests are dramatically expanded, for example, the price of forest products may go down —benefitting consumers but penalizing forest owners. There may also be cross-border effects, where one country's economic or environmental interests are served at the expense of additional costs or impacts imposed on another country. One country's decision to reduce or expand timber harvest may, for example, create international market changes with far-reaching effects.

6. TREE PLANTING PROJECTS

Tree planting projects appear to be one of the more cost-effective ways to increase terrestrial C storage, and many countries are considering or have implemented tree planting programs as one element of their climate change action plan (Sedjo et al., 1995). While the economic analysis is, in some regards, more straight-forward than is the case in some of the forest management or protection projects, there still remains the need for consistent analytic methods on these projects.

Foresters have developed reliable growth models for the commonly-planted tree species on most soils and sites, so establishing a future growth curve with reasonable reliability is fairly straight-forward. Converting forest growth models (which normally express growth in terms of marketable timber volumes) to total biomass weight, and then into C, can be done by using factors developed for each species. (For rough estimates, one m^3 of merchantable tree volume growth will result in the addition of about 0.5 tonne of total C to the site.)

In addition, soil C storage may be increased on some plantations, depending primarily on the condition of the soil at the time of planting (Sampson 1995). If the soil is C-depleted, such as might be found in an abandoned crop field, soil C levels equivalent to the normal forest soil C for the soil type will likely be restored in the first forest growth cycle of 15-25 years. Field measurements of soil C are more difficult than above-ground biomass measurements, however, so most project evaluations will depend on available soil survey data to estimate probable soil C change.

Other data needed for the C calculation include the planned uses of the wood to be grown in the project. C can be stored on-site, in protected forests, for a period of time, or wood can be harvested and placed in some long-term product usage. Alternatively, biomass can be used as a fuel to offset fossil fuel energy, resulting in no net C emissions provided the biomass is produced in sustainable management systems that provide replacement growth. For each of these pathways, estimates of the C fate need to be developed to provide the data needed to conduct the analysis.

Other benefits, both economic (e.g., value of wood products to be sold) and non-economic (e.g., improved air or water quality), may also be identified in association with the project and, to the extent they can be quantified, should be included in the calculation of the net benefit stream.

On the cost side, the direct costs of establishing a plantation may be fairly easy to identify, but care is needed to include future management, maintenance, and monitoring costs. Opportunity costs (the value of not using project resources in another use) should be included in the construction of a reference case. Where feasible, externality costs and/or benefits likely to be associated with the project and the reference case should be included.

For each project, a reference case should be developed to indicate what will occur, in terms of costs and benefits, if the project is not undertaken (Parks et al., 1995). In the case of abandoned agricultural land, this may be continued abandonment. Or, alternatively, it may be some other usage. In any case, it is critical to the analysis to establish an accurate idea of what is most likely to occur on the site in the absence of the project. Because the focus of these projects is on C sequestration or mitigation, it is critical to determine what will happen to the C balance in the absence of the project, being mindful of the fact that its change over time will be influenced by what occurs both naturally and under management influence on the site. Once the full range of costs and benefits for the project and the reference case have been identified, the comparison of the two, in terms of net benefits, can be accomplished.

7. SEQUESTERING C IN NATURAL FORESTS

Natural forests consist of a wide variety of opportunities, both in terms of the physical situation and the social changes that may be needed in order to sequester or store additional C. Here, the challenge to project planners may be less with the physical context of the forest, and more with the social and cultural changes needed to affect the C dynamics in the desired manner. These forests, while they are widely diverse, are usually the home of people who have cultural roots deep in the forest and its use. Where those uses have grown unsustainable, or where C stocks are being needlessly depleted, it may be difficult to cause real change to occur.

While the cost-benefit analysis of projects affecting natural forest management can follow the same general format as for a plantation — the net benefits of the proposed project minus the net benefits of the reference case — natural forests provide far more difficult measurement challenges (Binkley et al., 1995). A baseline estimate of how fast the forest is currently adding biomass and C requires a forest assessment which may, in remote areas or forest types where growth dynamics are not well known, be difficult enough in itself. Added to that difficulty are the problems of forecasting what kinds of biomass dynamics will occur under reference case assumptions, and predicting what kinds of changes may be possible with project action. Subsequent monitoring to evaluate project success will face similar measurement problems.

Part of the opportunity cost of increasing C storage in natural forests may be passing up the opportunity to harvest economically valuable timber. There may be widely different ways of viewing that case, however. Allowing forest stands to get older, and building C stocks above normal reference levels in a stand, may increase the biophysical risk of losing the stand to a disturbance or epidemic. On the other hand, older, larger trees may be scarcer and more valuable as timber sources than their smaller, younger counterparts, so delaying harvest by a few decades, if catastrophic risk can be avoided, may realize some economic gains or offsets that affect costs.

Complicating any assessment is the difficulty in determining the extent to which a project to protect a natural forest in one area will simply deflect forest disturbances elsewhere. Under such a situation, the net forest protection achieved, and the total C emissions mitigated, can be much less than suggested by focusing only on the project itself.

8. JOINT IMPLEMENTATION AND INTERNATIONAL COOPERATION

The First Conference of Parties, held in Berlin, Germany, in 1995 to consider implementation of the U.N. Framework Convention for Climate Change, recognized the opportunity for parties to work cooperatively in establishing greenhouse gas mitigation efforts. In working together (termed joint implementation), opportunities may exist that do not present themselves within the borders of a single nation. There are, obviously, significant complications created by such joint efforts which reach far beyond economic analysis. In addition to the difficulty in creating a credible technical analysis of the reference case and the project impacts, there is the need to address the political, economic, and environmental interests of many different parties. These difficulties were recognized during the Berlin Conference when it was agreed to initiate a pilot phase of joint implementation during which no official carbon credits would be recognized. The pilot phase is to undergo a comprehensive review, however, so that the parties can consider a conclusive decision about the fate of joint implementation by the year 2000. This provides a fairly brief time in which to establish internationally-accepted methods of evaluation and monitoring of mitigation projects (Jepma et al., 1995).

9. CONCLUSIONS

There are significant conceptual and methodological issues to be addressed before a consistent approach to economic analysis of forestry projects for carbon sequestration is achieved. While there is an emerging consensus on many of these issues, today's analyst needs to take extra precautions to assure that all assumptions and methods used in an analysis are clearly stated. Unless the underlying assumptions and concepts are consistent, very different conclusions can emerge from similar situations.

ACKNOWLEDGMENTS

The authors wish to thank Joe Wisniewski, Robert Dixon, John Kinsman, and Peter Parks for insightful reviews and comments on the manuscript. Much of the intellectual content of this overview was taken from discussions at the workshop led by the lead authors of the working group papers that are referenced below, and much credit goes to them, as well.

REFERENCES

Birdsey, R.A. (In Press). *Carbon Storage and Accumulation in Major Forest Types and Regions.* In Sampson, R.N. and Dwight Hair (eds.) Forests and Global Change, Volume 2: Opportunities for Managing Existing Forests. Washington, DC: American Forests.

Binkley, C.S., Apps, M.J., Dixon, R.K., Kauppi, P., and L-O. Nilsson. This volume. *Sequestering Carbon in Natural Forests.*

Jepma, C.J., S. Nilsson, M. Amano, Y. Bonduki, L. Lönnstedt, J. Sathaye, and T. Wilson. This volume. *Carbon Sequestration and Sustainable Forest Management: Common Aspects and Assessment Procedures.*

Parks, P.J., Hall, D.O., Kriström, Masera, O.R., Moulton, R.J., Plantinga, A.J., Swisher, J.N., and J.K. Winjum. This volume. *An Economic Approach to Planting Trees for Carbon Storage.*

Richards. K.R., Alig, R., Kinsman, J.D., Palo, M., and B. Sohngen. This volume. *Consideration of Country and Forestry/Land-use Characteristics in Choosing Forestry Instruments to Achieve Climate Mitigation Goals.*

Sampson, R.N. 1995. *The Role of Forest Management in Affecting Soil Carbon: Policy Considerations.* In Lal, R., J. Kimble, E. Levine and B.A. Stewart (eds), Soil Management and Greenhouse Effect, Boca Raton, FL:Lewis Publishers, pp. 339-350.

Sedjo, R.A., J. Wisniewski, A. Sample and J.D. Kinsman. 1994. *Managing Carbon via Forestry: Assessment of Some Economic Studies.* Washington, DC:Resources for the Future. 39 p.

van Kooten, G.C., Grainger, A., Ley.E., Marland, G., and B. Solberg. This volume. *Conceptual Isuues Related to Carbon Sequestration: Uncertainty and Time.*

Critical Reviews in Environmental Science and Technology, 27(Special): S9–S21 (1997)

AN ECONOMIC APPROACH TO PLANTING TREES FOR CARBON STORAGE

PETER J. PARKS,[1] DAVID O. HALL,[2] BENGT KRISTRÖM,[3] OMAR R. MASERA,[4] ROBERT J. MOULTON,[5] ANDREW J. PLANTINGA,[6] JOEL N. SWISHER,[7] and JACK K. WINJUM[8]

[1]*Department of Agricultural Economics and Marketing, Cook College, Rutgers University, Post Office Box 231, New Brunswick, NJ 08903-0231, U.S.A.,* [2]*King's College, University of London, London, U.K.,* [3]*Faculty of Forestry, Swedish University of Agricultural Sciences, Umeå, Sweden,* [4]*Centro de Ecologia, National University of Mexico, Michoacan, Mexico,* [5]*U.S.D.A. Forest Service, Washington, DC, U.S.A.,* [6]*Department of Resource Economics and Policy, University of Maine, Orono, ME, U.S.A.,* [7]*UNEP Collaborating Centre, Risø National Laboratory, Roskilde, Denmark,* [8]*NCASI/USEPA National Health and Environmental Effects Research Laboratory, Corvallis, OR, U.S.A.*

ABSTRACT: Methods are described for evaluating economic and carbon storage aspects of tree planting projects (e.g., plantations for restoration, roundwood, bioenergy, and nonwood products). Total carbon (C) stock is dynamic and comprises C in vegetation, decomposing matter, soil, products, and fuel substituted. An alternative (reference) case is essential for project evaluation.

KEY WORDS: Global change, forests, carbon, plantations, tree planting, project evaluation.

1. INTRODUCTION

Over 120 developed and developing countries met this year in Berlin[1] to discuss mitigating carbon dioxide emissions as part of their responsibilities under the Framework Convention for Climate Change.[2] The Berlin mandate obligates countries to reduce net emissions and requires policy makers and analysts to rapidly[3] become familiar with new mitigation alternatives (e.g., tree planting). Although planting trees for carbon storage appears to be a cost-effective option for many countries, consistent analytical methods are still needed to quantify costs, benefits, and carbon storage associated with planted forests.

The United Nations Food and Agricultural Organisation (FAO) (1995) estimates that 10.5 million ha of new plantations were annually established during the 1980s. The net gain in plantation area for this period was about 2 million ha per year since new plantations can (*i*) regenerate harvested plantations; (*ii*) reforest removed natural forests; or (*iii*) afforest lands previously in nonforest uses. The total area in forest plantations in 1990 was about 130 million ha (Allan and Lanly 1991), of which 14 percent was in boreal regions, 63 percent in temperate regions, and 23 percent in tropical regions (FAO 1995).

Like natural forests, planted forests take up and store carbon (C) at high rates compared to other world land covers. Storage rates commonly range from 1 to 8 MgC ha^{-1} yr^{-1} (1 Mg equals 10^6 g), and typical mean C storage over a rotation period is from 50 to 80 MgC ha^{-1}

1064-3389/97/$.50

(Winjum and Schroeder 1995). Key carbon stocks associated with planted trees may include *in situ* carbon (i.e., carbon stored at the forest location in the vegetation, decomposing matter, and soils), carbon in forest products, and carbon in substituted fuel. Two alternatives for claiming carbon storage from planting trees should be conceptually distinguished: *(i)* Carbon saved by fossil fuel substitution; *(ii)* Carbon stored *in situ* and in wood products. The duration of C storage (permanent or transitory) is of paramount importance.

In the first case, the carbon stored is the net amount of carbon saved during the tree planting project by substituting biomass fuels from the planting project for fossil fuels. This storage may be transitory if the saved fossil fuels are later used, but is otherwise permanent. In the second case, permanent carbon storage only occurs when the plantation is permanent: Storage in products is transitory (these eventually decay) and much of the *in situ* carbon stock may return to the atmosphere if the planted lands are ultimately converted to nonforest use.

Carbon storage strategies such as establishing planted trees must be considered amidst an array of emissions reduction alternatives. This paper presents an effective method for policy makers and analysts to make clear evaluations and comparisons before selecting tree planting to aid in C storage. After discussing definitions and scope, the paper describes carbon stocks, costs, and benefits associated with tree planting projects. This is followed by a discussion of project evaluation criteria, a summary, and conclusions.

2. SCOPE AND DEFINITIONS

In this discussion, forest plantations include contiguous areas of planted trees occupying areas greater than one hectare. Tree planting should be sufficient to provide stocking to at least 20 percent crown cover. This definition is consistent with FAO (1995) terminology for forest plantations and stocked forests. Planted tree areas such as in urban settings, windbreaks, or boundary strips are outside the scope of this definition. In addition, while agroforestry[4] must be acknowledged as an important class of land uses, treatment of selected agroforestry systems is beyond the scope of the present work.

Major plantation types include those planted primarily for restoration, roundwood, bioenergy, and nonwood purposes. It is clear that specific plantations may span more than one of these categories,[5] but some structure is useful for categorizing important net changes associated with components of carbon stocks (e.g., changes in fuel carbon stocks associated with bioenergy plantations).

Restoration plantations are planted primarily to obtain environmental services other than commodity products (e.g., reduced soil erosion, habitat for biologically diverse plant and animal communities). Roundwood plantations are planted primarily to obtain raw materials for timber products (e.g., pulpwood, sawnwood). Bioenergy plantations are planted primarily to obtain materials for use as energy sources (e.g., fuelwood, charcoal). Plantations for nonwood purposes are planted primarily to obtain products other than timber commodities (e.g., rubber, nuts, or fruit). Evaluations of forest plantations for C storage must consider the social and institutional settings of the land owners. In developed countries, these settings may include public, large-, and small-private classes of owners. Examples of forest plantations for these owners include government forests, industrial tree farms, and individually-owned forests, respectively. In developing countries, forests owned as commons may also be included. Examples of public, commons, and private plantations include government forests, community forests, and farm woodlots, respectively. In general, the methodology described here is applicable to all these categories, though unique consideration may be required in some cases.

3. CARBON STOCKS

A proper account of the carbon implications of a tree planting project must include the different carbon stocks that may be influenced by the project. These include carbon stored in vegetation (above and below ground), decomposing matter, soils, wood products, and the carbon substituted by burning wood for energy instead of fossil fuels. The empirically relevant stock components will vary with the type of plantation (restoration, roundwood, bioenergy, nonwood). For example, in restoration projects there may be no wood products, and in bioenergy projects all biomass may be used for fuel.

The total MgC ha^{-1} stock of carbon stored at time t (C_t) is the sum of *(i)* carbon stored in living vegetation above and below ground (C_{vt}), *(ii)* carbon stored in decomposing matter (e.g., litter and coarse woody debris, including dead trees, dead branches, and dead leaves) (C_{dt}), *(iii)* carbon stored in soils (including humus and soil organic matter) (C_{st}), *(iv)* carbon stored in wood products[6] (e.g., furniture, lumber, paper and paperboard products) (C_{pt}), c2d *(v)* cumulative carbon substituted by the replacement of fossil fuels with biomass fuel from the project (C_{ft}),

$$C_t \; C_{vt} + C_{dt} + C_{st} + C_{pt} + C_{ft} \; .$$

Similarly, the rate of change in total carbon stock (c_t) is comprised of rates of change in carbon stored in vegetation, decomposing matter, soils, products, and substituted fuel (i.e., $c_t \; c_{vt} + c_{dt} + c_{st} + c_{pt} + c_{ft}$). (These rates of change will subsequently be referred to as carbon flows.)

For ease of exposition, consider a project planning horizon where time is measured in discrete units (e.g., years). The total carbon flow c_t is defined as $C_{t+1} - C_t$; carbon flows for each component of carbon stock are similarly defined. These flows are useful in analysing carbon stored during finite periods. For example, the total carbon stored during the life of a tree planting project over T years is

$$C_T = \sum_{t=0}^{T} c_{vt} + c_{dt} + c_{st} + c_{pt} + c_{ft}$$

Each flow may be positive or negative, and may be influenced by management activity pursued during the planning horizon. Consequently, empirical studies frequently calculate carbon stocks for finite periods by multiplying an average flow by the length of the planning horizon. Ideally, the planning horizon should be long enough to assure that the carbon stock has reached a steady (equilibrium) state (i.e., $c_T = 0$).

Tree planting projects are often evaluated using mean carbon storage over the life of the project. Collapsing the dynamic stream of carbon storage into a single statistic will obscure the timing of carbon storage benefits; however, defining this statistic is often unavoidable when tree planting projects must be compared to other emissions reduction projects (e.g., source reduction) of different lengths. Mean carbon storage (MCS) for a plantation of T years is $1/T \_ \sum_{t=1,T} C_t$. When T coincides with the length of a rotation (e.g., for roundwood, bioenergy, and nonwood product plantations) and soil carbon reaches equilibrium within T years,[7] then MCS for a perpetual series of identical rotations is equivalent to MCS for a single rotation. When the life of the project is finite, specific assumptions are required about which components of MCS legitimately may be claimed as permanent (see above).

Defining MCS for restoration plantations (that once established may never be harvested) is not simple. For example, if the restoration plantation reaches a steady state stock of carbon C_T after some finite time T (i.e., $c_T = 0$ " $t > T$), then an infinite number of observations of the stock C_T will be made following time T. Under this interpretation, MCS for the restoration plantation is equivalent to the steady state carbon stock C_T.[8]

Not all authors use the carbon stock categories described above (cf. Dewar 1990). Depending on the approach and the type of information available, authors may collapse vegetation and decomposing matter in one pool (Swisher 1991) or may use a component of C_{vt} (e.g., merchantable wood) to estimate total carbon (Moulton and Richards 1990). Carbon stocks of substituted fuel and products are central to an ongoing debate about the magnitude of carbon storage associated with fuel and materials substitution. The latter is attributed to the substitution of wood products from the tree planting project for energy intensive products (e.g., cement, steel). The net carbon storage gain associated with these stocks clearly depends on the alternative (i.e., reference case) to which the planting project is compared (see below).

4. COSTS

Depending on the type of plantation being considered, the direct cost of a tree planting project may include the value of resources (e.g., land, labour, materials) needed to establish, maintain, manage, monitor, and produce energy from the project. Project funding must cover the project's development and the expenses and incentives for its on-going operation, including maintenance, management and monitoring (see Table I).

The cost of establishing a plantation may include, for example, the costs of seeds or seedlings and other materials; labour costs for site preparation, planting, and building access roads; and materials and labour for replanting trees that do not survive the first year (Sedjo 1983). In developing countries, the value of labour inputs may be difficult to assess, especially when alternative uses of workers' time are in unpaid household production activities. Establishment costs are highest for restoration projects and biomass energy projects and tend to be lower for production of non-wood products.

Management costs include the cost of overall administration and technical supervision, and the costs of training, technical assistance and extension services to provide for a sufficient level of technical competence on the part of the participants. These costs tend to be relatively high for small-holder plantations, such as for production of non-wood products, and lower for roundwood plantations.

Maintenance costs may include weeding and thinning, road maintenance, and fire protection, as well as production costs including harvesting and transport. These costs tend to be higher for roundwood plantations and lower for restoration projects. The costs for biomass energy projects may also include the capital and operating costs of bioenergy production. The cost for biomass energy production is often expressed net of the value of

TABLE I
Relative Magnitudes for Selected Components of Cost for Restoration, roundwood, Bioenergy, and Non-Wood Product Plantations.

Establishment	Management Maintenance Energy[a]
Restoration	++ [b]
	+
	0
	0
Roundwood	+
	0
	++
	0
Bioenergy	++
	+
	+
	+
Non-wood	+
	++
	+
	0

[a] Refers to net costs of producing biomass energy (see text). [b] '++' indicates high cost, '+' indicates significant cost, '0' indicates little or no cost.

energy savings[9] (i.e., substituted fuel that need no longer be purchased) (Hall et al. 1991, Swisher 1995).

In addition, monitoring of all project types is a necessary expense to verify project success and carbon storage. Monitoring costs include the costs of conducting and updating site surveys, before and after soil testing, destructive tree measurements, and other procedures. These costs are roughly equivalent for the four types of plantations considered here.

Some costs, particularly maintenance and monitoring costs, may recur over the life of the project. To ensure that these costs are covered, and to provide an on-going incentive for participation in the project, recurring costs may be anticipated and provided for in the initial project funding.[10] In addition to direct costs of the project, externality costs (e.g., offsite damages from soil erosion) may be associated with some planting practices. These costs are most often associated with practices that are poorly performed, and may be minimized with appropriate monitoring and management activities.

The opportunity costs of not employing project resources in alternative uses (e.g., the value of land in agricultural use) require the definition of an alternative to the project (a reference case). This important category of costs will be discussed under project evaluation, after a discussion of benefits.

5. BENEFITS

The benefits from tree planting projects include those that may or may not be included in markets. Market benefits include the value of commodities produced by the tree plantation. For roundwood, bioenergy, and plantations for nonwood products, these can include harvested trees marketed for wood product resources (e.g., lumber, pulp and fuelwood), and nonwood extractions (e.g., latex for producing natural rubber). It is clearly possible for plantations to provide benefits that may be of value outside markets (e.g., stored carbon, reduced soil erosion, biologically diverse plant and animal communities). Some important benefits (e.g., socio-cultural, infrastructure, and political benefits) may not readily be quantified.

5.1. Quantifiable Economic Benefits

The value of marketed commodities may depend on the location of the market relative to the planted trees. Prices used to value timber commodities (stumpage prices) are often reported net of production costs, including transportation. It is possible for some services to influence the costs for marketed products produced away from the planted forest. For example, benefits attributable to reduced soil erosion may include larger fish stocks and decreased costs for fishing. In developing countries, the value of marketable commodities may be difficult to evaluate if markets do not exist or are far away from the planted trees.

Individuals may directly or indirectly benefit from planted environments (e.g., use values, and nonuse or existence values, respectively). Estimates of use values may be obtained by observing related market transactions. For example, observations of recreation site visits or property values are often used to estimate the value of recreation or aesthetic values.

Nonuse values can be divided into two categories, user's nonuse value and existence value (Freeman 1993). For remote locations, introducing planted trees can create or change amounts of environmental services (e.g., habitat for biologically diverse plant and animal communities) that individuals may value but not directly partake through use. A variety of hypothetical valuation methods (e.g., contingent valuation) have been developed to gain insight into nonuse values.

Whether goods and services provided by the plantations accrue to producers (as products or inputs) or consumers (as use or nonuse values) they can be further classified based on their degree of publicness. Rivalry and excludability are used to distinguish between public and private goods (Dixon and Sherman 1990). Public goods are nonrival and nonexcludable (e.g., aesthetic or scenic values).

5.2. Socio-Cultural, Infrastructure, and Political Benefits

Benefits may also include those which cannot readily be assigned a monetary value, either because data are not available or reliable enough, or because it is not clear how to quantify changes in their levels (e.g., limits to economic measurement, Dixon et al. 1994). Some nonquantifiable benefits can be crucial to long-term success of tree planting projects. The list of these benefits may be long, but many could be grouped into socio-cultural (including employment and equity), infrastructure (both human and capital), and political (both domestic and international) categories.

While some of these benefits may be evident to project implementors, they may not directly accrue to local land owners, individuals, and communities (Dixon and Sherman 1990). When benefits are not readily apparent, community members may not manage or protect the project for C storage benefits. Examples of this harm include deliberate fires, damage by animals, neglect of the plantation, and degradation of the plantation via open access use.

Socio-Cultural Benefits

For projects to succeed at the local level, benefits must accrue to individuals and local communities, in addition to entrepreneurs and the government.[11] Local benefits may be evident in a number of ways, such as off-season employment and the employment of community members in the collection or use of forest products and services. Depending on the type of tree planting project, these products may include fruit, medicines, fuelwood, grazing, fodder, and grasses. Some projects may influence intra-and intergenerational equity by changing the distribution of income or environmental benefits within and between generations.

Infrastructure Benefits

The availability and training of planners, entrepreneurs, managers, and skilled labour may be enhanced by the tree planting project, and physical assets may accompany its implementation. For some projects (including many that are funded by international lending agencies), improvement of human and technical capital, within the area of the project and nationwide, is a prerequisite for obtaining the long-term commitment needed for the projects to succeed.

Political Benefits

The public's perception that environmental issues, such as C storage or rehabilitation of degraded lands, are being addressed by tree planting projects is important both in domestic and international political decision making (e.g., devoting publicly-owned lands to planted forests). Vested commercial and institutional interests may have agendas inconsistent with long-term carbon storage, and thus must be considered when trying to ensure success of a project.

The existing institutional structure relevant to the proposed project (e.g., the level at which forest resource management decisions are made) should be consistent with the assumptions used to evaluate the planting project. In particular, the economic conditions under which planted forest commons (e.g., social forestry) succeed or fail may be important (Wade 1988).

6. PROJECT EVALUATION

This section presumes that an economic decision-maker (e.g., land owner, policy analyst, policy maker) must determine whether a specific tract of land should be devoted to a tree planting project that may result in C storage. Tree planting projects are frequently among a set of several alternatives for emissions reduction or land use. Because resources such as project funds or land may be scarce, not all projects may be pursued. As a result, the decision-maker must select among projects, including tree planting projects, based on their costs, benefits, and carbon effects.

For simplicity, we consider that the decision to plant trees will be made after considering the present value of net quantifiable benefits (Dixon et al. 1994), carbon storage, and unquantified benefits; other decision rules (cf. 'safety first' rules for subsistence agricultural production, or safe minimum habitat standards for species survival) may be relevant for evaluating specific tree planting projects. The decision-maker's concerns are very likely to include considerations (i.e., constraints) other than available land and funds. Such considerations may include desired levels of socio-cultural, infrastructure, or political benefits (see above) and carbon storage goals that must be met by specified dates within the planning horizon.

By pursuing the tree planting project, the decision-maker accomplishes carbon storage of C. Economic aspects of the project include quantifiable benefits and costs, for which the present net value may be written $PVNB$,[12] and unquantified benefits B. Carbon storage and unquantified benefits may be considered matrices of outcomes accrued over the planning horizon for the project, that is $C [C_0 \ C_1 \ldots C_t \ldots C_T]$ and $B [B_0 \ B_1 \ldots B_t \ldots B_T]$. Whether the decision-maker should pursue the project crucially depends on the alternative(s)[13] to the project and the evaluation criterion.

The decision-maker's alternative to the project (the reference case) accomplishes carbon storage of $C^0 [C_0^0 \ C_1^0 \ldots C_t^0 \ldots C_T^0]$, net economic benefits of $PVNB^0$ and unquantified benefits of $B^0 [B_0 B_1^0 \ldots B_t^0 \ldots B_T^0]$. Possible reference cases include the most likely projection of land uses without the project (cf. 'business as usual' or *status quo* alternatives). For planted trees intended to replace natural stands, the reference case may be the standing forest. For planted trees intended to afforest agricultural lands, the reference case may be agricultural land use. As will be seen, the attributes of the reference case are as important to the decision-maker's decision as the attributes of the tree planting project. A correct decision cannot be made without some knowledge of both alternatives.

The decision-maker selects an alternative according to a criterion ('welfare') that incorporates the present value of net benefits, the carbon storage, and the unquantified benefits, $W=W(PVNB, C, B)$. For simplicity, consider the linear and separable function $W=w_1 PVNB + w_2 C + w_3 B$, where the weights w_1, w_2, and w_3 describe the relative importance of economic benefits and costs, carbon, and unquantified benefits in the decision-maker's decision.[14] Alternatives that provide acceptable levels of carbon storage and unquantified benefits (described by the matrices C and B, respectively) can be compared by the decision-maker using the criterion W.

Pursuing the tree planting project necessarily means that the decision-maker does not pursue its alternative. This fact defines the opportunity cost to the decision-maker of whichever alternative is selected. For example, if the project is pursued, W is obtained and W^0 is not obtained. The decision-maker's net welfare from pursuing the project is W minus W^0, and it is in this context that W^0 (welfare forgone) may be considered the opportunity cost of pursuing the tree planting project. The decision-maker's opportunity cost is comprised of welfare values of (i) economic net benefits (e.g., returns to land, labour, or materials in alternative uses), (ii) carbon (e.g., carbon sequestered in crop ecosystems), and (iii) unquantified benefits.[15] In developing countries, the value of forgone subsistence agricultural uses displaced by the project may be difficult to quantify; when this is the case, minimum subsistence production may be included in the constraints, B.

For example, the welfare derived from the reference case is $w_1 PVNB^0 + w_2 C^0 + w_3 B^0$. If the decision-maker chooses the tree planting project, then the welfare change is $W = W - W^0$. It is convenient to represent the welfare change as the sum of its components, for example, $W = w_1 PVNB + w_2 C + w_3 B$. The change in the carbon stock, $C = C - C^0$, must (by assumption) be compared to any obligated emissions reductions targets.

The decision-maker's choice to pursue the tree planting project includes determining the net economic gain from pursuing the project; if $w_1 PVNB$ exceeds zero, then the project contributes a net economic gain. In some circumstances (e.g., when carbon storage and unquantified benefits are roughly the same with and without the project), no further analysis may be needed. However, if $w_1 PVNB$ fails to exceed zero, then the project cannot be justified on economic grounds. If the decision-maker chooses to pursue the project under these circumstances an outside observer must conclude that the net welfare contributions from carbon storage and unquantified benefits, $w_2 C + w_3 B$, outweigh the economic costs of pursuing the project.[16]

Conversely, if added funds are required to induce the decision-maker to pursue the project, then the minimum amount required indicates a price for an outside transaction (e.g., a carbon offset trade) under the joint implementation provisions of the Framework Convention on Climate Change.[17] A potential buyer's maximum price for such an offset might be determined by the minimum marginal cost of emission reductions, applicable carbon tax or emissions charge rates, and the market price of other available carbon offsets.

When comparing a tree planting project to a reference case, the difference in carbon stock with and without the project is of crucial importance. As discussed earlier, $C = C - C^0$, which is comprised of changes in stock components. Omitting stock components (e.g., substituted fuel) in an analysis is equivalent to the assumption that $C_f = C_f - C_f^0 = 0$. Whether such assumptions are valid clearly depends on both the planting project and the reference case; however, the rationale and validity for omissions such as these should be discussed in all studies.

6.1. Monitoring

Once a project has been implemented it is essential to have an ongoing program of monitoring in place (and adequately funded). This is necessary to ensure the viability of the project for optimal long-term C storage. Thus the project management requires adequate data and feedbacks (both economic and non-economic) so that an adaptive management strategy (e.g., Walters 1986) can be used. This strategy should direct the project along the

C storage pathway (the main objective) while ensuring that the benefits are adequate to the investors and the local communities.

7. SUMMARY AND CONCLUSIONS

Economic evaluations of tree planting for C storage must consider the social and institutional settings of the land owners. Settings and institutions will vary by country, particularly between developed and developing countries and among boreal, temperate, and tropical forest regions. In general however, the methodology presented is applicable to tree planting throughout the world.

Major tree planting project categories include restoration plantations, roundwood plantations, bioenergy plantations, and plantations for nonwood products; combinations of these categories are common. Plantations include contiguous areas of planted trees greater than one hectare. Total carbon stock comprises carbon in vegetation, decomposing matter, soil, product, and fuel substituted. Carbon stored in each of these components is dynamic, and may change over time.

Economic costs of tree planting projects may include: The prices of all direct and indirect inputs (e.g., land, labour, management, materials) employed in the tree planting project; External costs of practices (if any) associated with tree planting. Benefits may include the values of goods and services attributable to tree planting (may or may not be exchanged in markets). Other unquantified benefits may include socio-cultural benefits, infrastructure benefits, and political-institutional benefits.

The carbon storage, benefits, and costs associated with an alternative to the project (a reference case) must be identified before the tree planting project may be evaluated. Carbon storage may be defined as the net change in total carbon storage between that stored in the project and in the reference case. A net carbon storage statistic exists for each time period in the planning horizon for the tree planting project.

If a summary statistic is needed, mean carbon storage may be defined as the cumulative total carbon storage over the planning horizon, divided by the length of the planning horizon. The assumptions implicit in this definition must be carefully considered when the planning horizon is of infinite length. Employing a summary statistic such as mean carbon storage masks some of the dynamics of the carbon storage process.

If an economic summary statistic is needed, the present value of net benefits may be obtained by discounting quantifiable benefits and costs. A present value calculation necessarily omits unquantified aspects (e.g., equity). Employing a summary statistic such as present net benefit masks some of the distribution of benefits and costs within and between periods in the planning horizon.

Economic evaluation of tree planting projects must consider *(i)* the perspective of the analysis (e.g., individual, community, societal) *(ii)* methods to account for nonmarket and unquantified benefits and costs *(iii)* the implications of risk, uncertainty, and project reversibility. Projects may require monitoring to inform adaptive management decisions when new information develops about project viability, economic or unquantified considerations, and changes in carbon stocks.

NOTES

[1]See, for example, Climate Talks Enter Harder Phase of Cutting Back Emissions', *New York Times* 11 April 1995.

[2]The Framework Convention is among several Conventions (cf. the Biodiversity Convention) signed by numerous United Nations member countries during the 1992 United Nations Conference on Environment and Development (UNCED) in Rio de Janiero. See Report of the Intergovernmental Negotiating Committee for a Framework Convention on Climate Change on the Work of the Second Part of Its Fifth Session, Held At New York from 30 April to 9 May 1992', United Nations Document A/AC.237/18 (Part II). New York.

[3]Some industrialized countries are obligated by the Framework Convention limit net emissions to 1990 levels as early as the year 2000.

[4]Agroforestry in this context is the cultivation of trees on the same land with agricultural crops and/or livestock. Note that several types of projects commonly described as agroforestry, such as small tree plantings (at least 1 ha) intended for nonwood products, and plantings of multiple tree species, are still within the scope of our definition of plantations.

[5]Evidence includes the fact that U.S. primary wood-using mills make use of 94 percent of all bark and wood residue at the millsite: 40 percent is used for fuel, 38 percent for wood chips and pulp, and 16 percent for such miscellaneous products as cooperage, bark mulch, and soil amendments (Powell et al. 1993).

[6]Depending on the goals of the study, carbon stored in products, C_p, may be subdivided into that stored in short-term, mid-term, and long-term product classes.

[7]Soil carbon does not reach equilibrium in rotations usually used for these types of plantations. Using MCS for a single rotation as a proxy for a perpetual series of rotations may therefore underestimate accumulated soil carbon.

[8]A similar method may also be employed to analyse the carbon stored in standing biomass for projects involving the protection of natural forests (Swisher 1995).

[9]The consumption of nonbiomass fuel in the reference case (see below) must clearly be known in order to calculate this savings.

[10]The reason for emphasizing this endowment approach is to ensure not only that all relevant costs are included, but to capture the assumption that local market and non-market benefits are retained as necessary incentives to keep the project operating over multiple rotation periods (Swisher 1991). Alternatively, these cash flow considerations might be incorporated into the project evaluation as constraints that revenues exceed costs for relevant periods.

[11]See, for example, Dixon and Sherman (1990 pp. 194-195) and Footnote 10.

[12]When the distribution of benefits and costs within and between generations is important and may be influenced by the project, then it may be desirable to avoid summarizing benefit and cost streams in a single *PVNB* index.

[13]The problem as presented is a choice between two mutually exclusive alternatives. The logic is readily applicable to choice scenarios with more than two alternatives and to continuous decisions (e.g., choosing the amount of land to allocate to the project and other uses).

[14]This might arise when allocating H hectares between the project and the alternative (reference) use. Allocating h hectares to the project and H-h hectares to its alternative based

on *PVNB* subject to minimum total acceptable levels of C_{min} (carbon) and B_{min} (unquantified benefits) gives rise to the Lagrangian $PVNB(h)+PVNB^0(H-h)+M_1[C(h) + C^0(H-h) - C_{min}] + M_2[B(h) + B^0(H-h) - B_{min}]$, for which M_1 and M_2 are multipliers. This is of the form $W=w_1PVNB + w_2C + w_3B$ when economic values are the units used to measure welfare. In the present value calculation, the weights w_1 could be interpreted to indicate rates of time preference for benefits, i.e., $w_1 [1 (1+r)^{-1} (1+r)^{-2} \ldots (1+r)^{-t} \ldots (1+r)^{-T}]$ where r is the rate of time preference.

[15]When welfare is only measured in economic terms, then the planner's opportunity cost will only consist of economic benefits forgone by pursuing the project (e.g. the value of land in its next best use).

[16]Alternatively, *PVNB* may be taken as a lower bound on the planner's willingness to exchange economic benefits for carbon and noneconomic benefits (e.g. Dixon et al. 1994 pp. 52-53, Krutilla and Fisher 1985 pp. 84-150).

[17]See Footnote 2.

REFERENCES

Allan, T. and J.P. Lanly. 1991. 'Overview of Status and Trends of World Forests', in D. Howlett and C. Sargent, eds., *Technical Workshop to Explore Options for Global Forestry Management,* London: International Institute for Environment and Development.

Dewar, R.C. 1990. 'A Model of Carbon Storage in Forests and Forest Products', *Tree Physiology* 6: 417-428.

Dixon, J.A., Scura, L.F., Carpenter, R.A. and P.B. Sherman. 1994. *Economic Analysis of Environmental Impacts,* second edition, London: Earthscan Publications.

Dixon, J.A. and P.B. Sherman. 1990. *Economics of Protected Areas: A New Look at Benefits and Costs,* Covelo: Island Press.

Freeman, A.M. III. 1993. *The Measurement of Environmental and Resource Values,* Washington: Resources for the Future.

Hall, D.O., Myrick, H.E. and R.H. Williams. 1991. 'Cooling the Greenhouse with Bioenergy', *Nature* 353: 11-12.

Krutilla, J.V. and A.C. Fisher. 1985. *The Economics of Natural Environments: Studies in the Valuation of Commodity and Amenity Resources,* Baltimore: Johns Hopkins University Press for Resources for the Future.

Moulton, R.J. and K.R. Richards. 1990. 'Costs of Sequestering Carbon Through Tree Planting and Forest Management in the United States', U.S. Department of Agriculture, Forest Service General Technical Report WO-58.

Powell, D.S., J.L. Faulkner, D.R. Darr, Z. Zhu, and D.W. MacCleery. 1993. 'Forest Resources of the United States', U.S. Department of Agriculture, Forest Service General Technical Report RM-234.

Sedjo, R.A. 1983. *The Comparative Economics of Plantation Forestry,* Washington: Resources for the Future.

Swisher, J. N. 1991. 'Cost and performance of CO_2 storage in forestry projects', *Biomass and Energy* 1(6): 317-328.

Swisher, J. N. 1995. 'The Incremental Costs of Terrestrial Carbon Storage in Forestry, Bioenergy and Land Use', *Environmental and Resource Economics*: In Press.

United Nations Food and Agricultural Organisation (FAO). 1995. 'Forest Resources Assessment 1990, Global Synthesis', FAO Forestry Paper 124.

Wade, R. 1988. *Village Republics: Economic Conditions for Collective Action in South India,* Cambridge: Cambridge University Press.

Walters, C.J. 1986. *Adaptive Management of Renewable Resources,* New York: MacMillan Publishing Company.

Winjum, J.K. and P.E. Schroeder 1995. 'Forest Plantations of the World: Their Extent, Ecological Attributes, and Carbon Storage', *Journal of World Forest Resource Management*: In Press.

Critical Reviews in Environmental Science and Technology, 27(Special): S23–S45 (1997)

SEQUESTERING CARBON IN NATURAL FORESTS

C.S. BINKLEY,[1] M. J. APPS,[2] R. K. DIXON,[3] P. E. KAUPPI,[4] and L.-O. NILSSON[5]

[1]*Faculty of Forestry, The University of British Columbia, Vancouver, B.C., Canada, V6T 1Z4;* [2]*Canadian Forest Service, Edmonton, Alberta, Canada T6H 3S5;* [3]*United States Environmental Protection Agency, Washington, D.C. 20585, U.S.A.;* [4]*Finnish Forest Research Institute, Helsinki, 00170, Finland;* [5]*Department of Ecology and Environmental Research, Swedish University of Agricultural Sciences, Uppsala, 75007, Sweden.*

ABSTRACT: Closed forests cover about 3 billion hectares, or 20% of the world's total land area (excluding Antarctica). Forest plantations comprise less than 1% of this area. Natural forests range from the intensively managed ones of Central Europe and Scandinavia to the wild boreal forests of Russia and Canada and the deep jungles and dry forests of the tropics. Numerous techniques—largely drawn from the ordinary repertoire of forest management—are available to enhance our ability of these forests to sequester and store C. Although the costs of sequestering additional C in these forests may be quite low (even in comparison with intensive plantation options), increased use of natural forests for this purpose raises a host of concerns about competing forest uses, biological risk, and the capacity to actually measure the incremental C sequestered. The problems of poverty, expanding populations, weak institutions, incomplete scientific knowledge, and climatic change itself will challenge the world's capacity to use natural forests as part of a CO_2 control strategy.

KEY WORDS: sequestering carbon, hectares, forest plantations.

1. INTRODUCTION

Forests serve humans in many ways. Indeed, some believe that forests have intrinsic values or even rights beyond those conferred by human use alone. This paper focuses on one of these uses: The capacity of natural forests to sequester CO_2 and store C emitted as a waste product of our industrial activities. This paper specifically discusses the mechanisms behind such C uptake and storage activities and the key points related to estimating the costs of accumulating C in natural forests. Most of these techniques are classical methods of forest management, albeit developed to achieve different objectives. While precise quantification of both the physical and economic amounts involved is possible in specific instances, it is beyond the scope of this paper. Instead, we provide a road map so the reader may pursue these important details independently. Dixon (this volume), Sedjo et al. (1995), Hoen and Solberg (1994 and this volume) and Richards and Stokes (1995) critically review some of the alternatives discussed in this paper and quantify some of the biophysical impacts and economic costs.

1064-3389/97/$.50

1.1. Definition of "Natural Forest"

For the purposes of this discussion the term "natural forests" refers to those lands currently occupied by closed forests or being regenerated to the same or similar species as removed from the site after logging. We specifically exclude agricultural and other non-forested lands which are planted to tree species. We do however, consider those cases where open areas become forested through the ordinary processes of forest succession (e.g., occupation of abandoned farm lands or lands laid bare by recent glaciation). We exclude non-forested peat lands, open woodlands and savannas. This definition probably encompasses virtually all of the world's forests (Dixon et al. 1994).

Human intervention in the development of natural forests ranges from intensive culture approximating the care lavished on agricultural crops to complete diffidence. Managed forests form one end of the spectrum and wild forests the other. Examples of the former include the industrial and multiple-use forests of Scandinavia and Central Europe; examples of the latter include the remote boreal wildernesses in Canada and Russia, and the highly diverse, inaccessible tropical jungles in Brazil and Zaire. Many parks and reserves lie towards the latter end of the spectrum, but require management intervention even if it is not for the purposes of producing industrial products. All of these landscapes share the quality that the species composition of the forest overstory has been largely determined by nature.

Globally, closed natural forests cover approximately 3 billion ha of the world's land (Sharma, 1992). Forty-three percent of the natural forests are found at low latitudes, 32% at the high latitudes and 25% at mid-latitudes (Dixon et al. 1994a). Low-latitude forests are currently shrinking at a rate of approximately 15 million ha annually, and thereby are a source of 1.6 Pg of C per year. Brazil, Indonesia, and Zaire are large CO_2 sources. If tropical deforestation and land-use change were slowed or eliminated, these forests would become a large sink of CO_2 (Brown et al. 1993). At present mid- and high-latitude natural forests sequester over 0.6 Pg C annually (Dixon et al. 1994a). Canada, Russia, USA, China, and the USA all have large natural forest CO_2 sinks.

1.2. How Do Natural Forests Sequester Carbon?

All natural forests store C in phytomass, forest floor litter and soils (Cole, 1995). Phytomass includes all above and below ground components of the dominant overstory tree species as well as all other vegetation comprising the overstory, understory, and forest floor cover (including lichens and mosses). In natural forests—particularly those at the wild end of the spectrum—the noncommercial components of the forest may represent a significant and even dominant component of the standing biomass. The forest floor ranges from a dominant pool in many northern forests to a negligible one in parts of the tropics (Brown, 1991). These facts make it extremely difficult to estimate C stocks from commercial inventories which—when they are available—often record only the merchantable bole volumes of the economic species. The lack of accurate allometric relationships to estimate root, branch and foliage biomass from stem measurements exacerbates the problem. The non-merchantable biomass components are important not only for their contribution to total biomass, but also in their role as inputs to the other C pools in the ecosystem.

Several factors distinguish C sequestration in natural forests from that in plantations, although the distinctions are not absolute. In natural forest systems where extensive forestry management is practiced, the fraction of Net Primary Production (NPP) that enters the forest floor as litter, dead snags or as dead-root turnover in the below-ground system is generally much higher than it is in plantation forests. Indeed, capturing this mortality as merchantable volume is one of the major objectives of plantation management. In this respect, the managed natural forests of Europe tend to resemble plantations, and the distinction between plantation and natural forests becomes very murky.

Life-cycle dynamics also distinguish natural forests and plantations. Plantation management usually seeks a 'normal' forest age-class distribution with an equal area in each age class up to the planned rotation age. Each cohort is harvested as it reaches the rotation age and is replaced with a new cohort which then becomes the youngest age class. Management of natural forests generally includes more chaotic, stochastic processes of mortality, disturbance and succession. Storing C in forests by increasing the standing above-ground biomass may increase the risk of devastating disturbances associated with fire, insects or diseases. Nonetheless, a C benefit still accrues because the increase in atmospheric concentrations of CO_2 is delayed.

The response and feedbacks of boreal, temperate and tropical forest systems to global climate change may be profound (Dixon et al. 1994b; Smith et al. 1991). Scenarios developed by coupling General Circulation Models (GCMs) of global climate with vegetation response models generally imply large shifts in the distribution and productivity of forest systems (Smith et al. 1991; Kauppi and Posch, 1988; (Songhen and Mendelsohn, this volume; Perez et al., this volume). Uncertainty regarding the potential redistribution of forest systems in response to global climate change complicates our ability to predict future CO_2 sequestration by natural forest systems (King 1993). Even if the GCMs are only partially correct, the productivity of existing natural forests will inevitably be affected, and the current C biogeochemistry, both pools and fluxes, will also change (Dixon et al. 1994a).

For the purposes of understanding the interaction between the C pools in the atmosphere, Net Ecosystem Productivity (NEP) may be a more useful concept than NPP. NEP refers to the net change in soil, litter and phytomass. Because different biogeochemical processes control each of these pools, it is possible for increases in phytomass to be negligible but for continued sequestration to occur in soil and litter pools. Such conditions may obtain in cool, moist micro-climatic conditions below the canopies of old-growth temperate forests where gap-phase replacement operates and stand-replacing disturbances are rare events (Harmon et al., 1990). In boreal forests, break-up of the overstory may be accompanied by a decrease in both the standing phytomass—and because of increased exposure and soil warmth—decreased forest floor and soil C.

Peat is formed in natural forests in waterlogged conditions. Trees can survive on large areas at least in the early phases of paludification (the peatland formation process). Such specialized peatland plants as *Sphaghum* mosses transfer organic material released from the overstory into the peat layer. Peatland forests differ from other natural forests because CO_2 fixation exceeds CO_2 release over time frames stretching from centuries to millennia. The boreal and subarctic peatlands comprise a C pool estimated at 455 Pg that has accumulated during the post-glacial period at an average net rate of 96 Tg/yr (Gorham, 1991). While this is insignificant in an overall global C budget where the flux from fossil-fuel combustion is at least two orders of magnitude greater, the relative contribution of peatlands to national C budgets is significant in some northern countries.

1.3. Sequestering C in the Economic System

Management of natural forests may require the removal of trees, either as a planned component of a larger industrial activity, or as a means of sustaining specific ecosystem conditions in a park or forest reserve. Such removals ordinarily enter the economic system. While the total amount of C fixed in forest products pools globally is less than 20 Pg over the last 50 years (Dixon, 1994a), it may be critical in the C budgets of those countries where forest products production is high and consumption is low (e.g., Finland, Sweden and Canada). This paper explicitly treats the impact of removing the tree from the forest, but excludes the impacts of that material on other parts of the economic system. This procedure obviously ignores the potential role of forests to substitute for fossil fuels in the energy production, or to substitute for such energy-intensive materials as concrete or steel in construction applications. Such calculations are beyond the scope of our analysis but are extensively treated elsewhere (e.g., Marland, this volume; Matthews, 1995).

In considering forest C sequestration, it is important to note that an effective option might be to optimize the capacity of forests to *remove* CO_2 from the atmosphere but to *store* the C in forms other than plant phytomass (i.e., as a substitute for fossil fuels, or in long-lived forest products). Such a strategy would involve harvesting forests at a comparatively young age, and resulting in a lower standing inventory of timber than would occur if the forests themselves were used to sequester C (Harmon et al., 1990). We discuss the impact of such strategies on the forests themselves, but do not expand this analysis to include the impact on C offset in the energy system or other economic activities.

1.4. Other Policies Affect C Sequestration in Natural Forests

Policies in spheres of public affairs other than forestry frequently have a profound impact on forests. For example, agricultural subsidies have long encouraged clearing of forested land in many countries—both developed and developing. Such land-use conversion release a pulse of C into the atmosphere, and may also reduce the annual rate of C sequestration. As another example, in Scandinavia and in parts of eastern North America it is well understood that N deposition associated with high levels of industrial air pollution fertilizes forests, increases forest growth, and induces a higher level of C sequestration (Eriksson and Johansson, 1993; Nilsson and Wiklund, 1992). As a third example, international agreements on the conservation of biological diversity may require countries to sustain conditions in forested ecosystems which do not optimize their capacity to sequester C (Victor and Salt, 1995). Policy makers in these areas, seemingly remote from forests, should consider such impacts on forests (Sharma, 1992).

2. CATALOG OF TECHNIQUES: TWELVE WAYS TO SEQUESTER CARBON

The remainder of this paper catalogs various techniques for sequestering C in natural forests. The first section below describes twelve possibilities which we believe collectively cover the range of alternatives. In each case we describe the biophysical response mechanism and the possible impact on C sequestration. The next section covers the problems of determining the economic cost of each practice. In dealing with both the biophysical responses and the economic costs, the discussion focuses on the appropriate conceptual approach for evaluating the technique with references to specific examples on the technique reported in the literature.

2.1. Biophysical Responses

We are specifically interested in the total stock of C held in natural forests at any point in time. Annual accumulation of C can be measured either as the change in C stocks from one year to the next, or as the sum of the changes in the stocks of the individual components of the ecosystem. In considering the different forest management techniques in different locales, one or the other method may be the more straightforward to apply, but they are conceptually identical. The total amount of C sequestered in natural forests simply equals the storage per hectare multiplied by the forested area. Sequestration strategies logically focus both on increasing the storage per hectare and on increasing the forested area over what it would otherwise be (Winjum et al., 1993).

2.1.1. Protect Against Fires

Fires, of both natural and anthropogenic origin, play an important role in the life cycle of many natural forests. From the point of view of C storage, fires may be broadly categorized into two types: non-stand-replacing and stand-replacing. The first are commonly associated with low-intensity but relatively frequent ground fires, and relatively open woodland structures. They often result in uneven-aged forests. Some high-latitude forest species (e.g. *Pinus ponderosa*) are well adapted to this fire regime but most low-latitude species are not. Such fires produce relatively low immediate C releases, and—by clearing understory debris—may both cleanse the forest floor of pathogens and remobilize nutrient for over-story utilization, thereby increasing subsequent C uptake. As a result, managers of some forests types prescribe fires as a regular practice. When fine fuels on the forest floor build to high enough levels, however, crown fires can be sustained, and stand-replacing fires may occur with very different consequences.

Stand-replacing fires are dramatic events having locally catastrophic effects leading to complete mortality of the overstory. Large, intense conflagrations are the dominant type of fire in many boreal systems (Apps and Kurz, 1993). As an example, in 1989, more than six million ha of boreal forest (an area 50% larger than Switzerland) burned in Northern Saskatchewan in a single fire season. The Great Black Dragon Fire in Northern China and the Russian Far East covered more than 10 million ha, and the resulting smoke plume was easily seen in satellite images extending far to the East for many days (Salisbury, 1987; Stocks, 1991). A large fire burned in Borneo in 1993 for many weeks before it was finally detected from satellite telemetry.

The effects of these fires are threefold: (i) C is redistributed amongst the various ecosystem pools, (ii) C is released to the atmosphere as CO_2 and other C compounds including CO and CH4 (Levine, 1991; Crutzen and Goldammer, 1993), and (iii) the forest structure is changed as the stand age is 'reset' and seral succession is restarted. While such fires entail major and immediate C released to the atmosphere (Cofer et al., 1991; Levine, 1991) the forests where such fires occur naturally are adapted to them and, indeed, are dependent on them for regeneration, removal of pests and disease vectors, and a host of other ecological relationships.

The C releases associated with fire go beyond the immediate pulse to include subsequent emissions from the non-combusted, decomposing, dead biomass left on site. For example, Auclair and Carter (1993) estimate that post-fire releases may be as high as three times the immediate release. Some of this material becomes coarse woody debris (Harmon et al., 1986), or other forest floor reservoirs (Apps and Kurz, 1993; Dixon and Krankina, 1993).

While eliminating these large C releases is a potentially attractive option, suppression of fire may also merely open the way for other pathological agents which would normally have been kept at negligible levels by the periodic cleansing action of wild fire. For example, in Alaska the increased incidence of bark beetle and other insects has been associated with the increased frequency of overmature forests (Dixon and Krankina, 1993). The large fires in the American West during the last decade suggest that the thorough suppression of fire may be counter-productive because it results in a build-up of fine fuels and the eventual, inevitable, occurrence of a much higher intensity fire. The resulting fires have significantly greater damage to site fertility due to changes in soil structure and organic and nutrient capital than would have otherwise occurred with a series of smaller, more frequent fires.

Although fires do periodically devastate low-latitude in some regions (e.g., in Indonesia), in most moist tropical forests neither fuel loading nor fuel condition are conducive to large or intense fires. Fires may play a role in these forests in unusually dry weather patterns or after large-scale mortality caused by agents such as hurricanes or typhoons. Anthropogenic fires occur throughout low-latitude forests, as forest burning is a common management tool for resource-poor farmers. Although the size of these controlled fires is relatively small, their large number produce a globally significant pulse of greenhouse gases (Crutzen and Goldammer, 1993). During peak burning season in some low latitude countries such as Brazil, satellite sensors detect many thousands of small fires on any given day.

2.1.2. Protecting Against Disease, Pest Insects, and Other Herbivores

As with fire, these types of forest disturbance can be either stand replacing or endemic. Stand replacing events have the same three characteristics as stand-replacing fires, except that direct releases of C to the atmosphere is generally smaller (it occurs as respiration of the pathogens rather than as combustion products), and the transfer of C to the forest floor correspondingly greater. Protection against these sorts of events is, like fire, problematic. For example, insect-induced stand mortality depends on present year weather conditions, previous year's insect populations and (of some insects) stand-age, type and health status (Volney, 1995; Galinsky and Witrowski, 1995). Avoidance of conditions which are conducive to these disturbances is the best protection. Aerial applications of chemical or biological pesticides are generally effective in sustaining forests in a living condition long enough to harvest wood, but are generally not effective in eliminating insect populations altogether.

Endemic impacts of insect, disease and herbivores, may result in direct reduction of NPP by reducing the net increment of forest phytomass. The effect on NEP and C sequestration is not clear and depends on many factors, including forest age and disturbance type (Kurz et al., 1995). In relatively mature stands, phytomass decrements associated with the endemic disturbances largely reappear as increased C on the forest floor and soils. In these cases the net C balance is determined by the relative rate of biomass replacement and forest floor decomposition. The type of litter created is also a factor. (For example, dead roots, snags and other coarse woody debris associated with individual tree mortality have longer turnover times in the cooler sheltered micro-climatic conditions below a fully developed crown than does leaf and branch litter fall following attacks of defoliators). In younger, regenerating stands, the effect of herbivory in particular may cause a regeneration delay which has a temporary, and small, impact on C storage, but may have important economic repercussions. Some level of endemic insect, disease, and/or herbivory disturbance affects all forests. The fraction of NPP consumed by these agents has not, to our knowledge, been estimated except in a few specific instances.

Laboratory studies have demonstrated that insect herbivory of plants grown in high CO_2 environments is dramatically increased (Drake 1992). As the proportion of N declines relative to C in plants grown in high CO_2 conditions, insects and other grazers increase their consumption to compensate for the loss of protein. In the future, management of natural forests as CO_2 sinks will need to be adjusted to compensate for this increase in defoliation. If insects and other forest pests adapt a more aggressive consumption pattern in a future global climate, net CO_2 sequestration of future forests may be less than it is in current conditions (Dixon et al. 1995).

2.1.3. Salvage Dead and Dying Trees

Natural disturbances leave dead or dying trees that have both positive aspects (e.g., animal habitat, landscape diversity) and negative ones (e.g., increased fuel loadings, loss of valuable timber, reduced NPP). In some cases, where both the infrastructure and the markets exist, it is possible to salvage some or all of this dead or dying timber to recover valuable wood, reduce the risk of subsequent contagious disturbance by insects, fire or pathogens associated with the accumulation of logging debris, and rehabilitate the site for subsequent regeneration.

By moving decomposing C from the natural system to the economic system, it may be possible to increase the net C storage associated with the given piece of land. The extent to which there is a net C benefit in salvage operations depends on several factors. First, is the turnover rate of C transferred to forest products (including all process wastes) greater than the C turnover rate on site (Hendrikson, 1990)? Second, to what extent is site regrowth increased relative to the untreated condition? Third, how much fossil fuel is used in the salvage and product manufacture and distribution operations?

Natural disturbances may affect very large areas. There are however two significant barriers to salvage operations: operability and marketability. Lack of railroads, highways and waterways limit the access to many natural forests. Hazardous conditions that often exist after larger scale disturbances also restrict salvage operations. Salvaged timber has limited uses. For example, pulp and paper mills are understandably reluctant to accept wood charred by fires. Concern for inadvertent importation of insect and disease greatly restricts the markets for timber salvaged from outbreaks of these pathogens. Bioenergy might be a potentially viable market opportunity, but access and infrastructure would appear to restrict this use to highly specific instances.

2.1.4. Change Rotations

Cooper (1983) demonstrated that converting a forest region of fully stocked mature stands into a maximum sustained yield forest decreases the standing stock by about two thirds, and thus releases CO_2 into the atmosphere. Harmon et al. (1990) simulated conversion of old-growth forests into young fast-growing forests and concluded that on-site C storage does not approach old-growth capacity for at least 200 years. They estimated that the conversion of five million hectares of old-growth forests to younger plantations in western Oregon and Washington in the last 100 years has added 1.5-1.8 Pg C into the atmosphere.

Natural forests in disturbance-dominated systems include immature as well as mature stands. Therefore, logging in natural forests does not always result in such a dramatic decrease of C pool as Harmon et al. (1990) demonstrated for the US Pacific Northwest. In some cases logging can merely substitute the natural disturbances as the mechanism by which CO_2 is released into the atmosphere. In the Nordic countries, for example, the pool of C in forest vegetation has increased this century despite continuous and steadily increasing logging (e.g. Kauppi et al. 1995). In contrast, logging of tropical forests in Peninsular Malaysia has resulted in a dramatic reduction in C storage over the past 30 years (Brown, 1991).

In a steady-state system according to Dewar's (1990) model, a very long rotation can promote C sequestration, if the products decay quickly. But a short rotation can also serve the same purpose given high biomass yield, and a long product life-time and/or effective replacement of the consumption of fossil fuels. Most short rotations systems are "plantations" (and therefore beyond the scope of this paper), yet some coppice systems can be classified as "natural forest".

Dewar (1990) formulated a model that describes C storage in a forest and its timber products as function of the forest growth curve, the rotation period and the C retention curves in timber products. He showed that, when the forest is managed for a maximum yield of biomass, the contribution of asymptotic C storage from timber products is about $2.5D/T^*$ times the contribution from living trees, where D is the characteristic decay time for conversion of timber products to C dioxide, and T^* is the normal rotation period for maximum yield. When $D/T^* < 1$, as the rotation period is increased indefinitely, the asymptotic level of C storage increases monotonically toward the value of C content of living trees at maturity. But when $D/T^* > 1$ there is a finite, optimal rotation period, greater than T^*, for which asymptotic C storage is greater than the C content of trees at maturity.

Either lengthening or shortening the rotation age will increase the C stocks, depending on the initial conditions (e.g. age structure) of the forest, harvesting methods, silviculture and, in particular, the fate of harvested C (Schlamadinger and Marland, 1995; Marland and Schlamadinger, this volume). In some cases, making long-lived products from wood grown in short-rotation forests can sequester C effectively. Short-rotation management of woody crops for production of bioenergy fuel stocks, can replace or offset the need for fossil fuels (Sampson et al., 1993)

Although a careful analysis is needed for each specific case, as a rule of thumb that prolonging rotations in natural forests will generally contribute to C sequestration. Converting natural forests to short rotation systems can make positive sequestration contributions mainly in the longer term, that is, in the period beyond 20 to 50 years from now.

2.1.5. Control Stand Density

Thinning is widely used to alter the size of individual stems in a stand, the timing of their availability, and the overall amount of merchantable timber available. In some regions, stands are thinned three to five times during the rotation in order to collect material which otherwise would decay in the forest. When thinnings provide sufficient space, the remaining trees grow to larger dimensions more quickly and thus more valuable per cubic meter harvested. The trend of decreasing thinnings and an increasing share of final cuttings has prevailed in Europe since the 1970s because logging costs are higher in thinnings than in clear cutting.

Thinning affects C sequestration and storage in several ways. Low thinnings utilize small stems which would otherwise decay and release C back to the atmosphere. High thinnings which reduce overall forest growth rates may also help sequester C by providing more of the total volume in longer-lived solid wood products such as lumber. Thinning-to-waste will generally be neutral or negative in terms of carbon sequestration.

2.1.6. Enhance Available Nutrients

Water and nutrient availability control forest growth in most parts of the world (Nambiar and Sands, 1993). As a result, fertilization and irrigation will generally enhance forest NPP. For example, in Portugal and Brazil experimental *Eucalyptus* plots have reached annual production of up to 40 Mg/ha/yr (Campinhos, 1991). In tropical Australia, Fife and Nambiar (1995) found clear responses of N fertilization on the growth (up to 99% growth increase) of *Pinus radiata* grown in a very dry environment. The authors also concluded that improved N status of the trees leads to more efficient water utilization. In Sweden a series of fertilization experiments have shown clear growth response due to N fertilization on both *Pinus* and *Picea abies* (Tamm 1992; Linder, 1995). In the most polluted parts of south west Sweden biomass production in *Picea abies* was doubled due to liquid fertilization with irrigation in 30 year old *Picea abies* (Nilsson L.O; this volume). Davey (1990) and Weetman et al. (1987) cite evidence for positive growth responses due to fertilization in both the US and Canada. Cole (1995) and Henry et al. (1994) showed a sustainable increase in basal area of 4-5 times after application of nutrient rich sewage sludge in a 45 year old Douglas fir forest. Allen (1995) has noted similar growth responses with liquid fertilization and irrigation. Fertilization may also increase C storage in the soil (Berdén 1994, Nilsson 1995).

In most cases N is the major nutrient element explaining the production increase. However, other nutrients—mostly P, K, Mg and such micronutrients as B, Cu, Mn or Zn—limit NPP in some areas. For example, in New Zealand, soil fertilization with P, B or Zn dramatically increases the productivity of *Pinus radiata* and various *Eucalyptus* species plantations (Nambiar and Sands, 1993). Similar vigorous responses to forest fertilization are observed in Africa and Southeast Asia in stands of *Tectona, Shorea* and *Leucaena* (Dixon et al. 1994b).

2.1.7. Control the Water Table

About 11.5 million hectares of boreal peat lands have been drained in the world in order to lower the water table and thereby promote forest growth (Gorham 1991). Almost all of this area can be classified as "natural forests" given the definition used in this paper. Although drainage of forested peatlands promotes forest biomass growth, but the net C benefit is partly offset by increased respiration in the exposed peat and (Zoltai and Martikainen, 1995). Undrained northern peat lands annually remove 76 Tg C from the atmosphere, where as the drained peat lands release 8.5 Tg C annually (Gorham 1991). Zoltai and Martikainen (1995) find that lowering water tables by 20–30 cm increases CO_2 emissions by 150–200%. They also point out that a more important, but difficult to predict, consequence of altering the water table is a change in methane (CH_4) production, an even more potent greenhouse gas than CO_2.

Irrigating dry forests would have a positive impact on C sequestration, by increasing storage in all of the subsystems—soil, vegetation, and products. Danish (Beier et al. 1995) and Swedish (Linder, 1995; Nilsson, 1995) experiments have shown a large growth increase due to irrigation alone despite a relatively humid climate. Water also limits NPP and NEP in many low-latitude forest systems (Jain et al., 1989). Loss of forest systems and desertification is a global problem of growing significance (Sharma, 1992). Irrigation and other silvicultural techniques can be employed to stimulate NEP, but the economic costs of these practices and the lack of appropriate infrastructure make such efforts virtually impossible. Many tree species are well adapted to dry conditions, and reclamation of substandard soils and desertified sites is possible systems (Jain et al., 1989). Biomass production on these low-latitude sites can be 15-20/hectare. Given the large area of dryland natural forests, stimulation of CO_2 production in these ecosystems is a globally significant opportunity (Dixon et al., 1994b).

2.1.8. Select Useful Species and Genotypes

The natural variation in NPP between provenances of the same tree species originating from a relatively small geographical area may be substantial,(e.g., Fife and Nambiar, 1995). Thus, selection of provenances well adapted to the site and expected climate may improve the NPP.

Tree improvement with some genera, including *Pinus, Populus, Picea, Eucalyptus, Leucaena, Tectona,* has dramatically improved the yield and other favourable commerical attributes over the last 50 years. However, it is not clear to what extent these increases in yield stem from increased NPP of individual trees, or simply reallocation of C from the noncommercial below-ground portions to stemwood.

A basic factor for improving NPP is a proper choice of tree species. Forest growth has declined in several locations responding to a poor choice of species (e.g. the decision to plant *Picea abies* on hardwood sites in central Europe). Because N limits NPP in many ecosystems, the introduction of such N fixing species like *Alnus*, at high latitudes, or *Lueccaena*, at low latitudes, would improve the nutrient status of the site, which would in turn improve (Jain et al., 1989). A large number of multipurpose tree species have been identified which can be employed to increase C sequestration and storage, or to improve site productivity on a sustained basis (Burley and Stewart, 1985).

Finally, enrichment planting below the main canopy can increase NEP in cases where one or more aspects of the growing sites are under utilized. This might occur in natural stands which are not fully stocked as a result of various non-stand-replacing disturbances. This technique could also be effective if shade tolerant species were planted below less tolerant ones. So, for example, mixed stands of *Tsuga heterophylla* and *Pseudotsuga mensiesii* have higher above-ground productivity than do pure stands of either species alone.

2.1.9. Reduce Regeneration Delays

Recovery of disturbed natural forests—whether by harvest, insect, fire, disease, blowdown, drought or any of a host of other factors—depends on both the availability of viable seed sources and suitability of site conditions. There is often a period of time—the regeneration delay—following disturbance where these criteria are not met. On some sites, particularly low quality ones, losses of organic material and nutrient capital by leaching or erosion can both degrade the site potential and delay its realization. In some high quality sites, more successful opportunistic species, such as grasses (e.g., *Calamagrostis sp.*, in many northern systems) may invade the site and temporarily prevent the establishment of trees and the longer-term accumulation of phytomass. In both cases, such delays may be exacerbated by such factors as increased herbivory, and changes in soil structure or water table.

Appropriate choice of harvest timing and method can prevent regeneration delays. If regeneration is not prompt, such mitigation techniques as in-planting, site preparation, and nutrient supplements can be effective in speeding full stand occupancy with tree species (Winjum et al., 1993).

In extreme cases the site may fail to recover to its original forest cover and remain in a non-regenerated state (Dixon et al., 1994b). The C consequences are the same as land-use changes. In less extreme cases, the regeneration delay introduces a corresponding delay in C accumulation.

2.1.10. Select an Appropriate Harvest Method

The current techniques for harvesting trees varies from the ancient practice of hand felling individual trees and yarding by gravity or draft animals to highly mechanized, capital intensive methods. The fossil C consumption per unit of C harvested is low even in the highly mechanized methods. Therefore, the fossil fuel use is not decisive in determining the impact of harvesting method on C sequestration.

Analogously to fires and pest outbreaks, logging methods can be divided into stand conserving and stand-replacing ones. Clearcutting is the most straight-forward technique for replacing a stand. However, seed-tree and shelterwood systems also remove most of the trees with a few left behind to provide the seeds and/or shelter for the new stand. Selective logging maintains forest cover while gradually replacing the stand. It can be used in some cases in a way that there never occurs a stand replacement phase: individual trees rather than the whole stand are replaced and rotated. Normally, trees are planted only in connection with stand-replacing harvest methods, but this practice is also appropriate as an enrichment technique as described above.

The choice of the harvesting method can have impacts on C sequestration, although such differences are poorly documented in literature. Olsson (1995) found that clear felling high-latitude forests results in significant losses of C and N from the humus layer down to 20 cm depth. Harvesting and the associated disturbance also influence C pools and flux in low-latitude forests (Brown et al., 1991). Techniques which have been employed to conserve and sequester C in Malaysian forests include (i) preservation of non-harvested trees and associated vegetation by selective logging (versus clearcutting), (ii) before harvesting takes place, cutting vines which link boles of trees and thereby reducing the felling losses of non-target trees, and (iii) using low-impact harvesting systems which reduce soil disturbance (Dixon et al., 1993). The key to analyzing the impact is to measure and monitor the C pool in vegetation and soils over a fairly long period of time.

2.1.11. Manage Logging Residues

In some places, logging residues from forest management are, to a large extent, already being used as bioenergy (Sampson et al., 1993). In Finland and Sweden, bioenergy represents as much as 17-18% of all energy consumption when pulpmill residuals are included (Hall et al., 1993). Bioenergy substitutes directly for fossil fuels, in essence storing the C that would have been released in oil, gas or coal reserves.

Whole-tree harvesting —where branches, needles and possibly root and stumps are removed—can increase the amount of the forest available for bioenergy. However, this treatment also reduces the pool of nutrients in the ecosystem, particularly in coniferous forests (Nilsson and Wiklund, 1994; 1995). On sites where nutrient availability limits NPP or NEP, this may result in reduced fixation of C (Olsson 1995). Fertilization can counteract this effect, but the carbon costs may not be favourable.

2.1.12. Establish, Maintain and Manage Reserves

Deforestation currently contributes about 2/7s of all net C emissions (Dixon et al., 1994a; Houghton, et al., 1992). As a consequence, slowing the rate of deforestation will reduce C emissions as well as increasing the capacity of the globe to sequester C. Two questions arise with this strategy: can suitable human intervention increase the stocks held in reserves? And what is the long-term fate of these stocks? Increases in the C stocks in reserves can be accomplished by increasing the NEP of existing reserves or by sustaining or increasing the areal extent of such reserves.

At a stand-level, a set-aside forest reserve may accumulate C while in a growth phase, but as maturity is reached, growth decreases, respiration increases and net C sequestration slows and may even decline as stand-breakup occurs. In many temperate and tropical forests, the forest stand may be maintained indefinitely in a closed canopy, mature state through some form of gap-phase replacement (Botkin 1992; Shugart, 1984; Leemans and Prentice, 1985). In the absence of stand-replacing disturbances, relatively smooth changes in C accumulation make take place in response to changes in nutrient or environmental conditions. In disturbance-regulated forests, such as the boreal forest, however, C stocks may rise and fall as large C pulses are associated with disturbance events, regeneration and the regrowth following it. As discussed previously, controlling these swings—particularly in isolated reserves—may be difficult.

The overriding point is this: C can retained by avoiding deforestation *only if* the C pools in the reserves are explicitly managed. Management of C pools may or may not be compatible with other desirable objectives for forest reserves such as preservation of biological diversity. Creation of a reserve might also simply displace the competing land users to a different location, with consequent impact on C stocks and sequestration rates. To compute the total impact of the reserve requires netting out these offsite effects.

2.2. Economic Costs

Most analysts agree that it is most useful to describe the costs of C sequestration in terms of real monetary units (e.g., US\$ fixed at a specific date) per unit of C removed from the atmosphere (and perhaps per unit time). However, some confusion surrounds the exact way to measure this quantity, both for the numerator and for the denominator. Furthermore, most of the alternatives discussed above are continuous, in that to sequester more C the alternative can be more intensively applied (e.g., more fertilizer means higher growth rates; more protection means less C lost to fire). The optimal scale of the alternative must also be determined.

2.2.1. Measuring Project Costs

The economist's concept of cost for a specific activity refers to the value of the resources used if applied to their next best use. Where markets exist, market prices provide a useful guidepost of costs, so, for example, local wage rates measure the cost of labour used in a project and local machinery costs measure the costs of mechanized activities. In cases where markets are not well developed—either as a matter of government policy or as a matter of weak property or other institutions necessary for markets—the analyst must estimate the shadow prices of resource inputs in order to quantify the costs of C sequestration.

Ideally the total cost of a project to sequester C should include both the direct costs of the project and the opportunity costs of resources used. A fail-safe approach to comprehensive cost analysis is the "with and without" approach. With this procedure the analyst determines the total economic return of an activity without the specific action targeted to sequester C, and then the total economic return of the action with the C sequestering activity included. The cost of C sequestration is then the difference in the returns. For example, the cost of sequestering C by extending forest rotations is the difference in the net present values of the shorter and longer rotations, while accounting for *all* costs and benefits. That is, the costs of a C sequestration project should be the *net* costs. Any incremental benefits associated with the project not related to C sequestration should be subtracted from the project's direct costs. Examples include the recreation or aesthetic benefits of natural reserves, or the value of timber produced when management practices are intensified.

Because most natural forest management options extend across a long time period, the choice of discount rate will generally be quite consequential in determining the cost of C sequestration. Ideally the discount rate should reflect society's preferences for consumption in one period compared with the next. One measure, albeit imperfect, is the real cost of long-term government debt. In fast-growing developing countries, the discount rate is logically higher than it is for more slowly growing developed countries. In the latter, discount rates of 2-4% are generally justifiable, while rates as high as 10% might be appropriate for the former. Because of the uncertainty surrounding the choice of discount rate—and its consequences for the analysis—good analytical practice always computes the present value of project costs for a range of values.

2.2.2. Measuring C Sequestration

Most studies measure the amount of C sequestered as an average annual figure, and one of two approaches can be followed. In some cases, it will be possible to estimate the actual amount of C fixed each year (as, for example, being proportional to a conventional timber yield table). Alternatively, just as in the "with and without" analysis of economic costs, the amount of C fixed as the difference between stock of C without the activity and the stock with the activity divided by the number of years the project is in place.

These approaches assume that the present value of the damage of a unit of C (the shadow price for C) is the same regardless of when it is emitted. This assumption is unlikely to be true. Abatement, adaptation and mitigation technologies will change over time. Damages may be nonlinear in the total amount of CO_2 in the atmosphere. All of these considerations suggest that, just as the economic costs of a C sequestration project must be discounted to determine their present value, the annual C emissions should be discounted (Richards, this volume; Richards and Stokes, 1995). Because of the complexities and ambiguities of this procedure, however, most analysis measure the costs of C sequestration as the present value of all costs (or their annualized equivalent) divided by the average annual amount of C sequestered over the project's life. Because this procedure does not account for the temporal pattern of C releases, we suggest that the discounted sum of C flows (using the same discount rate as for costs) should also be reported.

Finally, the extant literature focuses on *ex ante* forecasts of the amount of C which would be sequestered under various assumptions. As C sequestration policies are put in place, it will become necessary to measure the *ex post* quantities of C that actually are being sequestered. Such measurements will be required to monitor progress towards targets for

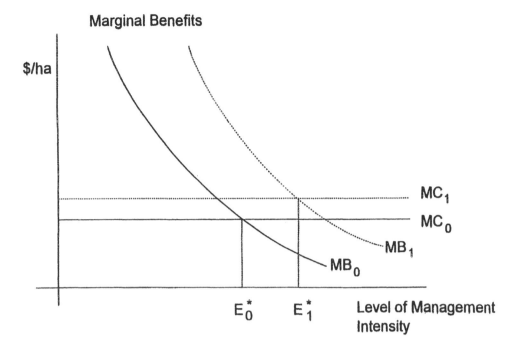

Marginal Benefits

$/ha

MC$_1$

MC$_0$

MB$_1$

MB$_0$

E$_0^*$ E$_1^*$ Level of Management Intensity

Protecting the Forest. Figure 1 shows the standard analysis of the optimal level of forest protection.

net emission reductions, and to verify any multiparty agreements about C offsets or "joint implementation projects". The level of precision needed for such measurements in natural forests currently exceeds that which ecosystem scientists can produce at a reasonable cost. We must either develop quick, inexpensive methods or be satisfied with indirect estimates from modeling procedures.

2.2.3. Optimal Project Scale

Because the biophysically feasible management alternatives discussed above are ordinary forest management practices, considerable past analysis has centred on their optimal application. In many cases, these analytical methods can be extended in a straight-forward way to incorporate the benefits of sequestering C and the cost of emitting it. To fix ideas, it is helpful to imagine that the C sequestration policy is implemented via a tax on C emissions and a subsidy for C sequestration, both levied on a per unit basis. The twelve alternatives discussed above fall into six general categories of economic concern.

Without considering the value of C sequestration, the optimal level occurs at P*0 where the marginal cost of another unit of protection (perhaps measured as total forest area which is managed under a protection program) just equals the marginal benefits of avoided losses (e.g. the value of the timber and recreational values protected). Now consider the case where releasing C to the atmosphere carries a cost. Protection against fires, insects and diseases produces additional benefits equal to the upward shift from MV0 to MV1. The optimal level of protection increases to P*1. Operationally this means that roads would be maintained further into the extensive margin of the forest, more people would be held on alert status during fire season, and so on. Finally, note that the marginal costs of protection might also rise in response both to a changed climate and to efforts to store more C in the forests themselves. While the optimal level of protection is likely to increase, it will still probably not be economically efficient to exclude devastating fires or pest events from all forested areas, even if were physically possible and ecologically desirable to do so.

Fire monitoring and management in natural forest systems is the most efficient near-term to conserve and sequester C in forest systems. Perhaps this point is best illustrated in boreal forest systems, where frequent occurrence of wildfire results in globally significant pulses of CO_2 to the atmosphere (Dixon and Krankina 1992). Within the Russian Federation, forest fire monitoring systems (e.g., aircraft, satellites) in East or West Siberia or the Far East have not proven sufficiently efficient to support a fire management system. C emissions from wildfire, including direct and post-fire release are estimated to be approximately 0.2 Pg C annually. Improvement in forest fire monitoring and management systems and its expansion to all forest land in Russia can potentially reduce the are burned annually by 20% and thus conserve 0.05 Pg C. Dixon and Krankina (1993) estimate that this C can be conserved in forest systems for much less than $1.00 per hectare.

Rotation Length and the Optimal Amount of Growing Stock. Policies designed to sequester C in the forests themselves will logically increase the optimal forest rotation (van Kooten et al., 1995; Plantinga and Birdsey, 1995, Hoen and Solberg, 1994) and the optimal amount of growing stock in uneven-aged management systems. The logic of this is captured in the first-order condition for the optimal rotation in the presence of a C eternality (van Kooten, et al. 1995). In comparison with the usual Faustmann case, the presence of the C eternality creates an additional benefit of holding the stand longer because of the value of the C sequestered. At the same time, the C externally increases the costs of harvesting the stand because of the C emitted both from the ecosystem and from the industrial activities involved with processing the harvested timber. Both factors work to lengthen rotation ages, and the impact may be quite large (van Kooten et al., 1995). By similar logic, the optimal level of growing stock in a selection forest would increase in the presence of a C externally.

Stocking Control. Thinning performs three economic functions—harvesting trees that would otherwise succumb to suppression, accelerating the timing of cash flows from the stand, and increasing the piece size of the remaining trees. These benefits are purchased at the cost of a possible reduction in final harvest volumes. While the solution to the optimal thinning problem is quite complex, the presence of C externalities should generally favour repeated light thinnings, with some anticipated loss in final harvest volumes. The repeated light thinnings would harvest trees whose death and decay would contribute directly to C emissions. As long as the thinning were used for fossil fuel substitution or products with a longer life than that of the dead trees left to decay in the forest, their harvest would reduce or delay total C emissions. Thinnings heavy enough to reduce final harvest volume should lead to higher piece sizes, and a larger fraction of the final harvest going into products with long useful lives. The tradeoff between the annual C reduction services of the unthinned stand and the higher amounts of C fixed when stand densities are lower must be computed for each specific case. Thinnings to waste should clearly be avoided (Hoen and Solberg, 1994), suggesting wider initial stockings than in situations where precommercial thinning is practiced.

Changes in Management Intensity. A number of management practices, including fertilization, and water control, are conceptually similar. The general problem of the

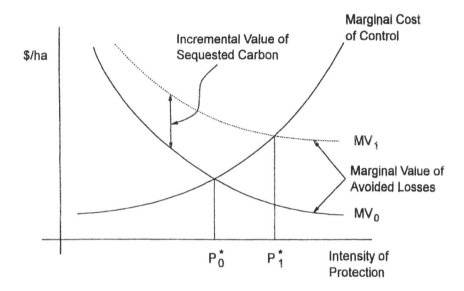

optimal level of management intensity is well understood (see, e.g., Hyde, 1980), and Figure 2 shows the general situation.

As with Figure 1, the subscripts 0 refer to the choice of management intensity in the absence of the C externally. The optimal level occurs where the cost of another unit of management intensity MC0 (e.g., the cost of a tonne of N fertilizer) just equals the marginal benefit of that unit of (e.g., the additional timber at the end of the rotation produced as a result of fertilization multiplied by the value of the timber). The C eternality will increase the marginal benefit of the management intensification, but may also increase the cost to the extent that C is released by the management intensification (e.g., in the production of fertilizers). The net effect is probably an increase in management intensity, although some specific circumstances might lead to the opposite result.

Utilization of Salvage and Logging Residues. The limit of utilization is defined by the smallest size and lowest quality of timber that just has enough value to "pay its way out of the woods." C sequestration policies will raise the value of small, low-quality logs either for fossil fuel offsets or as a store of C in the product sector. If such policies were implemented as a C tax on emissions and subsidy to sequestration, forest managers would utilize smaller stems rather than leave them as logging debris.

Land-use. The margins of land use occur where the value of one use (e.g. for agriculture) just equals the value for another (e.g., forestry). C sequestration policies will increase the economic value of land-uses which store more C over those which store less. Hence the intensive margin should shift inward, with more land dedicated to forestry and less to agriculture and other uses. At the same time, the benefits of managing remote sites will increase, both through the value of products stored in the economic system via harvesting and the increased storage made possible by forest management. As a consequence, the extensive margin of utilization will shift outward. Masera (this volume) provides an interesting example of these effects in Mexico.

Deforestation is, in effect, an outward shift of the intensive margin of land-use, where a more intensive use such as agriculture displaces the less intensive forest use. In some cases, policies such as agricultural subsidies drive these changes. Policies targeted on sequestering C will work to offset these margin shifts, and help halt deforestation. In these circumstances it is logical also to examine how policies outside the forest sector are influencing such margin shifts.

3. DISCUSSION AND CONCLUSIONS

"Natural forests" comprise those lands currently occupied by closed forests or being regenerated to the same or similar species as removed from the site. The vast majority of the world's forest fall within this definition (perhaps over 95%). They range from intensively managed natural forests of Central Europe and Scandinavia to the wild forests of the Russia and Canada and deep jungles of the tropics. Many indigenous people call these forests home. Much of the world's biological diversity is harboured within them.

There are numerous techniques for enhancing the capacity of these forests to sequester C. While the literature provides useful analysis of some of these techniques (e.g. changes in rotation length, fertilization, stocking control, forest reserves), we have identified a somewhat longer list of possibilities which merit attention.

These are ordinary techniques of forest management already used to achieve other societal objectives. Viewed in this context, CO_2 sequestration is but one of the multiple outputs of the forest. The C impact and cost of each of these techniques depends heavily on local conditions, perhaps even more so than is the case with afforestation and plantation management options because natural forest management begins by dealing with an existing growing stock. The literature suggests a range of costs for C sequestration in natural forests, from cheap to dear. Forest sector "joint implementation" projects, mostly established or sponsored by U.S. or European electric utilities, propose to conserve or sequester several Tg of C at initial project costs of less than $5.00 per Mg C (Dixon et al., 1993). Analysis of these projects reveals that the long-term costs of CO_2 sequestration in natural forests are generally low or negative (e.g., yield net benefits).

Despite the fact that the biophysical responses will differ widely as one moves among the various tropical, temperate and boreal forests, a few general principles seem clear:

- Accumulating higher levels of growing stock is apt to increase the biophysical risk of holding stands managed expressly with C sequestration in mind,
- Because of poor access and their extensive nature, natural forest are already subject to a host of biotic and abiotic stresses, and climatic change will to add to these;

- Increasing atmospheric concentrations of CO_2 and climatic change will affect the amount of C stored in these forests;
- While it is possible in many cases to estimate *ex ante* the amount of C that will be sequestered under various management strategies, it will be extremely difficult to measure and verify these quantities in the field.

It is conceptually straightforward to determine the incremental cost of adding C sequestration as an objective of managing natural forests. Similarly, the literature contains many reports of attempts to compute the present value of these incremental costs of C sequestered on an average annual basis. However, the temporal pattern of C sequestration in natural forests will generally be complex, so all of the problems of "C discounting" will occur. As is the case with plantations, correct estimates of C sequestration will include that held in the economic system, both as fossil-fuel offsets and as long-lived products. While the economic concepts for measuring the net costs of C sequestration programs are clear, application of these concepts to specific cases may face daunting empirical obstacles related to the quality and availability of data, and the high cost of studies needed to estimate the value of such elusive benefits as the option value of preservation.

Weak physical and institutional infrastructure will impede implementing C sequestration policies in many natural forests. In some places, traditional use of these forests by aboriginal people raises serious ethical questions about storing the offal from industrial society there. In other places, lack of physical access proscribes management intervention. Three general recommendations for policy attention seem appropriate:

- attend to how policies from other spheres of societal concern influence the capacity of natural forests to sequester C (e.g. agricultural subsidies encourage deforestation);
- look for opportunities to combine the solution of several problems into a single policy (e.g. use municipal waste water to fertilize forests);
- remember that the techniques for using natural forests to sequester C are quite traditional forest management practices. As a consequence, quite a lot is known about policy implementation, even if it is difficult.

Even in the absence of the problems associated with climatic change, the problems of poverty, weak political institutions, and expanding populations challenge our policies for the management of natural forests throughout the world. The new requirement for sequestering C and responding to altered climatic regimes in forests add to this burden. Given the relative complexity of forest-sector C offset projects, a number of biogeochemical, institutional and socioeconomic, monitoring and regulatory issues merit further analysis before the potential of this greenhouse gas stabilization approach can be fully understood.

LITERATURE CITED

Apps, M.J. and Kurz, W.A. (1993), The Role of Canadian Forests in the Global C Balance. *In* M. Kannien (Ed), *C Balance on World's Forested Ecosystems: Towards a Global Assessment*, Proc. Intergov. Panel on Climate Change Workshop, Joensuu, Finland, 11-15 May 1992, Publications of the Academy of Finland, Helsinki, 14-28.

Auclair, A.N.D. and T. B. Carter (1993) Forest wildfires as a recent source of CO_2 at nothern latitudes. *Canadian Journal of Forest Research* **23**:1528-1536.

Beier, C., P. Gundersen, K. Hansen, and L. Rasmussen (1995), Experimental manipulation of water and nutrient input to a Norway spruce plantation at Klosterhede, Denmark. II Effects on tree growth and nutrition. *Plant and Soil* 168-169: 613-622.

Berdén, M. (1994), Ion leaching and soil acidification in a forest Haplic Podzol: Effects of N application and clear-cutting. Doctoral Dissertation. Swedish University of Agricultural Sciences, Department of Ecology and Environmental Research, Rep 73. 21 pp. ISSN 0348-422x.

Botkin D.B., J.F. Janak, and J.R. Wallis (1972), Some ecological consequences of a computer model of forest growth. *Journal of Ecology* **60**: 840-873.

Brown, S., C.A. Hall, W. Knabe, J. Raich, M.C. Trexler, and P. Woomer (1993), Tropical forests: their past, present and potential role in the terrestrial C budget. *Water, Air and Soil Pollution* **70**:71-94.

Brown, S., J.R. Gillespie, and A.E. Lugo (1991), Biomass of tropical forests of south and southeast Asia. *Canadian Journal of Forest Research* **21**:111-117.

Burley, J. and J. Stewart (1985), Increasing productivity of multipurpose trees. International Union of Forest Research Organizations, Vienna, Austria.

Campinhos Jr. E. (1991), Plantation of fast-growing species for tropical areas. *Actes du 10e Congrès Forestier Mondial* Vol. **5**, Revue Forestière Française Hors Série No. 511-120.

Cofer III W.R., J.S. Levine, E.L. Winstead, and B.J. Stocks (1991), Chapter 27: Trace gas and particulate emissions from biomass burning in temperate ecosystems. in: Levine J.S. (Ed), *Global biomass burning; atmospheric, climatic, and biospheric implications*, MIT Press, Cambridge Mass., 203-208.

Cole, D.W. (1995), Soil nutrient supply in natural and managed forests. *Plant and Soil* 168-169: 45-53.

Crutzen P.J., and J.G. Goldammer, eds. (1993). *Fire in the environment: The ecological, atmospheric, and climatic importance of vegetation fires*, John Wiley & Sons, New York.

Davey, C.B. (1990), Forest fertilization in the americas. *In*: Nutrition of trees. The Marcus Wallenberg Foundation. Symposia proceedings:6. Lectures given at the 1989 Marcus Wallenberg Symposium in Falun, Sweden, on September 14, 1989. Strålins Tryckeri AB, Grycksbo, Sweden. ISSN 0282-4647. p.p. 33-56.

Dixon R.K., and O.N. Krankina (1993), Forest fires in Russia: C dioxide emission to the atmosphere, *Canadian Journal of Forest Research.* **23**:700-705.

Dixon, R.K, J.K. Winjum, K.J. Andrasko, J.J. Lee and P.E. Schroeder (1994b), Integrated land-use systems: Assessment of promising agroforest and alternative land-use practices to enhance C conservation and sequestration. *Climatic Change* **27**:71-92.

Dixon, R.K. (1995), Silvicultural options to conserve and sequester C in forest systems: Preliminary economic assessment. this volume.

Dixon, R.K., J.K. Winjum and P.E. Schroeder (1993), Conservation and sequestration of C: The potential of forest and agroforest management practices. *Global Environmental Change* **2**:159-173.

Dixon, R.K., K.J. Andrasko, F.G. Sussman, M.A. Lavinson, M.C. Trexler and T.S. Vinson (1993), Forest sector carbon offset projects: Near-term opportunities to mitigate greenhouse gas emissions. *Water, Air and Soil Pollution* **70**:561-578.

Dixon, R.K., O.N. Krankina and K.I. Kobak (1995), Global climate change adaptation: anecdotes from Russian boreal forests. *In* Adaptation to *Global Climate Change* (J. Smith et al.,eds.) Springer-Verlag, Berlin, in press.

Dixon, R.K., S. Brown, R.A. Houghton, A.M. Solomon, M.C. Trexler, and J. Wisniewski (1994a), C pools and flux of global forest systems. *Science* **263**:185-190.

Drake, B.G. (1992) The impact of rising CO_2 on ecosystem production. *Water, Air and Soil Pollution* **64**:25-44.

Eriksson, H. and U.T. Joahnsson (1993), Yields of Norway spruce (*Picea abies(l) Karst.*) in two subsequent rotations in southwestern *Sweden Plant and Soil* **154**:239-247.

Fife, D.N. and E.K.S. Nambiar (1995), Effect of N on growth and water relations of radiata pine families *Plant and Soil* **168-169**:279-285.

Galinski W., J. Witowski (1995), Chapter 15: The C pulse resulting from forest dieback related to insect outbreaks: Case study of a forest district in the Sudety Mountains (southwest Poland). in: Apps M.J. and Price D.T., eds. *Forest Ecosystems, Forest Management and the Global C Cycle* NATO ASI series, Springer-Verlag, Heidleberg, in press.

Hall, D.O., F. Rosillo-Calle, R.H. Williams, and J. Woods (1993), Biomass for Energy: Supply Prospects. In: B.J. Johansson, H. Kelly, A.K.N. Reddy & R.H. Williams eds., Renewables for fuels and electricity. Island Press, Washington D.C.

Harmon, Franklin J.F., F.J. Swanson, P. Sollins, S.V. Gregory, J.D. Lattin, N.H. Anderson, S.P. Cline, N.G. Aumen, J.R. Sedell, G.W. Lienkaemper, K. Cromack Jr., K.W. Cummins (1986), Ecology of coarse woody debris in temperate forest ecosystems. *Advances in Ecological Research* **15**: 133-299.

Harmon, M.E., W.K. Ferrell and J.F. Franklin (1990), Effects on C storage of conversion of old-growth forests to young forests. *Science* **247**:699-702.

Hendrickson, O.Q. (1990), How does forestry influence atmospheric C? *Forestry Chronicle* **66**:469-472.

Henry, C.L., D.W. Cole, T.M. Hinckley, and R.B. Harrison (1994), The use of municipal and pulp and paper sludges to increase production in forestry. IUFRO Centennial Conference, Berlin, 1992

Houghton J.T., B.A. Callendar, and S.K.Varney (1992), Climate Change 1992. The Supplementary Report to the IPPCC Scientific Assessment. Cambridge University Press, Cambridge and New York.

Ingestad, T. and G. Agren (1995), Plant nutrition and growth: Basical principles. *Plant and Soil* **168-169**: 15-20.

Jain, R.K., K. Paliwal, R.K. Dixon, and D.H. Gjerstad (1989), Improving productivity of multipurpose trees growing on substandard soils. *Journal of Forestry* **87**:38-40.

Johnsson, D.W. (1995), Effects of elevated CO_2 and N on nutrient uptake in ponderosa pine seedlings. *Plant and Soil* **168-169**: 535-545.

Kauppi, P., and M. Posch (1988), Chapter 5: A case study of the effects of CO_2-induced climatic warming on forest growth and the forest sector: A. Productivity reactions of northern boreal forests. in: M.L. Parry, T.R. Carter, and N.T. Konijn, eds. The Impact of Climatic Variations on Agriculture. Kluwer Academic Publishers. p.p. 183-195.

Kauppi, P.E., E. Tomppo, and A. Ferm (1995), C and N storage in living trees within Finland since 1950s. *Plant and Soil* **168-169**: 633-638.

King, G.A. (1993), Conceptual approaches for incorporating climate change into the development of forest management options for sequestering C. *Climate Research* **3**:61-78.

Krankina, O.N. and R.K. Dixon (1992), Forest management in Russia: challenges and opportunities in the era of perestroika. *Journal of Forestry* **90**:29-34.

Krankina, O.N. and R.K. Dixon (1994), Forest management options to conserve and sequester terrestrial C in the Russian Federation. *World Resource Review* **6**:88-101.

Kurz WA, M.J. Apps, S.J. Beukema, and T. Lekstrum (1995), 20th century C budget of Canadian forests. *Tellus* **47B**: 170-177.

Leemans R. and I.C. Prentice (1989), FORSKA: A general forest succession model. *Meddelanden* **2**: 1-45. Växtbiologiska institutionen, Upsalla, Sweden.

Levine J.S., ed., (1991), Global biomass burning; atmospheric, climatic, and biospheric implications, MIT Press, Cambridge Mass.

Marland G and S Marland (1992), Should we store C in trees?, *Water, Air and Soil Pollution* **64**: 181–195.

Matthews R. (1995), Chapter 19: The influence of C budget methodology on assessments of the impacts of forest management on the C balance. in: Apps M.J. and Price D.T., eds., *Forest Ecosystems, Forest Management and the Global C Cycle* NATO ASI series, Springer-Verlag, Heidleberg, in press.

Matthews R., G.-J. Nabuurs, V. Alexeyev, R. Birdsey, A.Fischlin, J.P. Maclaren, J.P. Marland, and D.T. Price (1995), Chapter 24: Evaluating effects of alternative forest management regimes on the role of forests and forest products in the C cycle *In*: Apps M.J. and D.T. Price eds., *Forest Ecosystems, Forest Management and the Global C Cycle* NATO ASI series, Springer-Verlag, Heidleberg, in press.

Nambiar, E.K.S. and R. Sands (1993), Competition for water and nutrients in forests. *Canadian Journal for Forestry Research* **23**:1955-1968.

Nilsson, L.O. and K. Wiklund (1992), Influence of nutrient and water stress on Norway spruce production in south Sweden—the role of air pollutants. *Plant and Soil*, 147: 251-265.

Nilsson, L.O. and K. Wiklund (1994), N accumulation in a Norway spruce stand following ammonium sulphate application, fertigation, irrigation, drought and nitogen-free-fertilisation. *Plant and Soil* **164**:221-229.

Nilsson, L.O. and K. Wiklund (1995), Nutrient balance and P, K, Ca, Mg, S and B accumulation in a Norway spruce stand following ammonium sulphate application, fertigation, irrigation, drought and N-free-fertilisation. *Plant and Soil* **168-169**: 437-446.

Nilsson, L.O. (1995), Manipulation of conventional forest management practices to increase forest growth - results from the Swedish Skogaby project. *Forest Ecology and Management* (in press).

Olsson, B. (1995), Soil and vegetation changes after clear-felling coniferous forests: effects of varying removal of logging residues. Doctoral dissertation. Department of Ecology and Environmental Research, Swedish University of Agricultural Sciences. Report No. 80. ISSN 0348-422x. 25 pp.

Sampson, R.N., L.L. Wright, J.K. Winjum, J.D. Kinsman, J. Benneman, B. Kursten, and J.M.O. Scurlock (1993), Biomass management and energy. *Water, Air and Soil Pollution* **70**:139-162.

Schlamadinger B.,and G. Marland (1995), Chapter 18: C implications of forest management strategies. In: Apps M.J. and Price D.T., eds. *Forest Ecosystems, Forest Management and the Global C Cycle* NATO ASI series, Springer-Verlag, Heidleberg, in press.

Sharma, N.P. ed., (1992), Managing the World's Forests. Kendall Hunt Publishing, Dubuque, IA, USA, 605p.

Schroeder, P.E. (1991), Can intensive management increase carbon storage in forests? *Environmental Management* **15**:475-481.

Shugart H.H. (1984), *A theory of forest dynamics: the ecological implications of forest succession models*. Springer-Verlag, New York.

Smith, J.B. and D.A. Tirpak (1989), The potential effects of global climate change on the United States. EPA-230-05-89-050, U.S. EPA, Washington, DC, USA.

Smith, T.M., H.H. Shugart, G.B. Bonan and J.B. Smith (1991), Modeling the potential response of vegetation to global climate change. Advances in Ecological Research 22:93-116.

Stocks B.J. (1991), Chapter 26: The extent and impact of forest fires in northern circumpolar countries. In Levine J.S. (Ed), Global biomass burning; atmospheric, climatic, and biospheric implications, MIT Press, Cambridge Mass., 197-202.

Stocks, B.J. (1991), The extent and impact of forest fires in northern circumpolar countries. *In* J.S. Levine, ed., *Global Biomass Burning: atmospheric, climatic and biospheric implications*,. The MIT Press, Cambridge, MA, USA.

Tamm, C.O. (1991), N in terrestrial ecosystems. Questions of productivity, vegetational changes, and ecosystem stability. Ecological studies 81. Springer-Verlag Berlin New York London. ISSN 3-540-51807x. 116 pp.

Van Wagner C.E. (1978), Age-class distribution and the forest fire cycle. *Canadian Journal of Forest Research* **8**: 2220-227.

Victor, D.G. and J.E. Salt (1995), Keeping the climate treaty relevant. *Nature* **373**:280-282.

Volney W.J.A. (1995), Chapter 7: Climate change and management of insect defoliators in boreal forest ecosystems. in: M.J. Apps and D.T. Price eds. *Forest Ecosystems, Forest Management and the Global C Cycle* NATO ASI series, Springer-Verlag, Heidleberg, in press.

Weetman, G.F., H.H. Krause, E. Koller, and J.-M. Villeau (1987), "Interprovincial forest fertilization trials, 5 and 10 year results". *Forestry Chronicle* **63**:184-192.

Winjum, J.K., R.K. Dixon, and P.E. Schroeder (1993), Forest management and carbon storage: an analysis of 12 key nations. *Water, Air and Soil Pollution* **70**:239-257.

Zoltai, S.C. and P.J. Martikainen (1985), Chapter 4: Estimated extent of forest peatlands and their role in the global C cycle. in: M.J. Apps and D.T. Price eds. *Forest Ecosystems, Forest Management and the Global C Cycle* NATO ASI series, Springer-Verlag, Heidleberg, in press.

Critical Reviews in Environmental Science and Technology, 27(Special): S47–S64 (1997)

CONSIDERATION OF COUNTRY AND FORESTRY/ LAND-USE CHARACTERISTICS IN CHOOSING FORESTRY INSTRUMENTS TO ACHIEVE CLIMATE MITIGATION GOALS

KENNETH R. RICHARDS,[1] RALPH ALIG,[2] JOHN D. KINSMAN,[3] MATTI PALO,[4] and BRENT SOHNGEN[5]

[1]Pacific Northwest Laboratory, Washington, DC; [2]U.S. Department of Agriculture, Forest Service, Corvallis, Oregon; [3]Edison Electric Institute, Washington, DC; [4]Finnish Forest Research Institute, Finland [5]Yale University, New Haven, Connecticut

ABSTRACT: To implement effective carbon sequestration policies policymakers must analyze key characteristics of the country (geographic, institutional, economic, and infrastructural factors) and forestry and land-use practices (the degree of risk associated with investment, the relative difficulty in measuring sequestration, and the degree of discretion allowed). Without careful analysis of this type, policies may have unintended negative effects.

KEY WORDS: carbon sequestration, policy instruments, land use, forestry.

1. INTRODUCTION

Greenhouse gas mitigation efforts can include not only activities that reduce or avoid emissions of these gases, but also activities that sequester carbon, particularly in trees and other biomass. Table 1 shows many of the forestry practices that can increase carbon storage, maintain existing carbon storage, or reduce energy carbon emissions. These range from tree planting on non-forested land (afforestation) and previously-forested land (refor-estation), to reducing deforestation by forest preservation, to silvicultural options to in-crease forest growth. Trees can be used to reduce fossil fuel combustion, by serving as fuel, or by lessening the need for fossil fuel use (e.g., shade trees). Many authors have described in detail these and other practices to manage forest carbon (Binkley et al., this volume; Parks et al., this volume; Dixon et al., 1993; Kinsman and Trexler, 1993). Some of these practices occur autonomously under certain circumstances for economic or social reasons, while incentives, disincentives or education are needed to implement others.

The technical potential for forest carbon management is substantial. Dixon et al. (1993) estimated that over 2 billion hectares of deforested or degraded land area are technically suitable worldwide for expansion of forest or other improved land management techniques using tree cover. During the 1980s, on the average, 3.2 Pg C accumulated in the atmosphere annually, according to the Intergovernmental Panel on Climate Change (IPCC, 1994). Atmospheric carbon dioxide would be still higher but for a terrestrial carbon sink of about 2 Pg C per year (IPCC, 1994). Numerous studies indicate that humans can substantially enhance the terrestrial carbon sink and reduce the terrestrial carbon sink. Sampson et al. (1993) estimated that, by the anticipated time of doubled atmospheric CO_2 (probably between the year 2050 and 2100), managing the world's vegetation could turn the terrestrial biosphere from a source of carbon (likely range: 4.2 Pg C/yr source to 1.3 Pg C/yr sink) to a carbon sink (likely range: 0.1 Pg/yr source to 3.0 Pg/yr sink). In addition, although not coupled to the previous estimates, biomass energy was projected to be able to displace 0.5 3.8 Pg C/yr of fossil fuel emissions in a doubled CO_2 world (Sampson et al, 1993).

Individual countries may choose among practices shown in Table 1 and others in their plans to manage greenhouse gases. To establish and implement such plans, they will need to consider and develop domestic policies. The primary objective of this paper is to identify and discuss factors that influence national-level forestry policy choices. These include types of policy instruments, characteristics of land-use and forestry practices, and characteristics of countries.

TABLE 1
Land-use and Forestry Practices to Manage Carbon

Practice	Increase C Storage	Maintain C Storage/Avoid C Emissions	Reduce Energy C Emissions
Afforestation	x		
Agroforestry	x	x	x
Breeding/genetics	x	x	
Biomass for energy			x
Disease control		x	
Drainage	x		
Fertilization	x		
Fire control		x	
Herbivore control		x	
Improved regeneration	x		
Increased forest products	x	x	x
Insect control		x	
Irrigation	x		
Longer rotation	x	x	
Preservation (reserves)		x	
Recycling		x	x
Reduced impact logging		x	
Reforestation	x		
Salvage dead biomass		x	
Shade trees			x
Shelterbelts		x	x
Soil management	x	x	
Stocking control			
thinnings	x		
enrichment plantings	x		

TABLE 2
Domestic Policy Mechanisms To Enhance Carbon
Sequestration in Forest Ecosystems

Direct Control	Production on government land
	Government production on leased land
	Input regulation
	Output regulation
Indirect Control	Economic incentives:
	Taxes
	Subsidies
	Contracts
	Carbon markets
	Institutional incentives:
	Private property rights
	Market reforms
	Community-based forestry
	Education and extension
	Research and development
	Volunteerism and encouragement

2. DOMESTIC POLICY TOOLS

Countries can adopt many different policy tools to store more carbon within forests. For many different reasons, such as the type of government or the type of forestry practices available in an area, there may be no universally accepted form of policy tool that can be applied. In this section, we discuss some of these different options and consider some issues involved with choosing policies.

The policy tools we consider involve either direct or indirect government control (Baumol and Oates, 1988), as shown in Table 2. With direct controls, the government either produces carbon itself or regulates carbon production. With indirect controls, government utilizes economic tools to induce private landowners to increase the stocks of carbon on their land and/or sets up programs that will encourage additional carbon storage through better management or better harvest scheduling.

2.1. Direct Control

The government can directly control carbon stocks either by planting additional trees on its own or leased land or by forcing private landowners to follow certain guidelines. By controlling how individual landowners operate, these policy proposals will require well-established property rights and enforcement mechanisms. Well-designed policy initiatives will incorporate elements related to establishing the programs as well as maintaining them.

(1) Production on government land: Governments that are large landowners could enhance carbon storage by planting trees or implementing other carbon seques-

tration practices on their own land, particularly land that may at present be utilized for another purpose.

(2) Government production on leased land: The government may lease land from private landowners to augment carbon sinks on those acres. Lease agreements could stipulate that the government rents the land to grow trees for a given amount of time.

(3) Input regulation: Governments can regulate directly the inputs into forest management, such as mandatory replanting after harvest, minimum harvest ages or sizes, and best management practices.

(4) Output regulation: Government may directly regulate outputs from forests, for example, by limiting the area of forestland that can be cut or by forcing forestland owners to sequester a certain amount of carbon per year, through cutting and replanting.

2.2. Indirect Control

Indirect controls include economic incentives and institutional controls. The former are principally financial; the latter involve support to programs.

2.2.1. Market-Based Incentives

Market-based incentives allow the free market system to respond to a governmental policy to capture and store more carbon in the forest. Economists generally prefer market-based incentives to direct regulation because they provide more discretion to private forest managers than direct controls, particularly regulation.

(1) Tax and subsidy: Landowners may be taxed for the carbon they release by harvesting their forests. More recent studies have incorporated subsidies into the mechanism to recognize the carbon that is sequestered by growing forests (Binkley and van Kooten, 1994; Hoen and Solberger, this volume). With combined taxes and subsidies, landowners will hold timber for longer periods of time.

(2) Contracts: Two types of contracts may be employed by government. First, the government may contract with an existing forest owner not to harvest a fast-growing stand; and second, the government may contract with someone who currently holds land in another use to convert that land to timber.

(3) Marketable permits: The government may place an implicit value on carbon stored in trees by setting the overall pool of carbon it wishes to store. Individual landowners then must bargain among themselves to determine harvest and replanting levels.

2.2.2. Institutional Incentives

Government-sponsored programs may encourage individuals to manage forests prudently for both economic gain as well as additional carbon storage. They may be most appropriate in countries where free market economies are not well developed.

(1) Establish private property rights: A strong link has been established between sustainably managing natural resources and establishing private property rights (Palmer and Synnott, 1989). Many other direct or indirect mechanisms would require first establishing private property rights.

(2) Market reforms: Basic market reforms may enhance management of timber and thereby increase carbon storage. Vincent and Binkley (1992) point out the importance of ensuring complete information in timber markets. For instance, in order for people to invest in some sort of sustainable management activity, prices must be well known and established competitively, financial markets must be available, and landowners must be assured access to markets when they wish to harvest.

(3) Community-based forestry: Rather than either central government or private ownership of land, this policy option emphasizes local government ownership of timberland. In many developing countries, local communities often have rights to use the surrounding land according to custom and tradition. This may mean harvesting for fuelwood or other products derived from forests. Because these communities have a stake in sustainable land management and future productivity, their practices may encourage sustainable land management activities, which in turn will encourage carbon sequestration. Examples of this are seen in India (Arnold, 1992) and in Thailand (Boonkird et al., 1989; Dixon et al., 1994).

(4) Extension and education: Through extension services, governments provide information and education to landowners on land management practices that will enhance carbon sequestration. Such activities may include teaching people how to select and plant trees for maximum survival, how to obtain optimal stocking conditions, and how to harvest with minimal impacts.

(5) Research and development: The government may research new techniques for sequestering carbon. Examples include adopting new tree species or genetically improved tree species for a particular region; developing new agroforestry techniques; or developing herbicide, pesticide and fertilization techniques, or other procedures that can be used to enhance productivity.

(6) Volunteerism and encouragement: Some industries pursue carbon sequestration options voluntarily, mainly in response to proposed rewards or recognition from government agencies. Dixon et al. (1993) and Kinsman and Trexler (1993) discuss carbon offset projects, ranging from establishing agroforestry programs in developing countries to planting trees on private lands with private funds in the United States.

2.3. Choosing The Right Tool

No one policy tool should be applied globally; countries will have to make many choices when it comes to increasing carbon storage. Two major factors will help a country determine which policy tool(s) it should choose. First, policy tools will depend on general forestry practices within a country, often relying on the technical abilities of local forestry professionals. Second, the choice of a particular policy tool depends on specific characteristics of that country, such as the land tenure history and the natural type of vegetation. For example, one would not expect a country with little or no history of private property rights to set up a taxation and subsidy scheme for carbon output from land where the government theoretically owns the land anyhow. Although we recognize that the choice of sequestration practices is related to the country characteristics, we consider the two issues separately in this paper.

3. KEY CHARACTERISTICS OF CARBON SEQUESTRATION AND LAND-USE PRACTICES

As countries develop domestic mitigation strategies, carbon sequestration practices will likely be treated as part of a larger domestic land-use plan. In some cases, carbon sequestration practices will be undertaken purely for the sake of expanding the carbon sinks. But more commonly, carbon sequestration will be an important byproduct of other land-use goals such as timber production, soil preservation, and environmental protection. The policies that will be most useful in expanding carbon sinks must be consistent with the larger land-use goals, modifying current management practices to reflect another dimension in an already multipurpose land-use policy.

The goal of a domestic carbon sequestration program is to select the policy instruments that will bring about expansion of the nation's carbon sink at the lowest cost. The program can promote the carbon sink expansion by modifying government controlled activities and influencing private land-use decisions. The instruments listed in Table 2 can bring about both of these ends, but the choice of instruments depends upon the nature of the program. For example, where a sequestration program is subject to significant ambiguity in terms of goals and direction, the program may be better served by choosing instruments under which the government retains control over day-to-day decisions. That way, adjustments are internal to the government's hierarchy rather than requiring renegotiations between the government and the private sector participants. The choice of instruments may also be constrained by legal considerations. In the United States, for example, there are some types of activities that cannot be compelled by regulatory fiat (Richards, this volume).

Three general categories of characteristics will influence the matching of instruments with forestry practices: the riskiness of the sequestration practice, the difficulty of measuring the sequestration effects of the practice, and the importance of discretion in implementing the practice (Table 3).

3.1. Riskiness of Investment

The carbon sequestration practices listed in Table 1 vary substantially with respect to the risk they present to landowners. Four aspects to the risk presented by carbon sequestration

TABLE 3
Key Characteristics of Carbon Sequestration Practices

Riskiness of Investment	Timing of carbon uptake
	Capital intensity
	Market risk
	Natural risk
Difficulty of Measurement	Need for/ease of establishing a baseline
	Cost of measuring on-site carbon
	Potential leakage
	Potential for innovation in monitoring/measurement
Importance of Discretion	Variability of application
	Potential for innovation in practice

practices are particularly important: timing of the carbon uptake, capital intensity, market risk, and natural risk.

The first aspect relates to the timing of the carbon uptake that results from the practice. For example, preserving a forest that is under imminent threat of clearing provides an immediate benefit — emissions that would have taken place in the near term are avoided. In contrast, the carbon uptake associated with afforestation can spread over several decades or even a century (Richards et al., 1993). If a government adopts a policy instrument that rewards the capture of carbon or avoidance of carbon release, the forest preservation project will provide more immediate, and therefore less risky, returns. The rewards for the tree planting activities will be delayed for many years. The landowner may worry that while he is waiting for the payments the program could be dismantled and the payments discontinued.

The capital intensity of a sequestration practice is also a significant factor in the overall risk of the investment. Landowners are apt to choose to invest in sequestration activities that minimize their exposure to losses from changes in the direction of the program. For example, reduced impact logging, lengthening rotations, and fire control involve relatively little investment or spread the investment over a long period of time. Furthermore, the cost of reversing these practices is low relative to the cost of reversing practices such as afforestation on agricultural land.

Because many of the forestry practices listed in Table 1 are related to the production of forestry goods, market risk is also a key determinant of the overall riskiness of the carbon sequestration investment. Landowners may be willing to extend their rotation lengths, if they are confident in short-term projections of market prices for timber. Promoting large-scale afforestation is more difficult if landowners are uncertain about long-term timber prices and demand. Similarly, the value of large biomass energy production depends very much on the prevailing price of energy, a factor that has varied widely over the past 20 years.

Natural losses also affect the riskiness of sequestration investments (Moulton and Kelly, this volume). Many of the practices listed in Table 1 — fire, insect, and disease control — are specifically designed to reduce the risks of natural losses. In contrast, other practices increase the potential for large natural losses. For example, while the cost of extending the rotation length of a forest is not large, this practice puts the entire resource at risk of sudden natural loss. Not only would the incremental investment (delayed harvesting) be lost in case of fire, but the preexisting resource would be lost as well.

3.2. Difficulty of Measurement

Sequestration practices differ with respect to how difficult it is to measure the effects of the practice as it is applied to a defined area. Perhaps the greatest difficulty in measuring the effects of forestry practices is establishing the reference case or baseline to which performance is compared. In many cases, construction of the baseline involves substantial speculation about the biological and social factors that would have determined the carbon sequestration levels in the absence of implementing the carbon sequestration practice. For example, development of a reference case for a forest preservation project involves demonstrating that the subject land area was in fact threatened with deforestation as well as speculating about the rate at which the deforestation would have occurred.

Measurement difficulties also arise when it is costly to measure on-site carbon after the forestry practice has been adopted, and when there are no acceptable input proxy measures for the amount of on-site carbon. Some forestry and land-use carbon models provide estimates of the effects of certain sequestration practices, but these tend to be limited in geographic range and the scope of forestry practices that they cover. The costs of measuring on-site carbon are apt to increase as the density of the carbon change decreases. In this sense, carbon sequestration is like a nonpoint source pollutant where the cost of monitoring the pollutant source is high because the pollutant is very dispersed.

Another barrier to measurement arises when the implementation of a forestry practice on one site affects the level of carbon sequestration on another site. For example, a sequestration program to increase the amount of forest plantations in a country may lead to a decrease in the amount of carbon stocks in preexisting forests (Alig et al., this volume). In that case it may be difficult to assess the true impact of the carbon sink program on net emissions.

Finally, the potential for innovation in the measurement of effects may also have a dynamic influence on the choice of policy tools. For example, domestic policy may rely on education and government initiatives in the early stages of a carbon sequestration program, and move to more incentive-based tools as the capacity to measure and monitor outputs improves.

3.3. Importance of Discretion

Carbon sequestration practices differ with respect to the importance of allowing the practitioner discretion in implementation and day-to-day management. Where forest practices are well established, where there is little variability in how the practice is implemented, and where it is simple to identify where and when a given practice should be applied, providing discretion to the landowner or forester may not be important. However, many forestry practices are not so clearly defined. Most of the practices listed in Table 1 require significant technical expertise, site-specific knowledge and discretion in application. Further, there may be a number of different sequestration practices that could be applied to any given piece of land.

The market-based policy tools listed in Table 2 provide incentives to adopt the best existing practices for carbon sequestration. They also encourage innovation of new technologies and practices (Richards, this volume). The incentive to innovate may be more important for some carbon sequestration practices than others. For example, there may be little need for innovation in how forest preservation is implemented. In contrast, agroforestry is evolving and dynamic, both in terms of the practices themselves and the scope of their adoption. Understanding the potential for innovation in the carbon sequestration practices is critical to matching the practices with the appropriate policy tool.

3.4. Matching Instruments to Forestry Practices

The carbon sequestration practices listed in Table 1 differ significantly with respect to the key characteristics discussed above. This indicates a need to carefully match the instruments employed in the implementation of a domestic sequestration program with the types of land-use practices that are meant to be encouraged. Landowners will generally try to avoid the practices they perceive as particularly risky. If the domestic policy is intended to encourage those riskier practices, then the sequestration program may need to either have the government undertake those practices or employ the instruments that compel involuntary adoption of the practices by private parties.

Some of the instruments listed in Table 2, such as taxes and marketable allowances, require direct measurement of, or a reasonable proxy for, the carbon sequestration that is actually accomplished by each private party. These instruments may not be suitable for practices where measurement is costly or very inaccurate. In other instances, however, on-site carbon sequestration can be monitored and estimated through a combination of field sampling and satellite imagery, for example, in the case of afforestation (Winrock, 1995). Establishment of the baseline can be relatively simple for afforestation, since the reference case is often considered to be static over time — i.e., in the absence of the afforestation practice the land would remain in cropland with virtually no accumulation of carbon.

Finally, some of the instruments provide substantial discretion to landowners — holding them accountable only for results, the amount of carbon actually sequestered. These instruments allow landowners to optimize their practices to local conditions and provide incentives to undertake that extra effort; for example, the care with which a forest plantation is established and maintained can have a significant effect on the amount of carbon that is sequestered. These instruments are well suited to practices where local control of the practice is key to optimizing the productivity.

4. KEY CHARACTERISTICS OF COUNTRIES

Choosing relevant policy instruments involves careful consideration of the policy tools available (Section 2), current carbon sequestration and land-use practices (Section 3) and

TABLE 4
Relevant Country Characteristics

Agricultural policy	Biodiversity
Capital situation	Competing uses for land
Culture	Currency stability
Economic policy	Energy system and policy
Forest policy	Forestry know-how
Geographic location	Institutions
Land degradation	Land ownership
Legal infrastructure	Market systems
Physical infrastructure	Political stability
Population	Road access to forests
Structure of existing forests	Timber production
Wealth and income	

characteristics of the particular country (this section). Country characteristics are critically important; a policy that will be effective in one country is not possible or not workable in another. Table 4 lists relevant country characteristics, which are then applied in case study discussions of the Philippines and Finland.

4.1. Case Study: The Philippines

The Philippine Islands, located in Southeast Asia, have had an average annual population growth of 2.6 percent during the last decade, the fourth highest among the 17 tropical Asian countries. The physical infrastructure (transportation, telecommunications, and energy supply) is hindered by the need to connect many islands, mountainous terrain, and frequent volcanic eruptions and earthquakes.

The Philippines' political instability has meant deterioration in the legal infrastructure, which in turn has had implications for unclear land tenure conditions and protection of land ownership rights. Under such conditions, institutional and market developments, wealth accumulation, and income generation have been low. GNP per capita is currently about $800.

During the 1980s the average annual pace of deforestation in the Philippines was 3.3 percent, the third highest among the 17 Asian tropical countries (FAO, 1993). Forestry faced significant market and government failures, including corruption. Under such conditions, public forests have in practice become open access. Uncertain tenure in the uplands has often effectively hindered agricultural intensification. With a stagnating economy, an unstable political situation, low income levels, a low level of industrialization, and increasing population pressure, deforestation has been accelerating (Horne and Palo, 1995).

Compared to other developing countries, the Philippines has a high level of forestry education and research. The effective implementation of this know-how has, however, been problematic under the policy failure of public forestry administrations. Also, timeliness, validity, and reliability in monitoring forest resources and deforestation have been serious problems. This deficiency is substantial, particularly with regard to carbon accounting.

The government is the dominant forest owner in the Philippians. However, a recent Forestry Master Plan supports changing land ownership and tenure rights from the government to local people.

These characteristics of the Philippines help us understand some of the institutional barriers that will affect the choice of forestry practices to manage carbon. Activities such as afforestation, fertilization, and drainage probably are not relevant to the Philippines because of the continued quest for expanded agricultural production through a larger land base and because of technical barriers. Other practices, however, such as rehabilitation of degraded forest lands, may be more helpful for the country, as well as agroforestry, reforestation planting, anti-erosion plantations, and fire control. The Philippines may also consider expanding forest preservation for biodiversity and watershed protection, as well as production of wood biomass for energy.

Government regulation may not be effective in the Philippines because there are so few private owners. Cost-sharing could instead play a role in promoting agroforestry, reforestation, and farm- and community-based forestry. One of the most important tools, however, is government-sponsored planning of forestry-related activities. In a country where land markets are not well defined, and the legal infrastructure is not well established, both bottom-up and top-down planning can be beneficial. This includes public support of research and development, training, extension, and other educational services.

Enhancing market systems plays perhaps the most important role in several areas. Broader economic questions relating to development exist there, and the policy options that may enhance the ability of forests to sequester carbon in this country sound strikingly similar. Many of the issues that could promote more sound forestry practices in a country where soil erosion, land degradation, and harmful agricultural practices are abundant could also play a role in sequestering carbon from the atmosphere.

4.2. Case Study: Finland

Finland, located between the 60th and 70th degrees of Northern latitude, is among the most sparsely populated countries in Europe. Currently, its GNP is estimated at $22,000 per capita, 11th highest in the world. Both the physical and legal infrastructures are well developed.

Finland, in the boreal coniferous zone, has 75 percent of its land in forests. Silviculture is simplified by the limited number of commercially important tree species; logging conditions are also comparatively easy. Buying and logging, however, are complicated by the pattern of forest ownership. About 80 percent of timber is supplied by nonindustrial owners, with an average holding of 30 hectares.

Finland was the first country in the world to complete a systematic field sample-based national forest inventory, with pilot results published in 1923. Since then, seven more inventories have helped to develop a timely, valid, and reliable monitoring system that takes into consideration biodiversity and carbon accounting.

Long-range national forestry planning methods have been developed beginning in the 1940s, and in the 1960s the Finnish Forest Research Institute began to develop a comprehensive system of forest statistics. The comprehensive and reliable forest sector statistics along with high-level forest research and education enabled intensive national level forestry planning. This planning activity was motivated by the indicated threat that sustainability of the future cut was at stake in a country whose economy and particularly foreign trade depend heavily on forest products.

Various multiple-use issues, forest conservation, biodiversity, climate change, and other environmental issues have appeared on the agenda of national Finnish forest policy. The principal forestry laws are currently under revision with respect to these other environmental concerns.

Other policy areas are in transition. In the economy, Finland has been fighting the most serious recession in its history. Foreign debt has grown, and unemployment is the largest social problem. In agriculture, Finland has been decreasing the level of regulation that enabled it to be self-sufficient in food production. Energy policy, too, is in transition, with consideration of a massive increase in use of various wood fuels.

These salient characteristics of Finland enable us to understand some of the institutional barriers that will affect the choice of forestry practices to manage carbon. Activities such as rehabilitation of degraded forest lands, agroforestry, anti-erosion plantations, and fire control intensification are not relevant to Finland because of socioeconomic and natural circumstances.

However, other practices such as forest preservation and conservation, biomass for energy, increased solid-wood forest products, longer rotation, and field afforestation may be more relevant to Finland. Public cost-sharing in promoting biodiversity, landscaping, and carbon sequestration could be applicable, along with public regulation. In general, cost-sharing and other financial incentives may be favored in comparison to mere regulation when we consider the recent changes in Finnish economic policy. Research and development, education, training, extension, and new models of forestry planning could be publicly supported.

5. THE ROLE OF INTERNATIONAL COOPERATION

The discussion above suggested that the choice of instruments for a domestic sequestration policy will be constrained by both the characteristics of the country itself (Table 4) and the nature of the sequestration practices it seeks to promote (Table 3). Countries with relatively poor data and modeling capacities are not likely to use data-intensive instruments (such as taxes or marketable allowances) to promote those activities that are most difficult to measure. Countries with limited financial resources are unlikely to use mechanisms such as subsidies and contracts that require the government to pay for the production of carbon sinks. Similarly, where markets are poorly understood or developed, marketable allowances are unlikely to be a useful instrument.

A country may, however, modify its choice of instruments if the constraints it faces can be eased by international cooperation (for a discussion of international joint implementation of greenhouse gas mitigation activities, see Nilsson et al., this volume). International cooperation may serve to relieve some of the financial constraints faced by a country. The United States government has included four forestry projects in its first round of the U.S. Initiative on Joint Implementation. These projects have attracted funds from the United States to promote forest preservation, afforestation, and other practices in countries that have lacked the funds to implement the necessary policy tools. Similarly, international cooperation may allow countries to employ some of the more data-intensive instruments and some of the more sophisticated forestry practices by providing technical expertise and data gathering equipment and services. In short, as countries develop domestic sequestration policies, they should consider the resources available through international cooperation.

6. OTHER CONSIDERATIONS: THE REAL WORLD

Implementation of forest policy instruments under real world conditions can sometimes lead to outcomes that differ significantly from those intended. Here, we attempt to draw broad insights from previous examples of direct and indirect policies involving large-scale forest-related activities, in order to guide future policies for forest carbon sequestration. We examine selected cases to illustrate considerations for policymakers addressing forest carbon sequestration strategies, but this is not meant to be an exhaustive treatment.

Below we discuss selected factors that need to be recognized. These factors arise as one considers broader scales, such as linkages of the forest sector with other sectors in the economy. They may also relate to complex interactions between ecological and economic systems, including influences on private landowner behavior, that are not well understood at the time of policy initiation.

6.1. Market Forces

Basic market forces may be distorted by government intervention. Unforeseen links occur because we do not understand every possible outcome of a tax, subsidy, or other policy in advance. These types of market forces may in some cases offset, at least partially, land base and forest biomass changes intended by forest carbon sequestration policies.

Timber markets are dynamic in both natural growing conditions (e.g., Perez-Garcia et al., this volume) and human management. Private owners may increase harvesting in the near term if they perceive future impacts on inventories and longer term supply from reforestation associated with carbon sequestration programs. Owner responses in a broader context are also likely to include some form of adaptation to climate change if it occurs, perhaps leading to significant increases in both terrestrial and economic system carbon storage under some scenarios (Sohngen and Mendelsohn, this volume).

Inter-sector linkages and intra-sector feedbacks involve complex interactions that can diminish the thrust of certain policies. There may be market-driven impacts on forest area when large-scale afforestation is planned (Alig et al., this volume). If large-scale conversion or planting of agricultural land leads to higher agricultural commodity prices, other convertible forest land could be converted to agricultural use. This phenomena will likely increase with the scale of the program, especially those implemented in a relatively short amount of time and in one geographical region, for example, the southeastern region of the United States (Moulton and Richards, 1990). Market responses to afforestation programs also will influence the financial attractiveness of timber investment opportunities on existing timberland. Lower stumpage prices may lead to less timber management intensification and likely to less forest carbon sequestration on existing hectares (Alig et al., this volume). Policy instruments such as taxes can also distort prices and affect market signals. The extent of the distortion depends on the type of intervention, consumer preferences, and supply conditions.

Basic market forces can also lead to increased forest biomass when certain government policies are eliminated or modified for other sectors. In New Zealand, when subsidies for sheep grazing were eliminated on certain lands, results included lower land prices, given that agricultural subsidies had been capitalized into land values. From the forestry perspective, this meant lower opportunity costs, and a number of pastureland hectares could be profitably planted to trees. The forest area base was thus expanded, not directly through forest-related policies, but because of policies aimed at another sector. An example involving government intervention in the northern part of the United States is the case of declining profitability for some dairy operations, including a reduction in milk price supports. The lower financial returns meant that lands were abandoned, in many cases reverting naturally to forests.

6.2. Government Intervention and Institutional Factors

Regulatory uncertainty can add to the riskiness of an investment. As a long-term endeavor, forestry can involve timber rotations long enough that policies may change over the course of a rotation. Landowners may invest based on a certain policy trajectory which the government may change abruptly or unexpectedly before a rotation or investment is brought to fruition.

Evaluation and monitoring of programs are often plagued by uncertainty in measurement, potentially high measurement costs, and long-term monitoring requirements. This may mean that needed mid-course attention or adjustments may not be forthcoming for projects that can span decades and require many years before one can judge the success. Studies indicate that forestry projects often do not receive much investment of time or capital after installment, and that government assistance also is shifted to installing new projects.

Political developments are often unpredictable and new governments can deviate markedly from previous ones in regard to forest-related policies. A prominent recent example is the encouragement of private forest companies, including foreign firms, in the former Soviet Union. Previously, many lands were locked up in reserves. Stability of governments can influence investments in many sectors, including forestry, and needs to be factored into long-range planning, especially for global issues such as mitigation practices for global warming.

6.3. Infrastructure

Forestry research and inventory programs, as part of a country's infrastructure, can enhance analyses supporting policy formulation. Needed research includes better estimates of the global area in agricultural uses on marginal lands that could be converted to forests and the possible impacts on market prices and spillover effects from implementation of afforestation on such areas.

An example is the Finnish Forestry 2000 program that had the primary objective of timber production. Outcomes diverged from the original plans because improved forest inventory data part way into the program revealed more available timber than was previously estimated. This led to a larger allowable cut to be pursued, and a reduction in the necessary cost-sharing that was planned.

6.4. Integrated Policies

Historically, forest policies have not been given high priority in many developing countries, where food and housing policies have dominated. Myers (1995) asserts that non-forestry agencies often decide what forms of government investment will predominate, sometimes to the detriment of forests, and that basic forest policy is seldom formulated by foresters. Furthermore, concerns over endangered species, biodiversity, and other forest-related services or goods to date have often overshadowed global warming concerns. As a result, policy analysts are not well acquainted with and are less attentive to the unique considerations of forest carbon sequestration when formulating comprehensive policies.

Integrating carbon sequestration goals with those of broader forest policies involves emphasizing complementary benefits and examining values of carbon sequestration. An example of a positive spillover effect across sectors in the United States involved an agricultural policy designed to address excess production. One of the largest U.S. tree planting programs, the Soil Bank Program, was designed primarily to divert excess land out of crop production, but led as well to the establishment of approximately one million hectares of trees over about five years. In general, the interaction is probably strongest in most countries for the agricultural and forestry sectors. Forestry has often been affected more by policies designed for other sectors than policies designed specifically for the forest sector.

Policymakers have experienced many challenges when attempting to move from single to multiple objectives in policy implementation. One of the most successful single objective policies was the elimination of wildfire on millions of hectares annually in the United States, a positive and significant development for forest carbon sequestration, but perhaps having some drawbacks for land management (Sampson, et al., 1993). The burned area was reduced by millions of hectares annually, leading to much more forest biomass stock. When multiple objectives are introduced, as in recent forest planning in many countries, the complexities and associated gaps in knowledge or openings to legal challenges can either slow or dramatically alter the policy outcomes and attainment.

7. CONCLUSIONS

Individual countries may choose among a diverse menu of forest and land-use practices to implement plans for managing greenhouse gases, based upon types of policy instruments, characteristics of land-use and forestry practices, and characteristics of countries. Each country must customize its own policies based upon its economic, legal, political and cultural conditions; its forestry practices; and its resources.

Examples of country characteristics that influence land use and forest management include population trends, land-use and agricultural policy, legal infrastructure such as land ownership, physical infrastructure such as roads, and forestry expertise and markets. In general, these are the primary drivers of policy choices by a nation. It would be unreasonable to expect a nation with little or no history of private property rights to set up a taxation and subsidy scheme for influencing carbon, since the government essentially controls the land itself.

Forestry practices vary substantially with respect to the risk they present to landowners. Some policies require a government to instill confidence in landowners that they will receive expected benefits decades later. Investors are reluctant if they anticipate changing government policies or the inability to make long-term commitments.

Ease, certainty and cost are key aspects of measuring and accounting for carbon benefits associated with various forestry activities — difficult measurements, costly measurements, and inaccuracy increase riskiness and reduce the attractiveness of certain practices.

The timeframe over which the practices are to produce benefits is important, too. One must determine the relative importance of maintaining carbon benefits for a long period of time, such as more than 50 years, as opposed to for only a couple of decades. Some question the usefulness of such a short, in some cases, reversible impact.

Tools to implement policy decisions include subsidies, taxes, market development, education, and research. Policy mechanisms can be direct (i.e., the government can undertake the activity itself by managing carbon on its own land) or indirect (i.e., the government can institute policies that affect private landowners through market incentives — such as taxes and subsidies — or through institutional incentives — such as research, education and establishing property rights). A "mixed instrument" approach (e.g., increased storage on government land coupled with subsidies for private landowners and command-and-control replanting) may hold the greatest promise in countries with diverse resources and economic characteristics.

Some countries might be able to overcome internal limitations through international cooperation, such as bilateral or multilateral programs with other nations, or through joint projects enacted with the private sector of other nations. Such international cooperation can improve finances, as well as provide expertise, equipment and data.

Policymakers need to be aware of possible divergences between intentions and outcomes, some resulting from real-world conditions that can involve complex interactions among market-based, cultural, political, and other factors within and across sectors of the economy. Some potential obstacles (e.g., political complications that affect funding stability) are more widespread than others that may be more country-specific (e.g., interactions between sectors of the economy in certain countries).

An externality such as forest carbon inherently involves market imperfections that exist for a number of long-standing reasons in many cases in the real world. Policy instruments should accordingly be chosen prudently, considering complications that can arise. Perceptions about the seriousness of global warming may increase or decrease as new scientific evidence becomes available (Richards, this volume), and this likewise could affect the efficacy of programs that are inherently long-term and that will be subject to many complications over the course of their useful lives.

REFERENCES

Alig, R., D. Adams, B. McCarl, J. Callaway and S. Winnett. (This volume). 'Assessing Effects of Mitigation Strategies for Global Climate Change with an Interterporal Model of the U.S. Forest and Agriculture Sectors.'

Arnold, J.E.M. (1992), 'Production of Forest Products in Agricultural and Common Land Systems: Economic and Policy Issues', in N. Sharma, ed., *Managing the World's Forests, Looking for Balance Between Conservation and Development*, Kendall/Hunt, Dubuque.

Baumol, W.J. and W.E. Oates (1988), *The Theory of Environmental Policy*, Cambridge University Press, Cambridge.

Binkley, C.S., M.J. Apps, R.K. Dixon, P.E. Kauppi, and L.O. Nilsson. (This volume.) 'Sequestering carbon in Natural Forests'

Binkley, C.S. and G.C. van Kooten (1994), 'Integrating Climatic Change and Forests: Economic and Ecologic Assessments', *Climatic Change* 28, 91-110.

Boonkird, S.A., E.C.M. Fernandes and P.K.R. Nair (1989), 'Forest Villages: An Agroforestry Approach to Rehabilitating Forest Land Degraded by Shifting Cultivation in Thailand', in P.K.R. Nair, ed., *Agroforestry Systems in the Tropics*, Kluwer, The Netherlands.

Dixon, R.K., K.J. Andrasko, F.G. Sussman, M.A. Lavinson, M.C. Trexler, and T.S. Vinson (1993), 'Forest Sector Carbon Offset Projects: Near-Term Opportunities to Mitigate Greenhouse Gas Emissions', *Water, Air and Soil Pollution* 70, 561-577.

Dixon, R.K., S. Brown, R.A. Houghton, A.M. Solomon, M.C. Trexler and J. Wisniewski (1994), 'Carbon Pools and Flux of Global Forest Ecosystems', *Science* 263, 185-190.

FAO (Food and Agricultural Organization) (1993), 'Forest Resources Assessment 1990', FAO Forestry Paper 112.

Hoen, H. F. and B. Solberg (This volume). 'CO$_2$-taxing, Timber Rotations, and Market Implications.'

Horne, P. and M. Palo (1995), 'Benefits and Costs of Deforestation: The Case of the Philippines', in O. Sandbukt, ed., *Management of Tropical Forests: Towards an Integrated Perspective*, Centre for Development and Environment, University of Oslo.

IPCC (Intergovernmental Panel on Climate Change) (1994), 'Radiative Forcing of Climate Change: The 1994 Report of the Scientific Assessment Working Group of IPCC', Summary for Policymakers.

Jepma, C.J., S. Nilsson, M. Amano, Y. Bonduki, L. Lonnstedt, J. Sathaye, and T. Wilson. (This volume.) Carbon Sequestration and Sustainable Forest Management: Common Aspects and Assessment Procedures.

Kinsman, J.D. and M.C. Trexler (1993), 'Terrestrial Carbon Management and Electric Utilities', *Water, Air and Soil Pollution* 70, 545-560.

Marland, G. and S. Marland (1992), 'Should We Store Carbon in Trees?', *Water, Air and Soil Pollution* 64, 181-195.

Moulton, R. and John F. Kelly. (This volume). 'The Physical Risks of Refrestation as a Strategy to Offset Global Climate Change.'

Moulton, R., and K. Richards (1990), 'Costs of Sequestering Carbon through Tree Planting and Forest Management in the United States', General Technical Report WO-58, USDA Forest Service, Washington, DC.

Myers, N. (1995), 'The World's Forests: Need for a Policy Appraisal', *Science* 268, 823-824.

Palmer, J. and T.J. Synnott (1992), 'The Management of Natural Forests', in N. Sharma, ed., *Managing the World's Forests, Looking for Balance Between Conservation and Development*, Kendall/Hunt, Dubuque.

Parks, P.J., D.O. Hall, B. Kristrom, O.R. Masera, R.J. Moulton, A.J. Plantinga, J. N. Swisher, and J. K. Winjum. (This volume.) An Economic Approach to Planting Trees for Carbon Storage.

Perez-Garcia, J., L.A. Joyce, A.D. McGuire and C.S. Binkley. (This volume). 'Economic Impact of Climate Change on the Global Forest Sector'.

Plantinga, A. (This volume), 'The Cost of Carbon Sequestration in Forests: A Positive Example',

Richards, K. (This volume), 'Coercion and Enterprise in the Provision of Environmental Public Goods'

Richards, K., R. Moulton and R. Birdsey (1993), 'Costs of Creating Carbon Sinks in the U.S.', in P. Riemer, ed., *Proceedings of the International Energy Agency Carbon Dioxide Disposal Symposium*, Pergamon Press, Oxford.

Sampson, R.N., M. Apps, S. Brown, C.V. Cole, J. Downing, L.S. Heath, D.S. Ojima, T.M. Smith, A.M. Solomon and J. Wisniewski (1993), 'Workshop Summary Statement: Terrestrial Biospheric Carbon Fluxes — Quantification of Sinks and Sources of CO_2', *Water, Air and Soil Pollution* 70, 3-15.

Sohngen, B. and R. Mendelsohn (This volume), 'A Dynamic Model of Forest Carbon Storage in the United States'

Vincent, J.R. and C.S. Binkley (1992), 'Forest-Based Industrialization: A Dynamic Perspective', in N. Sharma, ed., *Managing the World's Forests, Looking for Balance Between Conservation and Development*, Kendall/Hunt, Dubuque.

Winrock (1995), 'A Plan for the Monitoring of Carbon in Forest Plantations', Mimeo, Winrock International Institute for Agricultural Development, Arlington, VA.

Critical Reviews in Environmental Science and Technology, 27(Special): S65–S82 (1997)

CONCEPTUAL ISSUES RELATED TO CARBON SEQUESTRATION: UNCERTAINTY AND TIME

G. C. van KOOTEN,[1] A. GRAINGER,[2] E. LEY,[3] G. MARLAND,[4] and B. SOLBERG[5]

[1]Department of Agricultural Economics and Faculty of Forestry, University of British Columbia, Vancouver, Canada. [2]School of Geography, University of Leeds, Leeds, U.K. [3]Resources for the Future, Washington DC, U.S.A. [4]Oakridge National Laboratory, Oak Ridge TN, U.S.A. [5]European Forest Institute, Joensuu, Finland.

ABSTRACT: Global climate change is about uncertainty related to ecological and economic processes, and political responses. It is about fairness and income distribution among nations, both now and in the future. It is a dynamic problem that involves national carbon transition functions, damage functions and discount rates. These issues form the basis of the current paper, which examines them from a conceptual point of view.

KEY WORDS: Optimal control of carbon sequestration; forest economics; nonmarket values; political and distributional aspects of climate change.

1. INTRODUCTION

The scientific basis for anticipated global climate change is twofold: (1) observations that human activities are increasing atmospheric concentrations of greenhouse gases (GHGs), principally CO_2, and some (perhaps inconclusive) evidence that global temperatures have recently been increasing; and (2) knowledge of atmospheric chemistry and the interactions among the atmosphere, oceans and terrestrial systems, a knowledge that is embodied in global circulation models (GCMs). The global community is now contemplating strategies to mitigate climate change, and these strategies include changes in the management of forest and agricultural lands. However, there remains a great deal of uncertainty about the magnitude and distribution, and even the nature, of climate changes that might occur and about the damages these will inflict on human and natural systems. Confronted with this uncertainty, some believe that we should not invest in mitigation strategies until it is clear that their cost will be less than the damages they would avert. Others argue on the basis of the "precautionary principle" that we should act now to limit damages that are unknown, but could be large or even catastrophic.[1] Still others take a middle ground and advocate a "no regrets" strategy, one which would implement measures that have very low cost and/ or provide other societal benefits, such as cleaner air.

Any policy approach to global climate change needs to address difficult questions, of which two are particularly important. First, what are the risks of climate change and what are the risks of trying to mitigate climate change? Second, how much are we as a society and as individuals willing to spend on aversion or mitigation, or both? The answers to these

1064-3389/97/$.50

questions vary greatly among countries, and among individuals depending on their personal interests or stakes in global warming, their aversion to risk, their demands on other resources, and their attitudes toward other communities and future generations. Despite the great uncertainty and diversity of interests, we can elaborate some general principles that acknowledge our varied interests and guide our responses.

Early estimates of the costs and benefits of different mitigation strategies suggest that measures like reforestation and forest protection may meet both of the "no-regrets" criteria, while providing a significant net reduction in emissions of CO_2. Forest management measures can both sequester carbon and offset carbon (C) emissions by displacing fossil fuels, and Solberg (this volume) notes that forests can provide an additional valuable role by providing flexibility for dealing with an uncertain future. Since forest and other land policies seem to offer some of the low-cost options for mitigating climate change (e.g., Parks and Hardie 1995), it is appropriate to explore some of these issues in the context of forest managment. In fact, many of the papers in this volume address the specific issue of how much of net greenhouse gas emissions can be avoided through changes in land managment, and at what cost, but they do not raise the more general issues that are addressed in this paper.

Whether or not the international community takes collective action to mitigate global climate change depends on its assessment of the comparative feasibility of alternative mitigation strategies, as determined by their economic viability and the likely effectiveness of implementation. Undertaking such an assessment presents great challenges.

The purpose of the current paper is to examine conceptual issues related to global warming in a dynamic context. After all, global climate change does not occur instantaneously; rather, it is a process that takes place over a period of decades or more, with the possibility of a double-CO_2 climate perhaps already unavoidable. It also involves an interaction between human and natural systems, and this "coevolutionary development" (Norgaard 1984) is often not very well understood. To improve our understanding even a little bit, at least in the context of terrestrial carbon sequestration, we proceed by first considering how a country's carbon emissions, and its reliance on fossil fuels, change as the country develops. Then, in section 3, we discuss three types of uncertainty related to climate change and mitigation activities, concluding that there is a role for forest-sector mitigation strategies as a result of some forms of uncertainty. Distributional issues are examined in section 4, but not in the traditional inter-generational sense. Rather, the focus is on inter-nation equity from the standpoint of development. Finally, we focus our attention on the use of discounting, even though that subject has been discussed by Richards (this volume). The conclusions ensue.

2. CARBON TRANSITION AND DAMAGE FUNCTIONS: TWO USEFUL CONCEPTS?

2.1. Carbon Transition Functions

Countries contribute differently to GHG emissions because disparities in development have influenced patterns of energy use, and this, in turn, affects the speed at which the world can respond to the challenge of climate change. Developed countries historically have released larger quantities of CO_2 to the atmosphere because their economies have relied more on fossil fuels for longer, and in many cases economic development involved clearing of forest land for agricultural, urban and industrial uses. Developed countries gained developmental benefits from these historic emissions of CO_2, ignoring their impact on global climate—the externality imposed on all.

Developing countries, on the other hand, have contributed less to anthropogenic increases in atmospheric CO_2, and many have expanses of in-tact forest in addition to lower current dependence on fossil fuels. These countries have not yet experienced the development benefits that accompanied the developed countries' large historic discharge of CO_2. If developing countries follow the same development path as have the developed countries, there is large potential for accelerated discharge of CO_2 from both fossil fuels and land use changes. Annual per capita emissions of CO_2 from energy systems in the U.S. amount to 5.3 tonnes of C, whereas those in China are 0.6 tonnes. Global emissions would increase several-fold if all developing countries were to achieve, say, the per-capita emissions rate of Western Europe (2.3 tonnes of C per year). However, while these countries may not experience such high levels of per capita emissions in the future as currently in Europe or North America, they can be expected to increase several fold.

In order to aid in one's understanding of the emissions dilemma, we can postulate the existence of national "carbon transition functions" (Figure 1). These relate carbon emissions to the stage of a country's economic development. Countries (A, B, C, ...) face disparate carbon transition curves because these reflect differences in natural resource endowments, cultural and other national attributes, technological possibilities, and population size. However, carbon transition functions exhibit certain general characteristics. Overall energy use rises at least until a country has passed through its demographic transition and population has levelled off. Fuelwood is usually the main energy source in the early stages of development, but, later, fossil fuel use increases and per capita energy use rises. Fossil fuels eventually give way to renewable and nuclear sources of energy. Initially, carbon emissions tend to increase with per capita income, but they level off once income rises to the point where citizens demand more of public goods associated with reduced carbon emissions.

Carbon transition functions are illustrated in Figure 1. Net carbon emissions from energy use rise as population increases and the economy grows, and most energy is obtained from fuelwood (Stage I as illustrated for country A). Emissions peak during the next period (Stage II) when fossil fuels are dominant (Stage II), and then decline as the use of renewable and nuclear sources of energy increases (Stage III). Support for the general transition hypothesis of carbon transition functions comes from long-term trends in carbon emissions from developed countries, where emissions are tending to level off in the 1990s after rising steeply for several decades following World War II. The history of energy use in the U.S. shows an initial dependence upon fuelwood, followed by coal, and then oil and natural gas after 1950. Nuclear and renewable energy use also grew from that date, but still constitute only a minor share of all energy use. In theory, carbon emissions from energy generation fall to zero when a country depends solely on renewable energy and the national carbon transition has ended. A great challenge is to envision what this post-transition state looks like and how developing countries might reach Stage III without passing through the high-emissions intermediate states (I and II).

If true, the *C* transition hypothesis has major implications for attempts to mitigate global climate change. For there to be rapid mitigation, all countries should pass through their carbon transitions as soon as possible. Unfortunately, because many countries are still growing in population and economic activity, this could take some time without specific intervention. Carbon emissions in developed countries will level off soon even without intervention to mitigate global climate change, but those in developing countries will take much longer to level off. International intervention might accelerate the carbon transitions in both sets of countries, thereby shifting transition curves to the left, but there will still be a delay before the global curve of net carbon emissions peaks and begins to fall.

Because the world has such inertia with regard to energy use and to a lesser extent energy sources, energy use patterns are unlikely to change very rapidly. This provides justification for instituting a substantial forest-based mitigation programme sooner rather than later, so as to offset the considerable emissions expected in the first half of the next century. There remain doubts about whether countries actually follow carbon transitions as described above. The evidence is weak, based only on the experiences of some of the developed countries, and there are no observations of Stage III in the transition function.

2.2. Damage Functions

Another theoretical concept that is useful, but perhaps unmeasureable is that of the damage function, Although it is essential for determining the economic viability of mitigation. The damage function translates the biophysical impact of climate change on agricultural and timber production, ecosystem viability (e.g., loss or gain in biodiversity), sea level rise and so on into monetary values—the differences over time between the costs and benefits of climate change. Using economic theory, one would suspect that damages will rise with increases in atmospheric concentrations of GHGs, and at an increasing rate.

Global costs of climate change are still very uncertain owing to lack of knowledge of purely physical, chemical and biological phenomena. Thus, forecasts of the change in climate resulting from previous greenhouse gas emissions are of poor quality, because of the limitations of global climate models and lack of knowledge of world atmospheric processes, the link between climate and terrestrial (and ocean) ecosystems, *et cetera*. Further, climate change lags GHG emissions in a complex way that is not fully understood. Attempts to project future emission levels are fraught with uncertainty, which is exacerbated by the influence that mitigation programmes might have on both net emission rates and climate change. Calculating trends in net global costs is made more difficult by the way in which relative costs and benefits are expected to vary geographically with the nature and magnitude of climate change, and yet climate change at the sub-nation level is most difficult to forecast. Nonetheless, a large number of national trends must be combined to derive estimates of global costs.

Economists postulate a damage function in their models of climate change, much as atmospheric scientists and others postulate certain relationships in GCMs. There is a significant difference, however, because the damage function is determined from the interaction between human and natural systems, while the relationships in GCMs are based solely in the natural realm (even though emissions levels are human caused). Therefore, one might urge as great, or perhaps greater, research funding for economic projects related to climate change as is spent on GCM-oriented research.

3. UNCERTAINTY, IRREVERSIBILITY AND TIME

Before investing in climate change mitigation programmes, policy makers will want to have some notion of the potential costs and benefits of global warming, and how alternative mitigation strategies compare with one another. Some of the required information is set forth in this volume, but it is far from complete. Also important for policy makers is whether benefits exceed costs, particularly for their own country and, more generally, for the globe as a whole. In this section, we consider risk and uncertainty with respect to climate change, because it is a particular obstacle to potential mitigation agreements and their subsequent success both locally and internationally.

In evaluating changes in forest and land management for mitigating climate change, three types of uncertainty can be identified:[2] (1) the risk of a changing climate, or not spending enough on mitigation; (2) the risk of spending too much on mitigation; and (3) the risk that projects that are implemented will fail. It is not clear which, if any, of these risks might be greater or lesser than any other. In this section, we discuss these risks in more detail. We also consider the possibility of catastrophe and irreversibility, and the concept of a safe minimum standard of conservation.

3.1. Risks Posed by Spending too Little or too Much on Climate Change Mitigation

It has been customary to treat climate change as a chronic problem and assume that costs and benefits will change continuously over time. However, it is also possible that world climate could soon begin to change much more rapidly than hitherto, thereby becoming an acute rather than a chronic problem. This could happen if, for example, carbon sinks fill up, or there are large carbon pulses into the atmosphere owing to the long delays between when vegetation dies back and subsequent regrowth in areas affected by climate change. Such delays are possible given the long time taken for new tree species to colonise an area and allow regrowth. There could simply be an unanticipated biomass dieback with only minor regrowth as an ecosystem disappears due to climate change. The rate of atmospheric change could accelerate, or a mode change in the climate system could be encountered (e.g., in the circulation dynamics of the oceans). An enhanced greenhouse effect could also simply lead to greater climatic variability, which results in an increase in the frequency of extreme climatic events.

Catastrophes do not have to be only physical to have a major impact on humanity. The overall costs of global climate change will be determined by a combination of its physical, social and economic impacts. Greater climatic variability could well cause more environmental disasters (e.g., floods, droughts, wind storms) that will have such a large impact on world financial markets that a global economic catastrophe could occur even if a physical catastrophe does not. A succession of natural disasters in the United States in recent years has, for example, led to such large insurance claims that the Lloyds insurance market in London has been virtually brought to its knees. Pindyck (1995) distinguishes between this "economic uncertainty"—uncertainty over future costs and benefits of climate change and its abatement—and "ecological uncertainty—uncertainty over the evolution of climate variables and their impact on natural systems.

Some advocate that governments should do more to mitigate global warming at this time to avoid the risk of some unforseen catastrophe. The argument runs as follows: if the possibility exists that, once a threshold level of GHG concentration in the atmosphere is passed, a biophysical and/or financial catastrophe could occur, then it makes sense to put all of one's eggs in the mitigation basket. At the extreme, climate change should be avoided at all cost. But would we really want to stop climate change from occurring at all cost, assuming it is even possible for humans to prevent it from happening? Perhaps it is a moot argument as there are many who suggest that the natural system is sufficiently resilient to prevent catastrophe, and many governments behave as if a catstrophe is unlikely or, at least, that they are unwilling to prevent one at any cost.

Then there are practical realities that need to be taken into consideration. No matter what is done to prevent greenhouse gas emissions today, climate change is unavoidable due to current concentrations of GHGs in the atmosphere, which will take some time to dissipate or remove. There are constraints on how fast each country can reduce its dependency on fossil fuels, and this will lead to even more uncertainty about the relative merits of alternative mitigation strategies. Further, it has been recognized that, owing to organizational constraints, it would take some time for a large international tree planting programme to achieve its target annual planting rate. This will delay the speed at which carbon sequestration can increase and affect calculations of that programme's economic feasibility (Grainger 1991).

Then there is the risk that too much will be spent on mitigation. Maddison (this volume) argues that risks of the second type are greater than risks of the first type—we can do great damage to human systems by unnecessarily commiting current resources to mitigation. By impoverishing human systems (see below), they impact adversely on natural systems with the resulting damage to the natural systems being greater than that resulting from unrestrained global warming. Further, GHG mitigation imposes sunk costs on society (e.g., construction of a nuclear power plant or investment in a plantation forest), and these will bias future cost-benefit calculations.

3.2. Option Value and Climate Change

Economists are interested in maximizing the difference between expected benefits and expected costs, appropriately discounted (see section 5 below), in determining optimal climate change mitigation policies (including forest sector policies). The general approach is as follows. All possible mitigation options need to be identified. Then expected net present values are calculated for each subset of all possible options and their orderings (since the order in which calculations are made will matter), with the subset that leads to the highest positive expected net present value to be chosen. If no subset of projects results in a positive net present value, no mitigation strategy should be implemented.

This approach assumes that a decision about mitigation is to be made once-and-for-all; that is, the choice is either to mitigate now or never. However, the decision is not that simple; rather, the decision involves a choice of mitigating today, delaying the mitigation decision to the next period, and so on. Further, in each period, an optimal decision may require that different levels of mitigation (different policies) are chosen. In other words, there may be a benefit to delaying some or all mitigation options. This is true even if the original mitigate-now-or-never decision yielded a positive expected NPV. It may also be true that, if the original decision yielded a negative NPV, a decision to mitigate climate change at a future date yields a positive NPV (Pindyck 1995). Likewise, one might find that temporary emission-reduction policies dominate permanent ones (Kolstad 1994). (Further, as more C is sequestered, there is a feedback to the shadow price of C that makes the original decision, based on the discount rate, wrong, unless the feedback effect is explicitly accounted for.) The point is that any cost-benefit analysis needs to take into account the lost opportunity cost that comes from delay and resolution of uncertainty as part of the cost of any investment in mitigation. This opportunity cost is referred to as option value, or the value of keeping investment options open (Dixit and Pindyck 1994).

3.3. Safe Minimum Standard of Conservation

At the moment there is no evidence that a catastrophic change in the relationship between atmospheric GHGs and the climate is pending. Should it occur, however, it would then be too late to take any significant mitigating action, while the biophysical and/or financial cost may be great. It is important, therefore, to account for this possibility in economic analyses. Yet, one would not want to take action today to prevent an uncertain irreversibility if that action came at too high a cost. Ciriacy-Wantrup (1968) proposed the concept of a safe minimum standard (SMS) of conservation, arguing that, from a safety-first perspective, society would want to prevent irreversibility of any environmental good. However, there is a cost to maintaining a SMS. Thus, a reasonable approach is to maintain a SMS for any environmental good as long as the cost of doing so was not prohibitive. Presumably, prohibitive would be determined through the political process.

Cost-benefit analysis operates at the margin, but is unable to keep track of what is happening at the aggregate. This "murder at the margin" can be corrected for by incorporating the SMS of conservation as a constraint in cost-benefit analysis (Randall and Farmer 1995). Indeed, Randall and Farmer (1995) demonstrate that the same general rule— maximizing net present value subject to a SMS constraint— "is admissable under consequentialist, duty-based and contractarian reasoning" (p.36). In addition, they indicate how one might set the SMS constraint in practice. Although their analysis applies to the possibility of species extinction, the SMS could be set at a particular level of global CO_2 emissions or atmospheric concentration of GHGs.

Delaying the start of substantive action, or acting in only a gradual manner, is encouraged by the perception that climate change is a chronic problem. This is very different from ozone depletion, which was tackled promptly because it was seen as an acute if not catastrophic problem, capable of leading to a major decline in human welfare over a fairly short period of time. In this instance, a SMS appears to have been identified, although ease of technological control was likely a key factor in getting an agreement. What would a SMS of conservation be for climate change, if it is either a chronic or an acute problem? How important is the role of forest sector policies in these cases?

3.4. Risk of Project Failure

Risk of project failure is a real risk. The usual example that one might envision in light of terrestrial carbon sequestration is the risk that disease and/or insects attack a plantation forest established to sequester carbon (Moulton this volume). Such risks are generally knowable and, to some extent, can be partially mitigated against, although, if climate change is actually occurring at the same time, uncertainty increases and previous knowledge about risks may provide no guide as to future risks. Further, mitigating the risk of project failure in these circumstances might require expenditures that are not appropriately accounted for in a cost-benefit analysis of projects to sequester carbon.

But there is another, perhaps more important, risk of project failure and that is failure of government support, which occurs because of budget difficulties (e.g., debts and deficits) or simply lack of political will. We consider several aspects of political failure related to uncertainty.

There exist many conflicting pressures on policy makers (e.g., to balance budgets and reduce debts, to maintain economic growth), so that the choice of mitigation strategies will probably be determined as much by political expediency as by their relative economic feasibilities. Some governments might decide to act now to mitigate the causes of the problem; others might decide to take no action, in the hope that a technological solution will eventually be found. The greater the uncertainty, the more likely it is that individual governments will prevaricate and impede formulation of international policies.

Although implementation of international agreements to mitigate climate change are discussed by Jepma et al. (this volume), there is a particular aspect of implementation that appertains to uncertainty and is briefly examined here. It relates to the desires of politicians and citizens in various countries to reach an agreement that limits a country's powers over its domestic policy and environment. Assignment of regional powers and functions through a country's constitution has been examined previously by Musgrave (1959), Olson (1969), Oates (1972), Breton and Scott (1978), and Breton (1987), among others. While these authors considered assignment of powers from a theoretical and normative perspective, focusing on economic efficiency, Scott and van Kooten (1995) examined the assignment of powers and functions from an empirical and regional perspective, using Canada's current constitutional debate as a backdrop. For purposes of the current discussion, their findings concerning regional support for assignment of functions over the environment in the constitution are instructive since reaching an international accord is similar to constitution making.

Politicians participating in procedures for implementing and enforcing an international climate accord might focus on economic efficiency, seeking an agreement that enhances global welfare. We do not know that they do not attempt to do so. However, it seems obvious that their behaviour can be better described as a process of treaty-making, in which the participants are driven mainly by the attempt to capture or retain such powers and rents that benefit themselves and their civil advisors. Most such benefits will be derived from decisions that distribute global GNP toward citizens in their own countries.

To function on behalf of his/her country in these procedural circumstances, a representative bargainer *must* anticipate or project both the effect of global climate change on his/her country and determine how restrictive or costly implementation of a global accord is likely to be. In addition, attitudes toward risk differ between countries (Chichilnisky and

Heal 1994), usually because of wealth levels, but sometimes just because of idiosyncratic national differences. Even identical countries confronted with the same global climate change impacts will react differently if their attitudes towards risk are not the same. This could explain the divergence of positions at the recent Berlin summit that separated the European and American positions.[3]

Thus, countries can be loosely classified into those that will more than likely benefit from global climate change, those that will likely be net losers, and those who could be either net gainers or net losers—the uncertainty is simply too great to know either way. In addition, there is uncertainty about how countries will be affected, in the long term, by any global accord that they sign. As uncertainty about any of the risk categories identified above increases and countries are not sure into which of the loser-gainer categories they fall, the greater will be the ambiguity and vagueness of a global accord to which they will agree. Ambiguity and vagueness can be found in any of the elements of an accord, including targets, penalties, enforcement, and so on.

It is understandable, therefore, that taking no action at all, or taking no action to significantly cut fossil fuel use, can become a political priority. This makes forest-based mitigation strategies seem attractive, because by offsetting emissions from fossil fuels they allow a country to cut its net carbon emissions while maintaining politically acceptable rates of fossil fuel use. Many governments would find it politically impossible to make heavy cuts in fossil fuel use because of the impact on national economic growth rates, and thereby on budgets. Policies that conserve existing forests or result in tree planting to increase the country's annual carbon sequestration rate enable a country to "buy time", while the magnitude of climate change becomes clearer and the search continues for technological solutions that have less drastic effects on the national economy. Implementing forest-sector policies also enables countries to be seen as doing something about a global problem, while not relinquishing freedom to pursue domestic policies, including ones that might increase CO_2 emissions (e.g., regional subsidies to develop coal or oil deposits). Any additional cost of forest-based mitigation is offset by the huge opportunity costs of making too drastic a cut in fossil fuel use. Finally, while having climate benefits, protecting forest ecosystems, planting trees and investing in silviculture are politically attractive policies because they satisfy other objectives of governments, such as appeasing environmentalists and protecting forest sector jobs.

4. DISTRIBUTIONAL ISSUES

Income distributional issues associated with global climate change are generally couched in terms of intergenerational transfers of environmental and other resources. However, intergenerational equity is not our concern here as it has been addressed by Howarth and Norgaard (1995). Our concern is rather with the distribution of mitigation costs and climate change damages among nations, both today and in the future. We are also not interested in the distribution of costs and damages across individuals.[4] Individuals are subject to risks of the first kind (spending too little on mitigation), directly because changes in climate may affect their welfare and indirectly because climate changes may affect the prices of goods and services they purchase. They are also subject to risks of the second kind (spending too much on mitigation) because it is individuals that will ultimately bear the burden of mitigation costs.

No single country can solve the problem of global climate change by itself, but, as noted earlier, the extent to which all countries can agree on what action to take will depend on whether individual governments see climate change and/or mitigation as advantageous or disadvantageous to their national interests. Several distributional considerations will influence negotiations between countries about how much each will contribute to the costs of international action, including: (1) whether countries perceive themselves to be net beneficiaries from global climate change or not; (2) the relative contributions that different countries have made to the atmospheric concentration of CO_2 through their historic carbon emissions; and (3) the extent to which either reducing fossil fuel use or not exploiting forests will deprive developing countries of their chance to develop. Items 1 and 2 will affect a country's willingness to pay for climate change mitigation. Reducing fossil fuel use and/or not exploiting resources (item 3) likely reduces GDP growth and, hence, a country's ability to pay for future mitigation, adaptation or investment in technological solutions. The first item was discussed previously. The other two items raise major questions of internation equity and will be discussed in more detail.

Some countries have contributed more to global climate change than others. Developed countries have released a lot of carbon into the atmosphere, by clearing their forests and consuming vast quantities of fossil fuels. They have gained major developmental benefits as a result of this, but have ignored the external costs that these emissions would have on others through global warming. Developing countries, on the other hand, have contributed much less to climate change and its external costs. Large areas of forests are still intact,[5] dependence on fossil fuels is still relatively low, and they are still to realise the developmental benefits gained by developed countries. They could, therefore, quite properly object if they are called on to curtail dramatically future carbon emissions or to plant large areas of forests to offset carbon emissions from fossil fuels, since most of these emissions come from the developed world. If developing countries were also asked to set aside large areas of unexploited forests as part of an international mitigation programme, they would deserve considerable compensation for the resulting cut in their rate of economic development (Grainger this volume).

The atmosphere is still an open-access global commons. CO_2 containment efforts imply in one form or another managing this global commons with particular property-rights arrangements. These, in turn, translate into specific wealth redistributions. Burtraw and Toman (1992) review four of the cost-sharing rules that could, *a priori*, serve as focal points in an international CO_2-limitation agreement. These are: (1) ability to pay, (2) polluter pays, (3) equal percentage cuts, and (4) tradeable population-based emission rights. They conclude that these seemingly plausible alternative rules can have very different consequences with respect to burden-sharing. They suggest that the CO_2-containment issue could be usefully linked to other resource and development matters. However, using historic carbon release as a basis for fairness in the setting of future emission limits contentious, because: (i) at the time of development, the potential for climatic damage was still unkown; (ii) developing countries may have actually benefitted from development in richer countries (but some of these benefits may have been misappropriated by corrupt and incompetent governments); and (iii) it is unclear as to whether developing countries will actually be significant losers in the future as potential for climate change and its associated costs and benefits are uncertain.

In 1992, at the Rio summit, the United Nations Framework Convention on Climate Change (FCCC) recognized that the developed nations are responsible for most of the current stock of GHGs in the atmosphere. The FCCC basic principle is that the twenty-four countries that were members of the OECD in 1992 should support climate change activities in developing countries by providing financial support above and beyond any financial assistance they already provide to these countries. Specific commitments concerning efforts to limit greenhouse gas emissions and enhance natural sinks apply to the OECD countries as well as to the twelve "economies in transition" (Central and Eastern Europe and the former Soviet Union). Although the treaty language is vague, it is generally accepted that the OECD and transition countries should at a minimum seek to return by the year 2000 to the greenhouse gas emission levels they had in 1990 (IUCC 1994). In acting on these guidelines, the Global Environmental Facility of the World Bank provides resources to the developing countries for the incremental costs incurred by measures that result in global environmental benefits, including GHG reduction.

With regard to the occurrence of global climate change, it will not treat all nations uniformly—changes in precipitation, humidity, temperature, wind patterns and so on will occur locally, having distinct regional consequences. There might be winners and losers (or perhaps only losers if catastrophic changes occur globally), but, in any case, some nations will lose more than others. Countries also face different risks from global climate change and have different abilities to contribute towards mitigation. These problems are not only spatial, but are temporal: the abilities of countries will vary from one period to the next depending on where they are on their development and, hence, carbon transition paths. This will affect their desire to participate in international agreements and affects the compensation levels that will be needed.

It is often argued that less developed nations are more vulnerable to climate change because of their higher dependence on agriculture and other primary sectors, and their lesser ability to adapt to changes. On the other hand, wealthier nations have, by definition, more that can be lost when facing adverse climate change. While wealthier nations may have no particular obligation to the developing countries on the basis of their past emissions (see above), it is possible to argue that they have an obligation on other grounds. Economists have determined, from their theoretical models, that rich countries have a proclivity to move toward a cleaner steady-state environment, while poorer countries choose a lower level of abatement. However, this dichotomy disappears, on efficiency grounds alone, if the available technology permits the substitution of capital-intensive, lower-emission techniques as pollution accumulates (Toman et al. 1995). But there is a difference between efficient and sustainable outcomes. While efficient use of resources in developing countries is an important condition for sustainability, it may only be a necessary but not sufficient condition. Some amount of rich-country aid may make the difference between an environmentally poor country driving itself to environmental catastrophe or achieving an environmentally sustainable state (Toman et al. 1995, pp.156-57). While these models have not explicitly examined climate change, the connection between a clean environment and global warming should be clear. Thus, helping poor countries achieve a sustainable environment on the basis of local concerns only will provide global benefits.

5. TIME AND DISCOUNTING

Discounting is a source of controversy in many environmental debates, with some advocating the use of a social rate of time preference or, at the extreme, a zero rate of discount. (See Lind (1982) for a review of discounting.) In a review of carbon sequestration studies, Richards and Stokes (1995) found that both approaches—discounting and no discounting—were used to estimate the costs of carbon uptake on a per tonne basis, with the alternative approaches leading to large differences in cost estimates and subsequent confusion for policy makers. One argument is that the timing of carbon uptake should not matter (i.e., no discounting), but that costs of providing C uptake services should be discounted at an appropriate rate, although the rate is often left unspecified. Richards (this volume) has demonstrated that the use of a zero discount rate for timing of C uptake benefits (i.e., the value of the damages averted) can be justified when damages from atmospheric carbon increase at a rate that is equal to the discount rate. In this section, we extend Richards' analysis by further exploring the discount rate issue and optimal investment in carbon sequestration, using the damage function of section 2 and introducing the possibility of nonmarket benefits from tree planting.

5.1. Market Rate of Discount

The market rate of discount (i.e., the rate at which a country can invest capital for growth) is the appropriate one to use in thinking about climate change. Suppose that climate change results in $1.05 of expected damages in the next period. This damage can be avoided by investing $1 today as a mitigation strategy. Suppose, however, that the market rate of return over this period is 10%. Then, by investing $1 today in a market instrument, one can have $1.10 in hand in the next period. Since climate change was not avoided, $1.05 is needed to cover the damage caused by climate change, leaving $0.05 in the second period. Had the $1 available in the first period been spent on mitigation, nothing would have been left, but there would also have been no damage. The strategy that yields the greatest benefit ($0.05) is to suffer the consequences of (or adapt to) climate change. This is the approach taken by Madisson (this volume).

The arguments against this NPV approach are legion. One is that nonmarket values associated with such things as biodiversity are not generally taken into account, although the analysis can be extended (see below) to include compensatory payments for loss of such attributes. The magnitude of the payments could be determined from contingent valuation surveys or through *a priori* agreements among the various parties (e.g., governments, NGOs), as has been advocated by Knetsch (1995). In establishing any such compensation levels, it will be necessary to keep in mind global budget constraints and political realities.

Market rates of discount and inclusion of nonmarket amenities may be one explanation as to why developing countries are less keen on spending monies on mitigation than developed ones. Real discount rates in many developing countries are much higher than those in developed countries. While one would expect arbitrage to equate rates across countries via a flow of capital from developed towards developing countries, transactions costs (e.g., cultural differences) appear to prevent this from occurring on a sufficiently large scale. The result of these interest rate differentials is that governments in developing countries have a predilection for adaptation, while those in developed countries are more prone to favour mitigation.

Further, people in developed countries are better off and, as a result of their higher incomes, demand more public goods, such as clean water and air, biodiversity, scenic amenities and so on. Carbon reduction and uptake strategies, such as investments in non-fossil fuel energy, tree plantations and silviculture, also provide more of these public goods. As income levels in developed countries rise, more environmental amenities will be desired, and countries will rely less on carbon, as illustrated by the C transition functions.

5.2. Shadow Price of Carbon, Damage Function, Nonmarket Values and the Discount Rate: An Optimal Control Model of Sequestration

The forest carbon uptake problem can be represented mathematically as follows:

$$\min_{s(t)} \int_0^\infty \left\{ d[C(t)] + c[s(t)] - b[\varepsilon(t)] - n[s(t)] \right\} e^{-rt} dt \qquad (1)$$

$$\text{subject to} \quad \dot{C}(t) = \varepsilon(t) - s(t) - \gamma C(t) \qquad (2)$$

where $C(t)$ is the stock of atmospheric carbon at any instant, $\dot{C}(t) \equiv dC(t)/dt$ is the rate of change in the stock of atmospheric carbon, $s(t)$ is the amount of carbon sequestered at any instant through growth of forests, $\varepsilon(t)$ is the rate of carbon emissions, r is the discount rate, and γ is the atmospheric dissipation rate for C. The damage function, $d(C)$, increases with increasing atmospheric C ($d' > 0$), likely at an increasing rate ($d'' > 0$). As noted earlier, a major problem with climate change policy is the lack of knowledge about the damage function, and the possibility of catastrophe. The cost of sequestering carbon, $c(s)$, is a function of the amount of C that is sequestered, and is assumed to increase with C uptake at an increasing rate ($c' > 0$, $c'' > 0$). Unlike the damage function, there is some information available about the costs of sequestering carbon (e.g., Parks this volume).

The next to last term in objective function (1) is the benefit to society of emitting carbon, $b(\varepsilon)$; global benefits likely increase with emissions, but at a decreasing rate ($b' > 0$, $b'' < 0$). These are the benefits from consuming the goods and services produced as a result of processes that release C to the atmosphere. Since emissions are treated exogenously, the benefit function is included here only for completeness.

Planting trees and investing in silviculture yield nonmarket benefits (e.g., scenic amenities, habitat for some wildlife species, etc.) that are captured by the last term, $n(s)$. Nonmarket benefits are a function of the amount of carbon that is sequestered in forests, but they are unknown and difficult to measure. It is assumed that nonmarket benefits increase with the amount of carbon that is sequestered in forest ecosystems, but at a decreasing rate ($n' > 0$, $n'' < 0$).

The current-value Hamiltonian associated with the problem represented by (1) and (2) is:

$$H(C, \varepsilon, s, \mu) = d(C) + c(s) - b(\varepsilon) - n(s) + \mu(\varepsilon - s - \gamma C) \qquad (3)$$

where all of the variables are functions of t and μ is the current-value shadow price of a unit of carbon in the atmosphere. The first-order conditions for a minimum are:

$$\frac{\partial H}{\partial s} = 0 \quad \Rightarrow \quad \mu = c'(s) - n'(s) \tag{4}$$

$$-\frac{\partial H}{\partial C} = \dot{\mu} - r\mu \quad \Rightarrow \quad -d'(C) + \gamma\mu = \dot{\mu} - r\mu \tag{5}$$

Condition (4) states that, at any instant, the marginal benefit of sequestering an additional unit of carbon must equal its marginal cost. The marginal benefit is given by the shadow price of removing a unit of carbon from the atmosphere plus the marginal nonmarket benefits of "sinking" that C in trees—for example, the marginal benefits of scenic and wildlife habitat amenities associated with additional forest biomass.

Rearranging (5) gives:

$$\frac{\dot{\mu}}{\mu} - \mu + \frac{d'(C)}{\mu} = r \tag{6}$$

This is the standard result that the last dollar invested (in C sequestration) must earn a (social) return at any instant equal to what could be earned elsewhere, which is given by the discount rate r. The benefit is equal to the rate of change in the shadow price (which takes into account the effect of current C uptake on future atmospheric CO_2 concentration) minus the decay rate of atmospheric C plus the marginal damages averted.

Solving (4) and (5) gives:

$$\dot{s} = \frac{c'(s) - n'(s)}{c''(s) - n''(s)}(r + \gamma) - \frac{d'(C)}{c''(s) - n''(s)} \tag{7}$$

The phase-plane diagram for the differential equations (2) and (7) is provided in Figure 1. Optimal steady state values of atmospheric carbon and sequestration are $C^*(t)$ and $s^*(t)$, respectively; it is a closed-loop solution as C and s are constant, no longer depending on t. An increase in exogenous emissions (ε) will shift the $\dot{C} = 0$ function upwards to the right, so the optimal amount of C to be sequestered increases as does the amount that is permitted to reside in the atmosphere—it is optimal to put up with greater climate change. Likewise, an increase in either the discount rate (r) or the natural rate at which C is removed from the atmosphere (γ), or both, will reduce both C^* and s^*. An increase in damages per unit of C resident in the atmosphere affects the optimal state, shifting it upwards to the left, thereby making it worthwhile to sequester more carbon and reducing atmospheric CO_2.

Suppose damages are increasing over time at a rate $\partial > 0$, so that the damage function in the above equations is replaced by the term $d(C)e^{\partial t}$. In this case, the solution to the optimal control problem is open loop, meaning that it depends crucially on time. In

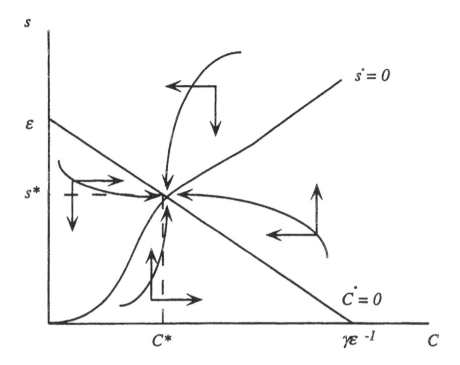

FIGURE 1. Phase-plane diagram for carbon and carbon sequestration.

equation (7), the last term needs to be multiplied by $e^{\partial t}$, so that $\dot{s} = 0$ is continually shifting upwards. Then it is necessary to continually increase C uptake, which is an impossibility.

By including $e^{\partial t}$ as a multiplicative term on $d(C)$ in the objective function, one is tempted to combine this with the discount term e^{-rt} (see Richards this volume). The argument is that, if $\partial = r$, then damages should be discounted at a zero rate, while sequestration costs (and other costs and benefits) are to be discounted at the "normal" rate r. There are a number of problems with this argument. First, ∂ can take on any value; since its value is unknown, any discount rate could be justified, including negative ones. Further, there is the possibility that itself is unstable, and that for some values damages exceed some threshold level. Finally, even if is known exactly, choosing a discount rate based on the difference $r - \partial$ (≥ 0) implies a one-time NPV calculation to determine whether or not to sequester C. As noted above, this ignores the timing of investments and option value.

Space considerations prevent us from considering uncertainty in the forgoing analysis, although the above model is not up to the task because it ignores emissions reductions as a strategy for reducing atmospheric CO_2. However, research by others offers a clue as to what optimal policy might be in the presence of uncertainty about damages and other variables related to climate change. Pindyck (1995) indicates that uncertainty makes it optimal to proceed slowly with action to mitigate global warming. The literature on a safe minimum standard of conservation suggests that, in the absence of any information about the potential for irreversibility, policies that mitigate climate change should be implemented if their cost is not too high. Since the costs of forest sequestration are generally known and efforts to store C in trees provide benefits over and above those related to climate change, it appears that such forest policies can be justified on economic efficiency grounds, apart from arguments about the appropriate rate for discounting benefits (i.e., damages averted). However, this conclusion does not justify implementation of silvicultural investments and reforestation/afforestation projects willy nilly; projects should still be ranked according to a cost-benefit criterion, even though the benefits may simply be in terms of the amount of carbon sequestered and its timing. Even if physical C is discounted to take into account the timing of C uptake, choosing a discount rate different from the market rate constitutes a statement about damages and suggests that the decision maker (or someone else) has some knowledge of the damage function. If so, this knowledge should be taken into account directly, and not via the discount rate.

5. CONCLUSIONS

What does all this mean for forest sequestration policies (efforts)? If one looks at things realistically—i.e., considering risk and uncertainty, the potential for all countries to agree to take effective action, the political will within a single country to actually implement policy, and the time lag between action and results—then we may be left with forest-sector (and other-sector) policies and/or recommendations that are rather limited. What we are left with might include recommendations for more research (including a focus on pilot projects), greater efficiency in the use of wood products (including development of more efficient wood-burning, biomass technologies), and so on. The potential for large-scale plantation forests designed solely for sequestering C may not be as great, and needs to be justified on the basis on other amenities (e.g., scenic amenities, stream flow enhancement, water quality, *etc.*) and the safe minimum standard of conservation.

Nonetheless, forest sector strategies have certain advantages over other strategies for mitigating global climate change. In addition to their nonmarket benefits, which exist also for some other strategies (e.g., reduced burning of fossil fules results in cleaner air), the costs of C uptake policies are generally well documented, moreso than for other policies. Second, forest sector strategies are easily reversible. Third, forest biomass can substitute for fossil fuels, while forest products can replace products made from cement, plastics and so on, which contribute to pollution and global warming. Fourth, increasing the amount of commercial timber growing on sites around the globe will have a positive impact on employment, which is a benefit that politicians cannot ignore.

In conclusion, the value of a varied portfolio for addressing an uncertain future suggests that we make investments now in a variety of options that might be implemented at low cost now or in anticipation for a larger role in the future if our perception of the damage function becomes more ominous. Such a portfolio includes encouraging economic growth in developing countries that is both rapid and employs the best available technology for environmental reasons, as economic growth will enable countries to adapt and the higher income levels lead their citizens to place higher value on the environment. In addition to improving energy efficiency and adopting non-fossil fuel forms of energy, it would include improving carbon storage in forests, the productivity of forest land and the efficiency with which forest products are produced and used (whether for durable goods or for energy). To implement these types of policies will require a high level of international cooperation, perhaps one that is unlikely to be achieved or one that might not be desirable as it could imply a degree of coercion that is not tolerable in free societies.

NOTES

1. Some atmospheric chemists are concerned that anthropogenic activities (and not only those related to GCC) can bring about a chemical imbalance in the atmosphere that is both unanticipated and catastrophic.
2. Risk and uncertainty are often considered to be different, with risk refering to situations where probabilities of occurrence can be based on past events and uncertainty to those cases where the past offers no guidance to the future. In this paper, we often use the terms interchangeably because scientists have some notion about the possibility of occurrence of some components of the climate change picture but not others.
3. Differences in attitudes toward risk are good news for the creation of insurance markets, but one must be cautious about the potential of insurance in this situation because solvency might be an issue, especially if climate change has significant undesirable effects globally.
4. Individual welfare changes are the fundamental microcomponents of any aggregate welfare change. These individual changes always receive explicit or implict weights in any aggregation process (e.g., Ley and Sedjo this volume). That nations sacrifice a sizable portion of their income to achieve internal and external redistributional goals is an indication that the efficiency-distribution tradeoff is not negligible. Therefore, this issue should be addressed, at least at a conceptual level, when alternative courses of action are considered. For convenience and space considerations, we shall continue to refer only to national welfare in what follows.
5. Among developed countries, several still have significant tracts of undeveloped forests (e.g., Russia and Canada), while others have increased forest regions and protected areas (see Wang and van Kooten 1994).

REFERENCES

Chichilnisky, G. and G. Heal (1993). "Global Environmental Risks," *Journal of Economic Perspectives* 7: 65-86.

Breton, A. (1987). "Toward a Theory of Competitive Federalism", *European Journal of Political Economy* 3(1&2): 263-331.

Breton, A. and A. Scott (1978). *The Economic Constitution of Federal States.* Toronto: Univ. of Toronto Press.

Burtraw, D. and M.A. Toman (1992): "Equity and International Agreements for CO2 Containment," *Journal of Energy Engineering* 118(Aug): 122-135.

Ciriacy-Wantrup, S. von (1968). *Resource Conservation: Economics and Policies.* 3rd ed. Berkeley CA: Univ. of California Division of Agricultural Sciences.

Dixit, A. and R.S. Pindyck (1994). *Investment Under Uncertainty.* Princeton, NJ: Princeton Univ. Press.

Grainger A. (1991). "Overcoming Constraints on Assessing Feasible Afforestation and Revegetation Rates to Combat Global Climate Change". In *Proceedings of Technical Workshop to Explore Options for Global Forest Management, Bangkok, April 24-30, 1990,* edited by C. Sargent C. Bangkok/ Washington: Office of the National Environment Board, Thailand and US E.P.A. pp. 196-208.

Howarth, R.B. and R.B. Norgaard (1995). "Intergenerational Choices under Global Environmental Change". Ch. 6 in *The Handbook of Environmental Economics* edited by D.W. Bromley. Cambridge MA: Basil Blackwell Ltd.

IUCC (Information Unit on Climate Change) (1994). *Understanding Climate Change: A Beginner's Guide to the UN Framework Convention.* Geneva: UNEP/WMO.

Knetsch, J. (1995). "Wetlands and Valuation of Non-Pecuniary Values". Paper presented at the Wetlands Valuation Workshop, Simon Fraser Univ., Vancouver, April 10-11.

Kolstad, C.D. (1994). "Learning and Irreversibilities in Environmental Regulation: The Case of Greenhouse Gas Emissions". Working Paper No. 94-001, Dep. of Economics, Univ. of Illinois at Urbana-Champaign. Jan.

Lind, C. (1982). *Discounting for Time and Risk in Energy Policy.* Baltimore: Johns Hopkins Univ. Press.

Manabe, S. and R.J. Stouffer (1993), 'Century-Scale Effects of Increased Atmospheric CO_2 on the Ocean-Atmospher System', *Nature, 364:* 215-218.

Musgrave, R. (1959). *The Theory of Public Finance.* (New York: McGraw-Hill).

Norgaard, R.B. (1984). "Coevolutionary Development Potential", *Land Economics* 60: 160-73.

Oates, W.E. (1972). *Fiscal Federalism.* (New York: Harcourt Brace Jovanovich).

Olson, M., Jr. (1969). "The Principle of "Fiscal Equivalence": The Division of Responsibilities among Different Levels of Government", *American Economic Review* 59: 479-87.

Pindyck, R.S. (1995). "Sunk Costs and Sunk Benefits in Environmental Policy: I. Basic Theory". Dep. of Economics, MIT, Cambridge, MA, mimeograph. February 2. pp43.

Randall, A. and M.C. Farmer (1995). "Benefits, Costs, and the Safe Minimum Standard of Conservation". Ch. 2 in *The Handbook of Environmental Economics* edited by D.W. Bromley. Cambridge MA: Basil Blackwell Ltd.

Richards, K.R. and C. Stokes 1995. "Regional Studies of Carbon Sequestration: A Review and Critique", *Forest Science* In press.

Parks, P.J. and I.W. Hardie (1995). "Least Cost Forest Carbon Reserves: Cost-Effective Subsidies to Convert Marginal Agricultural Land to Forests", *Land Economics* 71: 122-36.

Scott, A. and G.C. van Kooten (1995). "Constitutional Crisis and the Economics of Environment and Resource Development in Western Canada", *Canadian Public Policy* 21: In press.

Toman, M.A., J. Pezzey and J. Krautkraemer (1995). "Neoclassical Economic Growth Theory and `Sustainability' ". Ch. 7 in *The Handbook of Environmental Economics* edited by D.W. Bromley. Cambridge MA: Basil Blackwell Ltd.

Wang, S. and G.C. van Kooten (1994). "Canada's Protected Areas Program: An Assessment in the International Context". Paper prepared for Canadian Forest Service under FRDA contract. Vancouver: FEPA Research Unit, Mar., 31pp.

Critical Reviews in Environmental Science and Technology, 27(Special): S83–S96 (1997)

CARBON SEQUESTRATION AND SUSTAINABLE FOREST MANAGEMENT: COMMON ASPECTS AND ASSESSMENT PROCEDURES

CATRINUS J. JEPMA[1] and STEN NILSSON[2] (Chairs), MASAHIRO AMANO,[3] YAMIL, BONDUKI,[4] LARS LÖNNSTEDT,[5] JAYANT SATHAYE,[6] and TOM WILSON[7]

[1]Department of Economics, University of Groningen, WSN Building P.O. Box 800, 9700 AV Groningen, The Netherlands; [2]Forest Resources Project, International Institute for Applied Systems Analysis, Schlossplatz 1, A-2361 Laxenburg, Austria; [3]Forest Products Research Institute, P.O. Box 16, Tsukuba Norin Kenkyu, Danchi-Nai, Ibaraki, 305, Japan; [4]Venezuelan Case-Study to Address Climate Change, Torre Oeste-Parque Central, Piso 19, Ofic. 19-3, Caracas, Venezuela; [5]Department of Forest Economics, Swedish University of Agricultural Sciences, S-901 83 Umeå, Sweden; [6]Lawrence Berkeley Labs 90/4000, Cyclotron Rd., Berkeley, CA 94720, U.S.A.; [7]Electric Power Research Institute, 3412 Hillview Ave., Palo Alto, CA 94333, U.S.A.

1. INTRODUCTION

Recently, two major developments have moved forestry-related research, and studies dealing with tropical forestry in particular, up on the environmental policy research agenda.

1. The recognition that reforestation and afforestation, as well as combating deforestation, can not only make a contribution to the local socioeconomic and physical conditions and climate, as well as to the preservation of biodiversity, but can also serve the purpose of acting as a carbon store. Thus the forestry option is recognized as belonging to the set of well-accepted greenhouse mitigation options, and therefore also as a potential candidate for joint implementation (JI) projects.

2. The insight that all forests — including tropical forests — should be managed sustainably to preserve them, and that policy measures must be taken and coordinated and monitored internationally to enforce this, especially as long as timber prices do not reflect all social costs involved in forest exploitation. This insight led to the policy target accepted within the International Tropical Timber Organization (ITTO) framework (recently reinforced by a resolution on certification of tropical timber by the European Parliament in June 1995), to allow only international trade in sustainably managed timber as of the year 2000.

1064-3389/97/$.50

2. CARBON SEQUESTRATION FOR SUSTAINABLE FOREST MANAGEMENT ASSESSMENT: COMMON CHARACTERISTICS

Both objectives — to use forestry for carbon sequestration purposes and to manage forests sustainably — stand on their own, because the first relates to the greenhouse problematique and the second essentially to the issue of the global commons. However, there are some interesting characteristics that are common to both developments that have not yet been recognized in policy debates, but should be further explored.

2.1. Time Pressure on Policy Decision Making

First, both developments require conceptual clarification in the near future. Insofar as carbon sequestration through forestry projects is carried out in the next few years either in Framework Convention on Climate Change (FCCC) Annex I countries or through JI pilot projects in non-Annex countries, numerous aspects must be explained before one can agree on a procedure that enables quantification and crediting of carbon sequestration at the level of individual projects in a manner that is acceptable to the various parties involved. On the one hand, the decision made during the Berlin Conference not to start crediting during the JI pilot phase reduces pressure on this exploratory phase; on the other hand, another resolution agreed upon during the Berlin Conference states that it is necessary to come up with 'a comprehensive review of the pilot phase in order to take a final decision on the pilot phase and the progression beyond that, no later than end of the present decade' — namely, the year 2000. The latter decision puts time pressure on the JI pilot phase, and suggests that the preparation of a feasible carbon sequestration crediting system is something that must be explored over the next few years.

A similar time pressure holds for the ITTO target that was adopted for the year 2000, a target and time frame set by the EU and reinforced during UNCED. Given that the year 2000 has been chosen as the target year by which timber trade between RITO members should effectively be restricted to sustainably produced timber, one must come to grips with the highly complicated sustainability concept that is to be applied, as well as with an assessment, monitoring, and possibly certification system, in an extremely short period of time. In an equally limited period of time, namely, the next couple of years, the issues of both crediting through carbon sequestration and certification for internationally traded timber from sustainably managed forests must be effectively solved, if these concepts are to be implemented through policy measures.

2.2. Compatibility with Forest Exploitation

Second, the two developments have similar conceptual problems. For example, if forests are used for carbon sequestration purposes and crediting based thereupon, one issue is to what extent this is compatible with forest *exploitation*. The advantages of additional exploitation are positive economic side effects, better chances that the local population and host country administration will support the project, lower average and marginal sequestration costs, and the possibility of increasing the carbon store if the timber produced remains stored and intact for long periods of time after being harvested. The disadvantages of carbon sequestration cum forest exploitation, however, are that it may become more difficult to keep illegal intruders out once inroads for exploitation have been made, fire risks increase, the chances of overexploitation due to the structure of incentives for the timber exploiter escalate, and control over the carbon stored in the timber harvested diminishes. Moreover, carbon storage cum exploitation may lead to inferior forest quality from a biological and biodiversity points of view, which may explain why biologists and ecologists seem to support the option of cemetery forests.

A similar discussion relates to the sustainable management concept; with regard to this point, however, there is a need to distinguish between forest categories — primary forests, secondary forests, plantations, conversion forests, agroforestry, etc. Clearly, the degree to which sustainability is compatible with forest exploitation differs depending on the forest type: in the case of primary forests one can often defend that any exploitation already conflicts with the sustainability concept, whereas some or even systematic exploitation respecting some clear rules for fallow periods, stumpage size, replanting, felling techniques, etc. would clearly be an acceptable aspect of sustainable management of secondary forests and plantations. So, for both developments one must differentiate the concept depending on the forest type, and carefully determine how quantifying the carbon sequestered/assessing sustainable management is affected as soon as forest exploitation is recognized as an acceptable complementary activity.

2.3. Establishing A Baseline Scenario

Third, the much-discussed baseline problematique — that is, the baseline emission patterns — enters the stage both in the measurement of carbon sequestration and in the sustainable forest management debate. Crediting of carbon sequestration requires a clear baseline concept; otherwise crediting becomes pointless. However, if credits are to be gained through forestry activities *in situ*, it is to the advantage of the organization that is responsible for carrying out the project and that expects to receive the credits to underestimate the reference level to increase the asserted net contribution of the project of the carbon being sequestered, and thereby the credits to be received. This is basically also true in JI projects if credits or other positive incentives to sequester carbon are somehow shared between the investor and the host country. In fact, if the host country government provides some incentives (say, subsidies or tax relief) for organizations to set up these types of projects, there are two sets of incentives to underestimate the reference scenario: the project participant vis-á-vis the investor country government receives more subsidies/tax relief, etc., and the investor country government vis-á-vis the other parties to the FCCC enlarges the amount of *international* credits built up through the project. Given that it is in the various parties' interest to underestimate systematically the reference case, much effort must be made in designing a procedure that is both manageable and objective.

Basically, one could argue that a fully objective assessment of the reference case cannot possibly be made because constructing the reference scenario theoretically involves not only the extrapolation of physical and economic trends (including population increase, food intake per person, GDP growth, etc.), but also guesstimates about policy trends. The latter cannot possibly be carried out without accepting various rather arbitrary assumptions based on available official and unofficial information or information from comparable cases. It, therefore, seems that attention and efforts should primarily be focused, not on the technical issue of how to determine the reference case, but rather on designing a reliable, acceptable, and objective *procedure* based on the professional judgment of internationally recognized experts who determine the reference case for a particular project or project type.

The argument must once again address the need to differentiate the concept of sustainable forest management depending on the socioeconomic context wherein forest management takes place. One example to clarify this relates to timber produced from conversion forests, i.e., from forests whose economic destination has been altered through government policy (e.g., allocated for agricultural use). Suppose, the timber produced by actually clearing the conversion forest to prepare the area for agricultural use is put on the international market. Could it then be certified as sustainably managed if the official conversion plans had already been adopted by the government before the sustainable forest management concept was even introduced into the international policy discussion? If not, the producing country would be penalized for a decision taken before introduction of the sustainability concept. Or, more generally, how should government plans that alter the forest destination, or that change the classification of a particular forest, be addressed in the interpretation of sustainable forest management?

Another example illustrating the need to clarify which baseline scenario should be accepted to establish if and when forests being managed sustainably refers to the need to differentiate between timber-producing countries — or even regions — depending on the overall economic context wherein timber production takes place. The ITTO guidelines on sustainable timber management distinguish between management at the country level and at the management unit level. Concerning sustainability at the country level, obviously the national forest laws and national land use planning are important ingredients for any assessment on sustainability. However, is it fair to use the same stringent screening procedures in an underdeveloped, densely populated, land-locked country and a developed, market economy without population pressures. The answer is most likely no. The next question then is, How should various socioeconomic development patterns be incorporated and what reasonable forest exploitation patterns might be compatible with sustainable forest management at a particular country level? The question boils down to determining a reference scenario which the country and any independent body charged with sustainability assessment can agree upon.

2.4. A Procedure for Monitoring

A final aspect that carbon sequestration and sustainable management assessment have in common is not only the need to take final decisions on concepts at rather short notice (see Section 2.1), but also the need to establish the right procedures to deal with the assessment. Because the conceptual debate cannot possibly exclude various arbitrary and subjective choices made in the final decision-making process, it is imperative that the procedure through which the decisions are made is objective and acceptable to all parties involved. This is true for carbon sequestration assessment as well as for timber certification.

Various discussions on how to set up an international body of surveillance have already taken place. The common denominator seems to be that such a body should consist of independent, internationally recognized teams of experts, possibly supported by local specialists in the various countries, applying uniform procedures under international supervision. Furthermore, there must be a solution for dispute settlement. Field inspections could be carried out by independent private units, and reporting for each management unit could be prepared by the forest user according to a predetermined format.

These basic requirements for a surveillance system are necessary for both timber certification and carbon sequestration assessment. This is one reason for arguing that these tasks should be carried out by one body of surveillance. Another argument is a major part of carbon sequestration forestry projects contains an element of forest exploitation. Wouldn't it be logical to assume that carbon crediting can only be provided under the provision that this exploitation be carried out sustainably? If so, two types of assessment must be performed in the framework of one carbon sequestration project: crediting and certification. This is another reason for authorizing the same surveillance unit to perform the dual task of determining both net carbon uptake and sustainable management conditions.

3. JOINT IMPLEMENTATION THROUGH FORESTRY: A COMPLICATED DECISION-MAKING PROCESS

In recent years, few North-South topics have attracted as much attention and led to as much confusion as JI. What makes the concept complex is not the underlying conviction to reach a certain goal in the most cost-effective manner, but the actual implementation that requires consent of and cooperation between at two governments, and potentially two private parties — one from each country. Given the number of parties involved in the negotiating process preceding the acceptance and eventual application of a JI project, a situation can easily unfold that can best be ascribed as a complex game. What makes matters even more complex is that the various participants not only have their own targets, but also have some convening and conflicting interests. To this a third element may be added that negotiations on JI are carried out in a multilateral framework that allows countries to form alliances to increase their chances of winning the game.

The overall situation becomes so complex — with three different games at different levels - that one runs the risk that playing the games absorbs a major part of the participants' energy and that sharing information to evaluate the patterns of the win-win situations for all parties involved, as well as for the environment, which is the ultimate goal, becomes secondary. The situation may be clarified with the help of examples.

Game 1

Suppose country A has accepted an obligation under an international treaty to reduce greenhouse gas emissions in a particular year by 10%, vis-á-vis its emissions in a base year. The application of the optimal mix of options - given the best available technology - to achieve the target costs the country, say, 0.5% of its GDP if applied at home. If, however, JI would be allowed for, say, 50% of the targeted emission reduction, the ensuing costs could be lowered to, say, 0.3% of the GDP. Now, if the savings — 0.2% of GDP — would be equally shared between country A and the JI host country, both parties involved in JI would benefit from the possibility of carrying out JI projects, but the environment would not, If, however, part of the net savings would be used to fulfill an extra greenhouse gas reduction commitment, both parties as well as the environment would win. So far, the interests of both parties coincide.

However, given the potential gains from JI, one has to agree upon how to divide the benefits between the investor and the host country (assumed 50-50 thus far). The investor will point out that it has made a commitment, and that the host country has not;[1] the host will point out the investor's past performance and its inability to contribute. The participant who plays the game hard has a good chance at achieving a better deal. However, if both participants play too hard, negotiations may fail and both parties may lose. The positions of both parties are symmetric; the game is overseeable and comparable to a duopoly.

Game 2

In reality, this game is not symmetric. One reason for this is that the initiative for JI is with the investor. The investor chooses from the various potential JI projects which one it is going to adopt.[2] If the investor sets the competing host countries against one another, theoretically almost all the JI benefits would accrue to the investor (possibly in conjunction with the environment), making the solution of the game rather predictable and the additionality principle meaningless.

However, because negotiations on JI, *in general,* are conducted in a multilateral framework, host countries can try to form a coalition to try to restore the power balance in negotiations. In that case, however, we are back to the situation described in Game 1.

Game 3

It is a popular misconception that, since it was decided during the March/April 1995 Berlin Conference that there shall be no crediting during the JI pilot phase (scheduled to end before 2000), no incentives can be provided to induce private institutions to participate in or start up JI projects. This misconception results from not distinguishing between the international credits to be gained through JI that could be used to claim fulfillment of FCCC commitments vis-á-vis other parties to the Convention and the *national incentives* every investor country is free to provide to enable or promote one of its private businesses or other private institutions to carry out JI activities. In the latter case, the only restrictions for the governments are that the incentives are not financed from regular aid budgets (nor GEF)

---

[1]   Some host countries even argue that the most cost-effective projects are 'occupied' by the donors, so that by the time they are expected to make their own contribution to the greenhouse problem, the best projects have been assigned to other investors. This argument, however, presupposes a precommitment from the host countries and could therefore easily backfire.

[2]   Theoretically in a fully transparent global JI system, JI crediting can be carried out through a tradable permit system, where the market determines the most cost-effective project, rather than the donor; such a system is, as yet, still far from reality.

- the additionality principle, albeit hard to prove - and that the incentives do not support exports in such a manner that they conflict with rules to establish an international playing field, but this area is ruled by rather soft law.

If we assume that the host country stands to gain from any type of joint project, the game is now between the investor country government (which may or may not provide these incentives depending on whose interests it intends to serve more, those of protectionists or those of free traders) and the potential private JI investor in the same country. Private businesses will try to convince the government that they should receive JI support (through subsidies, tax exemption, or otherwise) backed up by counterparts from the host country, once the distribution game has been settled.

In reality, the three games are played at the same time; sometimes the same individuals play in C, different games simultaneously. It is no wonder that decision making on JI is slow.

4. RECENT DECISION-MAKING ON JOINT IMPLEMENTATION

4.1. Joint Implementation Preceding and during the Berlin Conference

The common argument for a JI pilot phase usually is that time is required to solve technical issues such as establishing a baseline scenario, monitoring procedures, and building up local capacity to assist the projects and to gain experience with actual project application. Although this argument certainly makes sense — even if a fairly transparent and simple professional judgment approach through expert panels (see Section 5) might provide a considerable shortcut — the underlying argument to buy time and not to start crediting right from the beginning is that the games are still going on and that the parties feel insecure that the way in which they are going to benefit from joining joint activities. This may also explain why prior to the Convention of the Parties 1 (CoP 1) many participants had expressed their skepticism about the feasibility of progress. After all, the results of the International Negotiating Committee on the Framework Convention on Climate Change NI (INC XI) had not given rise to great optimism; especially with respect to JI, the points of view expressed officially and unofficially seem to differ so widely that, according to some pessimists, no *communis opinion* about a JI pilot phase would emerge during CoP 1. However, they turned out to be wrong.

Indeed, looking back at Berlin, it is fair to say that a breakthrough has been achieved in various respects, including JI. Indeed, one of the main results of CoP 1 - in addition to the acceptance of the Berlin Mandate to establish by 1997 a protocol to strengthen the FCCC for the period after 2000 — was the decision 'to establish a pilot phase for activities implemented jointly among Annex I Parties and, on a voluntary basis, with non-Annex Parties that so request.'

This result has not been achieved easily. One must acknowledge that during a large-scale negotiating process such as the Berlin Conference all issues are interpreted. Moreover, one must keep in mind the elements of the various games which may explain why positions taken by countries during a process of multilateral negotiations are based on tactics. Notwithstanding this, during the first part of the CoP I negotiations the divergent points of view on JI seemed almost unbridgeable. At first, only the G-77 and China were willing to accept JI projects between Annex I parties; others stressed the need to open the possibility of carrying out JI projects with non-Annex countries. Moreover, many countries objected against credits failing to any party as a result of greenhouse gas emission reductions or sequestration during the pilot phase from activities implemented jointly, whereas others, e.g., US representatives, stressed the important role of incentives also during the pilot phase.

However, in the course of CoP 1, the point of view that JI could ultimately provide a win-win situation for all participants increasingly gained adherence. India, as well as other developing countries, played a crucial role in achieving the final compromise. This compromise consisted of starting a JI pilot phase in Annex I parties and, on a voluntary basis, extending it to non-Annex parties.

4.2. Joint Implementation after Berlin: Some Procedural Aspects

As of July, 1995 16 projects between investors and host countries can be regarded as official JI projects if the criterion is employed that the projects have been agreed upon by the governments of the host and investing countries and have been (or will be) reported to the FCCC secretariat. Seven of these can be classified as forestry projects (see Box I for a short description): Rio Bravo conservation and forest management, Belize; CARFIX, Costa Rica; ECOLAND, Costa Rica; and RUSAFOR-SAP, Russian Federation (all approved for implementation in the framework of the USIJI program initiated in 1994); Krkonose, Czech Republic; and Profafor, Ecuador (both in the framework of the Dutch JI Pilot Project Program, through FACE Foundation); and Vologda, Russian Federation (recently initiated by the US). All projects have undergone some type of national screening procedure in the investing country but, in the absence of an internationally established procedure for approval, they have not gone through an international screening procedure.

Nevertheless, the variety of ongoing projects shows the types of projects, the way they are phrased, the levels of precision, and the complexity associated with success cases. Flexibility should be kept in mind during the evaluation process of proposals from host countries, as strict requirements for approval may underestimate the potential for success of many projects that may only meet adequate guidelines. From the developing countries' perspective, national capacities to put together a proposal within the JI initiative may need to be strengthened. Making general guidelines available is of critical importance to ensure that the requirements are effectively met from the beginning of the process.

Precisely because JI is in the early stages of its pilot phase, much discussion is still needed before the final criteria for JI projects and the procedure for JI project classification are established and widely endorsed by the various participants. Some examples of criteria for JI projects had been suggested prior to the Berlin Conference (in addition to draft criteria suggested by the UN interim secretariat): for example, the criteria for JI groundrules for the US initiative on JI and similar criteria outlined on behalf of the Australian and Canadian JI pilot initiative.

During the Berlin Conference the criteria for acceptance became clear. The following conditions with respect to JI activities were specified:

Box 1: JI forestry projects

Project: *Rio Bravo conservation and forest management, Belize*
Participants: Wisconsin Electric Power Company, Nature Conservancy of the US, and the Program for Belize.
Objective: Protection of 6,000 ha. from conversion to farmland.

Project: *CARFIX, Costa Rica*
Participants: FUNDECOR, the Costa Rican Ministry of Natural Resources, Energy and Mines, and Wachovia Timberland Investment Management.
Objective: Sustainable forest management in a major national park (71,000 ha.) and a buffer zone (20,000 ha.).

Project: *ECOLAND, Costa Rica*
Participants: Tenaska Washington Partners, Ltd., Trexler and Associates, the National Fish and Wildlife Foundation, COMBOS, the Costa Rican Ministry of Natural Resources, Energy, and Mines, the Council of the OSA Conservation Area, and Austrian NGOs.
Objective: Preservation of the Esquinas National Forest in Costa Rica.

Project: *RUSAFOR-SAP, Russian Federation*
Participants: Oregon State University, the Russian Federal Forest Service, the International Forestry Institute, and the US EPA.
Objective: Sequestration of greenhouse gases, prevention of soil erosion, and fostering of public participation in JI.

Project: *Krkonose, Czech Republic*
Participants: FACE Foundation and Krkonose National Park.
Objective: Reforestation and maintenance of 16,000 ha.

Project: *Profafor, Ecuador*
Participants: FACE Foundation and INEFAN.
Objectives: Reforestation and maintenance of 75,000 ha.

Project: *Vologda, Russian Federation*
Participants: Russian Federal Ministry of Environment, Vologda regional government, Center for Environmental Economics, Moscow State University, Commission for Productive and Natural Resources, Insitute.for Problems of Market Economy, Institute for the Economy in Transition, US EPA, and US EDF.
Objective: Assessment of several institutional issues related to JI in the Russian Federation and preparation of reforestation project (site: appr. 2,000 ha, mainly hayfields on collective farms, but originally spruce-pine forest) proposal in Vologda region.

- These activities must be compatible with and supportive of national and development priorities.
- The activities must be accepted, approved, or endorsed beforehand by the parties' governments.
- The environmental benefits expected as well as the financing of activities must be additional.

As was mentioned already it was also decided that no credits shall accrue to any party from JI activities during the pilot phase.

The pilot phase is essentially a teaming process. It must demonstrate how the many conceptual and practical questions with regard to JI will be answered in actual practice. The pilot phase should also make clear what the various parties can gain from participating in JI projects. Therefore, the Subsidiary Body for Scientific and Technological Advice (SUBSTA), in coordination with the Subsidiary Body on Implementation (SUBIM), is requested to establish a framework for reporting and to prepare a synthesis report for the CoP to consider at its annual sessions. On the basis of this, the CoP should take decisions on the pilot phase before 2000 and its progression beyond that.

5. A PROCEDURE FOR CARBON SEQUESTRATION ASSESSMENT

Assuming that a workable compromise on the criteria and procedures for determining which projects can be officially accepted as JI projects (and crediting thereupon) will be reached during the pilot phase, the next issue to address is how to measure the net carbon sequestered in JI forestry projects, and how to assess sustainability of forest management. With regard to greenhouse gas reduction through JI projects, the issue is not only to estimate and calculate the net contribution of the project, but also to verify it. The task of estimating the greenhouse gas reduction occurs during the project preparation stage. It is a necessary step in order to attract potential investors and to gain the approval of the relevant agencies. The task of calculating the greenhouse gas reduction occurs during or after project implementation. It is part of monitoring the progress of project implementation. Its objective is to calculate the greenhouse gas reduction of the project and to quantify the number of greenhouse gas credits that should be transferred between the host and investor countries, if the countries have adopted a greenhouse gas crediting system. The objective of the third task, verifying the greenhouse gas reduction, which may occur once or several times during and after the project implementation, is to establish whether the calculated greenhouse gas reduction actually occurred. Although the three tasks are different, the institutions required to perform them are the same.

Two questions now arise: How are these various tasks of carbon sequestration measurements to be carried out? What rules and procedures are to be used? To answer these questions two extreme positions can be taken, whereby criteria and indicators are approached along *inductive* lines, on the one hand, and along *mathematical* lines, on the other hand. By the inductive method a carbon sequestration is arrived at by a procedure that explicitly recognizes its subjective and judgmental elements. By contrast, the mathematical method assumes that judgmental factors can be more or less rules out if a sufficiently detailed set of norms is specified, which then can be quantified and added to the final objective assessment.

For various reasons we argue that it is a waste of time and money to try to design a detailed carbon sequestration procedure to reduce the maneuvers and the opinion-making role of the experts in assessing the size of net carbon sequestration at the project level.

First, there seems to be no procedure for answering the question about the difference between gross and net carbon sequestration via a particular JI forestry project in an indisputable way, because establishing the baseline cannot be carried out without making crude assumptions and guesstimates. For this reason alone consensus about the procedure seems much more important than about the precise application of the criteria.

Second, in forestry projects one must anticipate how the forest will be developed and exploited over a very long period of time, which is by nature often well beyond long-term legal, institutional, and political control. In fact, in all these respects one must assume that the long-term contract terms will not be breached, let alone that the forest will not be degraded by unforeseen natural causes such as fire.

Third, much of the long-term net carbon sequestration effect depends on the treatment of timber after it is harvested. Also, one can only make assumptions about this effect. In short, the assessment of net carbon sequestration through forestry can only be derived from professional estimates based on fair, reliable, and internationally accepted professional judgment: it is more important to ensure the fairness of the judge than to overemphasize the details of the rules. Overburdening international negotiations with the latter aspect may easily complicate the process in such manner that final decision making will be delayed.

In actual practice, a mixture of both extreme approaches seems to be the most likely outcome: a procedure that is based on professional judgment, recognizing the important role of experts who will support the final decision, on the one hand, and that recognizes the need for an overseeable and practical set of generally accepted criteria that experts should base their assessment on, on the other hand.

The institutional structure that is required for carbon sequestration assessment should include: a project-level team, a team of technical consultants, a verification team, and the UN JI secretariat.

The *project-level teams* can provide much of the project-specific data, and the *technical consultants* can provide the appropriate methodologies and technical expertise to perform the estimation and calculation of greenhouse gas reduction. These technical consultants could come from host and investor country research institutions, governmental agencies, nongovernmental organizations, and universities. Over time, project parties may themselves be able to estimate and calculate the greenhouse gas reduction. The task of the *verification team* requires the ability to check the data sources and check the methodologies used for calculating the greenhouse gas reduction. Thus, it may require a large team of experts working together to verify a project's performance. Furthermore, a verifier must be a trusted individual or firm whose credibility cannot be questioned.

The assessment tasks thus call for information exchange, training, and verification: activities where governmental intervention is appropriate and can succeed, but which can also to a large extent be delegated to the private sector. A *UN JI secretariat* might play a coordinative role, and focus on standardizing and disseminating assessment methodologies, training assessors in the use of the standardized methodologies, certifying verification teams, and resolving challenges to the verification or referring disputes to a tribunal.

Assessors should be trained by sector (e.g., forestry or energy) since data sources, methodologies, and the technical skills required vary greatly across sectors.

6. A PROCEDURE FOR SUSTAINABLE FOREST MANAGEMENT ASSESSMENT

In 1991 the ITTO council adopted the following definition of sustainable forest management:

> Sustainable forest management is the process of managing permanent forest land to achieve one or more clearly specified objectives of management with regard to the production of a continuous flow of desired forest products and services without undue reduction of its inherent values and future productivity and without undue undesirable effects on the physical and social environment.

This definition has been worked out in the ITTO framework and in a set of criteria that have been specified at both the national level and the management unit level. The further specification of the sustainability concept notwithstanding, it seems clear that a comprehensive scientifically justifiable assessment system of forest management is not possible due to lack of sufficient and reliable data on the exact norms to be applied and lack of any objective weighting system of the various criteria. In the end, both at the national and more notably at the management unit level, the assessment can only be quasi objective.

Obviously, various conditions must be fulfilled to arrive at a situation where the actual sustainability of forest management can be assessed in detail:

- The following minimum internationally recognized requirements must be met:
    - Endorse and implement international treaties and conventions, such as the Convention on Biodiversity and the Forest Principles,
    - Draw up an adequate system of laws and regulations ut) holding the system,
    - Draft rules for implementation,
    - Maintain adequate institutional capacity.
- Actual forest management policies and practices should be in line with the ITTO guidelines and criteria.
- Sufficient data and information must be available to enable the best professional judgments to be made.
- A proper institutional framework should exist allowing for correct and objective professional judgments.
- A proper institutional setting should be available in the country concerned to ensure that forest management practices are carried out sustainably and that the measures and regulations designed to achieve sustainability can be executed effectively.
- Adequate resources must be available to perform the sustainability assessment.

The ITTO and various NGOs have developed initiatives to design criteria with relevant indicators and norms for sustainable timber management assessment. These criteria might provide a basis for a timber certification system. In fact, various types of unofficial green labels for timber have already been introduced. These actions have met with different degrees of acceptance by consumers. If no international agreement is reached soon on the sustainable timber management concept and a certification system based thereupon, the initiative may well be taken over by one or a few well-organized NGOs in setting up a labeling system which is not completely under the control of the governments involved. The obvious disadvantage of such a development would be not only that the processes may well become unmanageable, but also that the system will not reflect the proper balance of the interests of both the timber-producing and timber-consuming countries, This provides another argument why one has to think hard about the fundamentals of an internationally agreed set of criteria for sustainable management assessment and a certification procedure based thereupon.

A procedure of certification could be carried out by different bodies varying in composition and tasks, very much analogous to the carbon sequestration assessment procedures. The certification procedure for sustainable timber management could include a monitoring committee, a joint working group, a multidiscipline@ inspection team, and a certifying body.

The monitoring committee should consist of an independent international body to draw up guidelines and procedures and to randomly monitor the certification process. The *joint working group* should be set up on a bilateral basis, and should consist of experts from various disciplines, if necessary joined by international experts; its task should include worldng. out procedures to be followed at the national level, establishing the minimum requirements, criteria, indicators, and norms for testing a country's forest policy for sustainability, and establishing whether a country's forest policy meets the conditions for sustainability. *Multidisciplinary inspection teams* should cuq out the actual inspection in the field on the basis of a checklist drawn up by the joint working group. In the inspection teams the required disciplines should be represented (e.g., forestry, ecology, anthropology); the teams should operate on the basis of information provided by the management authority, and produce advisory reports on the situation at the management unit level. An internationally monitored *certifying body* should take a final decision if a certificate of sustainability can be provided to the management unit or the acting forest manager considered, based on the reports received from the joint working group (on forest policy pursued) and from the inspection teams (on forest management). The certificate should be valid for a limited period of time and for the timber harvested during that period.

Before such a procedure can be put into operation many questions will have to be addressed:

- How is the responsibility for sustainable management divided among the forest manager, the concessionaire, the operator, and the forest users?
- How can the norms, with respect to sustainable forest management, vary depending on the situation in a country or region?
- How are the aspects that make up the sustainable timber management concept given their relevant importance and weighed against one another?

- How can consumers be convinced that sustainably managed timber production can ultimately lead to a certified supply of timber for the international market?
- How can sustainability (or the lack of it) at the national level be made compatible with sustainability at the management unit level?
- How will the sustainability concept be differentiated across the various forest types?
- How is timber from conversion forests dealt with in this respect?
- How is the sustainable management concept interpreted if forest activities in a particular management unit are also part of a carbon sequestration program?

7. CONCLUSION

There is an urgent need to deal with two distinct forestry-related concepts, both of which require assessment for international negotiating purposes: carbon sequestration via forestry (through JI) and sustainable forest management. In this paper we argue that the two concepts have many similarities, the main ones being that professional judgment must determine the net amount of carbon sequestration and that timber management must be labeled sustainable on an individual basis. It, therefore, is more important in international negotiations to focus on the design of a *fair, reliable, and internationally recognized procedure,* than to try to specify and quantify the various criteria in detail. In view of the many similarities in the potential procedures for the assessment of net carbon sequestration and of sustainable management (sketched in the paper), it seems logical to try to determine how far both procedures can be coordinated, if not carried out, by the same bodies.

Critical Reviews in Environmental Science and Technology, 27(Special): S97–S111 (1997)

ASSESSING EFFECTS OF MITIGATION STRATEGIES FOR GLOBAL CLIMATE CHANGE WITH AN INTERTEMPORAL MODEL OF THE U.S. FOREST AND AGRICULTURE SECTORS

RALPH ALIG,[1] DARIUS ADAMS,[2] BRUCE MCCARL,[3] J. M. CALLAWAY,[4] and STEVEN WINNETT[5]

[1]USDA Forest Service, Corvallis, OR 97331; [2]University of Montana, Missoula, MT 59812; [3]Texas A&M University, College Station, TX 77843; [4]RCG/Hagler Bailly Inc., Boulder, CO 80306; [5]U.S. Environmental Protection Agency, Washington, DC 20460, U.S.A.

ABSTRACT: A model of product and land markets in U.S. forest and agricultural sectors is used to examine the private forest management, land use, and market implications of carbon sequestration policies implemented in a "least social cost" fashion. Results suggest: policy-induced land use changes may generate compensating land use shifts through markets; land use shifts to meet policy targets need not be permanent; implementation of land use and management changes in a smooth or regular fashion over time may not be optimal; land use changes account for the largest part of adjustments to meet policy targets; and forest management changes involve higher intensity and less forest type conversion.

KEY WORDS: land-use change, afforestation, intersectoral, climate change, carbon.

1. INTRODUCTION

Increasing forest area and enhancing productivity of existing forests are two options being considered by U.S. policy makers to mitigate global climate change through the sequestration of carbon (C) in forests and forest products. Forests are an attractive vehicle for policy action, in part because the stocks themselves have values beyond that of the C sequestered and because programs for expanding stocks have been used for a variety of other objectives over many decades in the U.S. Thus, increments of C sequestration may be achievable in conjunction with other objectives of forest policy. Since forests sequester C from the atmosphere as part of the growth process, any increase in forest biomass constitutes a sink that will reduce the build-up of atmospheric carbon dioxide. However, anthropogenic activities involving forests, such as land use change and timber harvest, can alter the amount and temporal distribution of C storage. In past analyses of C sequestration in forests, land market interactions, timber harvest, and forest investment have received far less attention than the biophysical relationships between forest biomass and C sequestered (Sedjo et al., 1995). The potential impacts on C storage and fluxes of adjustments in

1064-3389/97/$.50

agricultural and timber markets have only been addressed in limited ways (Adams et al., 1993; Haynes et al., 1994). Opportunities to sequester more C in U.S. forests include both increasing net growth on existing timberland (Alig et al., 1992) and the conversion of suitable agricultural land to forest (Adams et al., 1994). In the latter case, programs designed to secure the afforestation of these agricultural lands could stimulate adjustments in both product and land markets that would partially offset land use shifts to forestry. The effect would be higher program costs and lower net increments in C sequestered relative to that suggested in static or single sector studies.

We apply a linked model of the U.S. forest and agriculture sectors, with endogenous land use and forest management investment decisions, to examine the consequences of intersectoral market forces on forest C storage, C fluxes, and costs. We investigate: 1) whether the effects of forest C sequestration programs differ significantly from those suggested in previous studies using static or single-sector approaches; and 2) the costs and the mixture of land base adjustments obtained when meeting a specific C sequestration target so as to minimize net social welfare costs in the intersectoral model. We briefly review previous studies that have estimated costs of terrestrial C sequestration. We then describe base case projections from our intersectoral model, results from simulation of an afforestation C program (Parks and Hardie, 1995), and projected impacts of meeting specific C sequestration "output" targets in a least social cost fashion.

2. METHODS OF PAST STUDIES

Previous studies of C sequestration through afforestation of marginal agricultural land can be differentiated (in part) by their treatment of management investment decisions and land prices in both the forest and agricultural sectors. A group of single sector studies, such as Haynes et al. (1994), focus on the impacts of various exogenous levels of land transfers on the forest sector and on forest C storage, based on econometric models centered on historical land use relationships (Alig, 1986; USDA Forest Service, 1990). Modeling of land transfers in these studies do not involve intertemporal feedbacks that recognize developments in agriculture or implications for forest land prices and the intensity of forest management on private ownerships. Moulton and Richards (1990) and Parks and Hardie (1995) explicitly consider the opportunity cost of the current use of land in deriving supply schedules for C sequestered in trees planted on marginal agricultural land. In Parks and Hardie (1995), for example, supply curves for transferable land are positively sloped (unit costs rise as more land is shifted) but are fixed over time with no feedback from changes in forest and agricultural markets. Moulton and Richards (1990) also provide supply curves for sequestering C by intensifying timber management on extant timberland, but do not consider interactions between the afforestation and timber management supply schedules. Adams et al. (1993) link a highly simplified model of forest sector markets (assuming exogenous management investment decisions) with a detailed model of agriculture that includes agricultural land—the Agricultural Sector Model (ASM) (Chang et al., 1992; McCarl et al. 1993). In a long-run equilibrium analysis, they establish the costs of land transfers for afforestation with (partial) adjustment in the agricultural sector, but with no feedback from, or management adjustment in, the forest sector.

3. MODEL OF FORESTRY AND AGRICULTURE

To allow land exchange and land price equilibration between the forest and agriculture sectors, we developed a linked intertemporal model of the two sectors. The Forest and Agricultural Sector Optimization Model (**FASOM**) is constructed as a multi-period, price-endogenous, spatial equilibrium market model (Takayama and Judge, 1971). The objective function maximizes the discounted economic welfare of producers and consumers in the U.S. agriculture and forest sectors over a finite time horizon. Quantity integrals of demand functions provide total willingness to pay, and the difference between total willingness to pay and production and processing costs is the sum of producer and consumer surpluses. The model operates on a decadal time step, with projections made for 9 decades to accommodate treatment of terminal inventories. Policy analysis is limited to results for the 50 year period from 1990 to 2040. Exogenous model elements in the forest sector component, drawn from the Timber Assessment Market Model (Adams and Haynes, 1995), are held constant after the fifth decade. Terminal inventories (at the end of the finite projection period) are valued in both sectors, assuming perpetual, steady state management following the last year of the time horizon (Adams et al., 1994). Solutions are obtained using a separable programming approach.

3.1. Forest and Agriculture Sectors

FASOM treats only the log market portion of the forest sector. Log demand is derived from the markets for processed products such as lumber, plywood, and paper. Logs are differentiated by six product classes: hardwood and softwood sawlogs, pulpwood, and fuelwood (Adams et al., 1995).[1] FASOM describes private timberland in terms of strata that are differentiated by: nine geographic regions, two classes of private ownership (industrial and nonindustrial), four forest types (describing species composition—softwoods or hardwoods—in the current and immediately preceding rotation), three site productivities (potential for wood volume growth), four management intensities (timber management regimes applied to the area),[2] suitability for transfer to or from agricultural use (four classes for crop or pasture plus a "forest only" class that can not shift use), and ten 10-year age classes.[3] Harvest age, management intensity, and forest type decisions are endogenous for private owners. FASOM simulates the growth of existing and regenerated stands by means of timber yield tables that give the net wood volume per hectare in unharvested stands by age class for each stratum. Public timber harvests are taken as exogenous.

The agriculture sector component in FASOM is adapted from the equilibrium price-endogenous ASM, aggregated to regions matching those used in forest assessments (Chang et al., 1992; McCarl et al., 1993). To convert ASM to a (disequilibrium) decade time step, the basic relations in ASM were treated as if they represented an annual period and repeated ten times each decade. Demand and supply components were updated between decades by means of projected growth rates in yield, domestic demand, exports, and imports. ASM simulates the production of 36 primary crop and livestock commodities and 39 secondary, or processed, commodities. Crops compete for land, labor, and irrigation water at the regional level. The cost of such inputs are included in budgets for regional production variables. More than 200 production possibilities (budgets) represent agricultural production in each decade, including field crop, livestock, and tree production. Farm programs are included only for the first decade in the model (Adams et al., 1994).

FASOM links the land inventories in the agricultural and forest sectors. Suitable nonindustrial private land can move, after timber harvest in the case of timberland conversion, between agricultural and forest uses based on considerations of inter-temporal profitability and subject to the availability of resources and the provisions of particular policies.[4] Estimates of the area of convertible forestland are derived from USDA studies and inventories of forestland with medium or high potential for conversion to crop or pasture use (USDA SCS, 1989, 1994). Area estimates for convertible agricultural land are drawn largely from Moulton and Richards' (1990) study of land suitable for tree planting. Land balances within sectors and exchanges between sectors are controlled by constraints described by Alig et al. (1995).

3.2. Carbon Accounting

FASOM accounts for forest C and stock changes for lands moving between sectors. The total C stored in the forest ecosystem of an unharvested stand is composed of five C pools: 1) tree; 2) woody debris; 3) soil; 4) forest floor; and 5) understory. FASOM accounts for C losses in nonmerchantable C pools from stands that are harvested, displaced fossil fuel from the burning of wood for fuel, and C decay in products derived from harvested timber. FASOM also accounts for differences in soil and understory C as land shifts between forest and agricultural uses. Soil and vegetation C on agricultural land is assumed to be a constant that varies according to land type (cropland and pastureland) and region. When a hectare of land is afforested, the C account is credited by an amount equal to the difference between the agricultural level and the higher, steady-state level of soil plus understory C for forests. As the forest stand matures, this soil C difference lessens as the soil and understory C move toward a steady state condition. When forested hectares revert back to the agricultural sector, the net loss of C associated with the soil, understory, and timber inventory, is accounted in reverse fashion.

4. INTERSECTORAL PROJECTIONS

With the linked model, we project a base case, a simulated afforestation program assuming a 4.9 million hectare transfer from agriculture to forest plantations in the 1990s, and three C output target scenarios.

4.1. Base Case: Large Potential for Intensifying Timber Management

We first consider the base case and then compare those projections to model results under the area transfer and C target scenarios. The base case projection entails an array of land use reallocations and land management shifts, leading to improvements in social welfare.

Base case assumptions for the forest sector derive from the USDA Forest Service's 1993 RPA Update (Haynes et al., 1995) and agriculture sector assumptions from Chang et al. (1992) and McCarl et al. (1993).[5]

4.1.1. Economic Welfare

Projected welfare measures for the base case and changes associated with the alternative simulations are shown in Table 1. Disparities in the size of base welfare measures between sectors, while striking, are unimportant to our analysis since we are interested in changes in the welfare measures at the margin, which are not necessarily proportional in size to the components of the overall welfare estimates.

4.1.2. Land Transfers

Projections of intersectoral land transfers for all scenarios are provided in Table 2. In the base case, approximately 9.3 million hectares are transferred between sectors from 1990 to 2039. Agriculture realizes a net gain from forestry of 5.7 million hectares (Figure 1), of which

TABLE 1
Distribution Of Welfare in the Base Case and Changes in the Net Present Value[a] of Selected Welfare Components for Afforestation and Carbon Target Scenarios ($1990 Billion)

Surplus Measure	Base Case	Afforest. Input	Carbon Target 1	Carbon Target 2	Carbon Target 3
		Change from the base case			
Forest Sector					
Domestic Consumers	2294.8	0.2	−16.0	−12.6	−26.8
Domestic Producers	108.7	0.6	11.4	11.4	19.3
Govt. For. Program Costs	—	1.9	—	—	—
Total	2541.5	−0.6	−7.0	−1.6	−12.0
Agriculture					
Domestic Consumers	638125.9	−4.6	−45.9	−60.4	−88.4
Domestic Producers	1008.8	2.9	38.5	48.1	78.6
Govt. Farm Prog. Costs	45.8	0.2	+	−2.4	+
Total	642481.0	−2.1	−10.2	−18.7	−19.5
Total (Both Sectors)	645022.5	−2.7	−17.2	−20.3	−31.5

[a] Calculated using a discount rate of 4% over 90 years. +/− signs indicate direction of changes for amounts less than $0.1 billion.

TABLE 2

FASOM Projections for Net Transfer of Land to Nonindustrial Private Forest from Agricultural uses for the Base Case and Four Scenarios, 1990–2039 ('000 hectares)[a]

Decade(s)	Base	Afforest. Input	Carbon Target 1	Carbon Target 2	Carbon Target 3
1990	−1096	869	546	18609	693
2000	−1841	−2623	−323	−6618	−625
1990–2039	−5713	−4120	5726	−2796	11312

[a] Positive number indicates net afforestation.

1.8 million hectares shift to pasture use and 3.9 million hectares to crop use. Base case land transfers fall within the range of historical experience, representing less than ten percent of the existing agricultural land base. Indicative of regional differences in land capabilities, transfers are concentrated in the East. Less than 0.4 million hectares change use in regions west of the Great Plains. Approximately half of the land transfers occur in the first decade, including the only transfers to forest use. Reallocation is concentrated in the initial periods due in part to the present value maximizing objective, the existence of farm programs only during the 1990s, and reduced availability of forest land (through exogenous losses) for conversion in later decades.

4.1.3. Forest Management Investment

In addition to land transfers, the base case involves significant changes in the intensity of private timber management. One indicator of the intensity of management is area in plantations, including conversion of hardwood types to softwoods. Relative to the 1990 level, the base case adds 12.6 million hectares of private forest plantations in the first decade (Figure 2), leading to reduced area of naturally regenerated stands, i.e., fewer low intensity and passively managed hectares. This exceeds the 9.7 million private hectares planted to forests over the past decade in the U.S., with the largest differences on nonindustrial private lands. In the second decade, private owners plant an additional 8.1 million hectares, then maintain the size of the total planted area for a decade before reducing it by 22 percent by 2039. A higher proportion of investments on industry ownerships are in relatively intensive plantation management (e.g., precommercial thinning, fertilization, and commercial thinning) compared to nonindustrial lands.

Most of the projected intensification of timber management is in the South and the Pacific Northwest. The area in commercially-preferred softwood types rises, with conversion of hardwood types to planted softwood stands. Conversion of hardwoods is driven by a relative softwood shortage (and rising softwood sawtimber prices). This, in turn, causes declining harvest volume levels and rising prices for hardwoods starting around 2020. Despite type changes, less than one-fifth of the future private forest landscape would

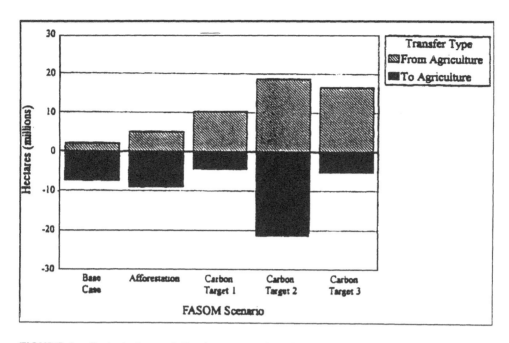

FIGURE 1. Projected cumulative intersectoral land transfers for the base case, an afforestation scenario, and three carbon target scenarios, 1900–2039.

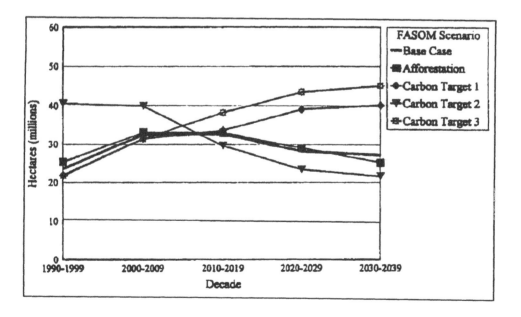

FIGURE 2. Projected private softwood plantation area for the base case, an afforestation scenario, and three carbon target scenarios, 1990–2039.

consist of plantations, and most timberland area in nonindustrial private ownership would still be concentrated in the two lowest management intensity classes that involve naturally regenerated stands.

4.1.4 Carbon Projections

Table 3 shows the projected time path of the C inventory on private timberland. The initial estimate for the C inventory (1990) is 23.71 gigatonnes (billions of metric tonnes), similar to Turner et al.'s (1993) estimate. C storage in the base case rises at an increasing rate between 1990-2009 and at a decreasing rate thereafter. In the 1990s, the C stock grows by 4 percent, equal to a "flux" (or average change in C stock over the decade) of 0.9 gigatonnes per decade. C fluxes remain positive over the projection, but drop to 0.5 gigatonnes by the 2030s decade.

4.2. Alternative Scenarios: Afforestation Area Input and Carbon Targets

4.2.1. Afforestation Program

Afforestation through tree planting has been proposed as one of the less expensive means of C sequestration in tree biomass (e.g., Moulton and Richards, 1990). Here we simulate a "forced" land use shift from agriculture through afforestation of 4.9 million hectares of pastureland between 1990-1999. The model is constrained to transfer at least the specified amount of land. Once forested, however, we do not require that these hectares remain in forestry indefinitely, nor do we restrict any contemporaneous or ensuing shifts in forest or agricultural land that might occur in response to this initial transfer.

In response to the pasture land transfer to forest plantations, the agricultural sector increases conversion of other forest land to agricultural use (Table 2), demonstrating that intersectoral responses can lead to outcomes that diverge from those intended in policy formulation. Compensating adjustments lead to more forest land being converted to agriculture than in the base case. In the South, the primary region for the afforestation, the conversion of forest to pasture use almost doubles compared to the base case, and most of the block of afforested land returns to agriculture when it is first eligible to transfer after timber harvest (once it reaches minimum harvest age). Thus, the net effect on land area in forest versus agriculture is significantly smaller than suggested in previous studies using static afforestation cost schedules (Moulton and Richards, 1990; Parks and Hardie, 1995). As a result, projected differences in timber harvest levels and timber inventory volume are relatively small between the base case and the afforestation scenario.

4.2.2. Carbon Sequestration Targets

Projections in the base case and those of several other studies (Birdsey, 1992; Turner et al., 1993) show a decreasing rate of C uptake by forests over time. Policy makers are interested,

TABLE 3
Projected Carbon Stocks for the Base Case, Afforestation Scenario, and Three Carbon Target Scenarios (with Targets in Left-Hand Columns and Projections in Right-Hand Columns), 1990–2039 (Gigatonnes, 10^9 mt)

Start of Decade	Base	Base Decadal Flux[a]	Afforest. Input	Fixed Increment (Carbon Target 1)			Fixed Increment Relative to Base (Carbon Target 2)			Growing Increment Relative to Base (Carbon Target 3)		
				TARG	PROJ	PROJ FLUX	TARG	PROJ	PROJ FLUX	TARG	PROJ	PROJ FLUX
<1990>	23.71	—	23.71		23.71			23.71			23.71	
2000	24.64	0.93	24.50	24.53	24.78	1.07	26.24	26.24	2.53	24.03	24.74	1.03
2010	25.77	1.13	25.67	26.13	26.36	1.58	27.37	28.13	1.89	25.33	26.24	1.50
2020	26.63	0.86	26.57	27.73	27.84	1.48	28.23	28.44	0.31	26.83	27.79	1.55
2030	27.28	0.65	27.30	29.33	29.33	1.49	28.88	28.88	0.44	28.53	29.51	1.72
2040	27.81	0.53	27.70	30.93	30.98	1.65	29.41	29.41	0.53	30.43	31.59	2.08

a   Decadal flux is computed as the difference in carbon stock levels at the start of successive decades.

however, in the costs and implications of achieving constant or increasing rates of forest C sequestration in the face of greenhouse gas emissions that are likely to grow due to rising population and energy use. We developed three scenarios in which C goals are expressed as a series of decadal C flux or inventory targets (Table 3 shows C inventory levels). Scenarios 1 and 3 give flux patterns that rise at varying rates from the base and remain at high levels, while scenario 2 has a high initial flux increase, then declines to base or lower levels. Projections were constrained to meet the following targets: 1) C **flux** at least 1.6 gigatonnes per decade beginning in the 2000-2009 period and all subsequent periods; 2) C **inventory** 1.6 gigatonnes larger than base levels by 2000 and all subsequent decades; and 3) C **flux** at least 1.3 gigatonnes per decade by 2000-2009, with growth accelerating by 0.2 gigatonnes each subsequent decade (1.3, 1.5, etc.). No restrictions were placed on how these C targets could be met, and resulting solutions can be considered least social cost allocations of land and investments to meet the targets. Least social cost is defined as the minimum loss in the NPV of the welfare of producers and consumers in the agriculture and forest sectors.

4.2.2.1. Projected Carbon Stocks and Fluxes

Figure 3 shows projections of changes in C fluxes relative to the base. Scenario 2 required the C stock to be 1.6 gigatonnes higher than base levels by the year 2000 and thereafter, a substantial increase in the size of the C stock over a short period. To reach this objective, first decade net afforestation is some 16.2 million hectares more than in the base case, roughly a ten-fold increase. This leads to some "overshooting" of the C target in the early part of the projection. Forest area then declines in subsequent decades, and by 2039 there is a net loss of land to the agricultural sector. Average harvest age of stands in some regions is also reduced in the first two decades of the projection, creating more area in younger reforested age classes with higher growth and C flux. By 2030, the requested C stock difference of 1.6 gigatonnes is met and maintained, or just barely exceeded, over the rest of the projection.

C stocks in scenarios 1 and 3 are constrained to increase over time without regard to the base case. By 2040, C stocks in scenarios 3 and 1 are about 3.8 gigatonnes and 3.2 gigatonnes higher, respectively, than in the base case. In both cases, net afforested land area is higher than in the base case, but the differences are not as dramatic as in scenario 2 (see figure 1).

4.2.2.4. Welfare Effects of Carbon Sequestration

Total net social welfare costs, representing the opportunity costs to society, of the three C target scenarios range from $17 billion for target 1 to $31 billion for target 3 (Table 1). Program costs reflect both the size of the C stock/flux targets and their temporal specifications. Scenario 3 has the highest costs and relatively high targets in the last three or four decades of the projection period, including the highest terminal C targets. The optimal solution exceeded targets in all of the intermediate years in order to achieve the terminal stock requirement, and attained much higher long-term levels of intensively managed

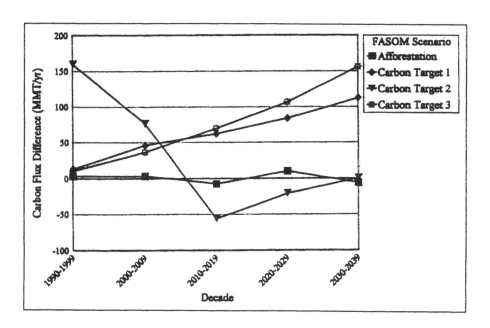

FIGURE 3. Projected differences in average annual carbon flux between the base case, and an afforestation scenario, and three carbon output target scenarios, 1990–2039.

softwood area than in the base case. Scenario 1, in contrast, involves neither the near-term land transfer extremes of scenario 2 nor the long-term inventory requirements of scenario 3.

The impacts of the C target scenarios on welfare in the agricultural sector are consistent with expectations. Consumers' surplus falls in all of the scenarios. Losses are largest in scenario 3 at some $88 billion (Table 1). The higher long-term C targets require net afforestation (Table 2) over most of the projection period. Reducing the amount of agricultural land lowers agricultural production and drives up land rents and agricultural commodity prices. Higher agricultural commodity prices translate into losses in consumers' surplus. Producer surpluses increase because of the inelastic nature of agricultural commodity demand curves. The largest increment is $79 billion in scenario 3.

Government costs are not significantly affected except for scenario 2, the only scenario in which more afforestation is projected in the 1990s decade while farm programs are still in effect. The 18.6 million hectares of afforestation acts to reduce the net present value of the deficiency payments to farmers by reducing agricultural land availability. This increases agricultural commodity prices, raises the prices consumers pay, and narrows the gap between target prices and market prices (Callaway and McCarl, 1995), leading to reduced government payments of more than $2 billion. Unknown amounts of government expenditures to induce the necessary timber management intensification and afforestation, however, are not considered in this accounting. For example, just for the fixed afforestation case, total expenditures to induce transfer of the 4.9 million hectares are estimated to be least $2 billion based on analysis of costs estimates by Parks and Hardie (1995) and comparisons with FASOM intersectoral projections. Under scenario 2, a substantial additional amount is likely to be required to induce about four times that level of net afforestation, given that opportunity costs of alternative land uses and program costs are likely to increase at higher levels of afforestation.

In the forest sector, consumers' surplus consistently falls across target scenarios and producers' surplus rises. The largest changes are in scenario 3, at -$27 billion and $19 billion, respectively. Given constraints to increase C storage, product harvest falls and prices rise (in varying patterns) across the scenarios. For scenario 2 short-lived price increases in the first period outweigh price reductions in later periods. The reverse is true for scenarios 1 and 3.

5. DISCUSSION AND CONCLUSIONS

Policy makers are currently debating the role of forest C sinks in achieving reductions in atmospheric carbon dioxide. Candidate programs to promote terrestrial C sequestration can produce complex interactions between the forest and agricultural sectors as a result of land use changes and alter inter-owner and inter-regional patterns of investment in forest management. Past studies have examined the impacts of potential policies with models that: (i) ignore spillovers in the other sector, or (ii) simply "add up" impacts across the two sectors, ignoring feedbacks or interactions through the markets for land, or (iii) treat forest management investment as exogenous. The present study applies a linked model of the U.S. forest and agriculture sectors that treats both land use and forest management investment decisions as endogenous.

In the three C target scenarios, afforestation of agriculture land provides the basis for most of the C inventory increases. These land use shifts were not, however, "once and for all" shifts. In some cases, large areas shifted from agriculture to forestry early in the projection, but then many afforested hectares shifted back to agriculture in later years. This suggests that minimum social cost strategies need not involve smooth or gradual trajectories of change in C fluxes or in patterns of implementing actions. For certain policies, an optimal strategy may be to overshoot flux targets in some periods (by increasing afforestation and/or forest management intensity) to meet targets in all periods. Such a strategy has more complex intersectoral welfare implications than schemes commonly considered.

The structure of forest management investment is also modified (relative to the base case) to grow or retain more wood to meet the C flux targets. Land shifting from agriculture must first be planted (which limits changes to softwood forest types in either the medium or high management intensity classes). In subsequent periods when these lands are harvested, they are regenerated into a mixture of the medium and passive management classes. Areas harvested from the passive class then shift back into agriculture. Softwood areas in the original forest inventory are managed in a modestly more intense fashion, with less area shifted to the passive management class. There is little change in the general pattern of management intensity allocations within hardwoods, but substantially fewer hectares are converted from hardwood to softwood types compared to the base. Over the projection period of 1990 to 2039, average softwood and hardwood rotation ages are significantly longer.

In total, the present value of surpluses in the forest sector falls in all of the scenarios, with the largest reduction approximately $12 billion in scenario 3. Losses in consumers' surplus in most cases are not offset by producers' surplus gains. This finding stands in contrast to most previous thinking on the welfare impacts of such programs as well as applied research using static models.

NOTES

1. Milling residues generated in sawlog processing can substitute for pulpwood.
2. The four management intensity classes are: "passive"—no management intervention between harvests of naturally regenerated aggregates; "low"—custodial management of naturally regenerated aggregates; "medium"—minimal management in planted aggregates; and "high"—genetically improved stock, fertilization and/or other intermediate treatments in planted aggregates. Specific practices vary by region, site quality, forest type, and agricultural suitability.
3. FASOM and the Timber Supply Model (TSM) (Sedjo and Lyon 1990) both model the forest inventory in an even-age format, with discrete age classes and endogenous decisions on management intensity made at time of planting. The TSM does not model land transfers with agriculture and is solved using methods of optimal control with an annual time increment.
4. Land is also exogenously shifted from both forest and agricultural uses of nonindustrial ownerships to urban/developed uses along with some timberland reclassified to reserved uses (Adams et al. 1994, Alig and Wear 1992).
5. Future climate conditions in the base case are assumed to approximate those of recent decades. Using FASOM, Burton et al. (1994) projected relatively small national and regional economic impacts for timber growth changes of less than 10 percent, as might occur under some global climate change scenarios.
6. None of the parameters in the scenarios are tied specifically to a projected rate of increase in emissions since there is no consensus on that projected rate of increase.

ACKNOWLEDGMENTS

The study was funded by the U.S. Environmental Protection Agency, Office of Policy Analysis, Climate Change Division, and the USDA Forest Service, Pacific Northwest Research Station. We wish to acknowledge the assistance of Pete Bettinger, USDA Forest Service, Corvallis, Oregon and John Chmelik and Eric Jensen, University of Montana.

REFERENCES

Adams, D. M. and R. W. Haynes. 1995. *The 1993 Timber Assessment Market Model: Structure, Projections, and Policy Simulations*. USDA Forest Service, Pacific Northwest Res. Stn., Gen. Tech. Rep., GTR-PNW-in press, Portland, OR.

Adams, R., D. Adams, J. Callaway, C. Chang, and B. McCarl. 1993. Sequestering carbon on agricultural land: Social costs and impacts on timber markets. *Contemporary Policy Issues* XI: 76-87.

Adams, D., R. Alig, J.M. Callaway, and B. McCarl. 1994. Forest and Agricultural Sector Optimization Model: Model Description. Final report to the Environ. Prot. Agency, Climate Change Division, Washington, DC. Variously numbered.

Adams, D., R. Alig, B.A. McCarl, J.M. Callaway, and S. Winnett. 1995. An analysis of the impacts of public timber harvest policies on private forest management in the U.S. *Forest Science* (in process).

Alig, R. 1986. Econometric analysis of the factors influencing forest acreage trends in the Southeast. *Forest Science* 32(1):119-134.

Alig, R., J. Vasievich, and K. Lee. 1992. Economic opportunities to increase timber growth on timberland. In *Forests in a Changing Climate*, A. Qureshi ed., Climate Institute, Washington, D.C. pp. 115-125.

Alig, R. and D. Wear. 1992. U.S. private timberlands, 1952-2040. *Journal of Forestry* 90(5): 31-36.

Alig, R., D. Adams, B. McCarl, J. Callaway, and S. Winnett. 1995. Land market interactions between the agriculture and forest sectors: Model development and policy projections. (In process).

Birdsey, R. 1992. Carbon storage in trees and forests. In *Forests and Global Change: Volume I - Opportunities for Increasing Forest Cover*, N. Sampson, and D. Hair eds., American Forestry Association. pp. 23-40.

Burton, D., B. McCarl, D. Adams, R. Alig, J. Callaway, and S. Winnett. 1994. An exploratory study of the economic impacts of climate change on southern forests. In proc., *1994 Southern Forest Economics Workshop*, Savannah, GA. March.

Callaway, J. and B. McCarl. 1995. The economic consequences of substituting carbon payments for crop subsidies in U.S. agriculture. *Environmental and Resource Economics* 5: 1-29.

Chang, C., B. McCarl, J. Mjelde, and J. Richardson. 1992. Sectoral implications of farm program modifications. *American Journal of Agricultural Economics* 74(1992): 38-49.

Haynes, R., R. Alig, and E. Moore. 1994. *Alternative simulations of forestry scenarios involving carbon sequestration options*. USDA Forest Service Gen. Tech. Rep. PNW-GTR-335. Pacific Northwest Res. Stn. Portland, Oregon. 66p.

Haynes, R., D. Adams, and J. Mills. 1995. *The 1993 RPA Timber Assessment Update*. USDA Forest Service Gen. Tech. Rep. RM-GTR-259. Ft. Collins, CO., Rocky Mountain Forest and Range Experiment Station. 66p.

McCarl, B.; C. Chang; J. Atwood; and W. Nayda. 1993. The U.S. Agricultural Sector Model. Texas A&M University, College Station, TX. (Draft).

Moulton, Robert J. and Kenneth R. Richards. 1990. *Costs of Sequestering Carbon Through Tree Planting and Forest Management in the United States*. USDA Forest Service Gen. Tech. Rep. WO-58. Washington, D.C. 44p.

Parks, P. J., and I. W. Hardie. 1995. Least-cost forest carbon reserves: Cost effective subsidies to convert marginal agricultural land to forests. *Land Economics* 71 (1): 122-36.

Sedjo, R. and K. Lyon. 1990. *Long-Term Adequacy of World Timber Supply*. Resources for the Future, Wash., DC. 230p.

Sedjo, R., J. Wisniewski, A.V. Sample, and J. Kinsman. 1995. The economics of managing carbon via forestry: Assessment of existing studies. *Journal of Environmental and Resource Economics* 6: 139–165.

Takayama, T. and G.G. Judge. 1971. *Spatial and temporal price and allocation models*. North-Holland Press, Amsterdam.

Turner, D. and others. 1993. The forestland carbon budget of the U.S. Current status and evaluation. Final rep., Environ. Protection Agency. ManTech, Corvallis, Oregon.

USDA Forest Service. 1990. *An analysis of the timber situation in the United States: 1989-2040.* USDA Forest Service RM GTR 199. Rocky Mountain Forest and Range Experiment Station, Ft. Collins. 268p.

USDA Soil Conservation Service. 1989. *The second RCA appraisal: Soil, water, and related resources on nonfederal land in the United States;analysis of conditions and trends.* Washington, DC. 280p.

USDA Soil Conservation Service. 1994. The 1992 National Resources Inventory: Preliminary Results. (Draft). Wash., DC.

Critical Reviews in Environmental Science and Technology, 27(Special): S113–S122 (1997)

INCORPORATING CLIMATE CONSIDERATIONS INTO THE NATIONAL FOREST BASIC PLAN IN JAPAN

MASAHIRO AMANO

Forestry and Forest Products Research Institute, Tsukuba 305, Japan

ABSTRACT: A computer simulation model developed for the Japanese Forest Resource Basic Plan which is the basis of the Japanese forest policies is used to assess mitigation policies under the three scenarios. These are a) stabilizing timber supply at the present level (case 1); b) adopting long rotation (case 2); and c) aggressively promoting domestic supply (case 3). These results indicate that each of the three scenarios contribute to storage carbon into the forest sector as shown in case 1, 2, and 3.

KEY WORDS: carbon storage, growing stock of forests, wood products, simulation, Japan, timber supply.

1. INTRODUCTION

Whenever the situation surrounding the forest sector changes greatly, the forest sector master plan of Japan (the Forest Resources Basic Plan) is revised. The last Basic Plan in Japan was reviewed in 1987, because wood prices had become stagnant, the growth rate of Japanese economy was stable, and domestic wood production activity was low. Seven years after the last Basic Plan was made public, the situation surrounding forestry has changed drastically. Wood prices in Japan became less expensive, and the ability of the domestic industry to compete internationally has been weakened by a remarkable rise of the yen in the currency exchange rate.

The present Basic Plan considered only the economic side of the forest sector. But another change is the increasing importance of environment protection at the global scale which is now becoming more recognized among the general citizens. In current review work, the plan has considered the environmental side of the forest sector. For example, in the new Basic Plan, a quantity of supply and effects derived from forest roles of non-timber use are to be evaluated. Of course, the mitigation effects on the global warming phenomenon by sequestration of carbon in the atmosphere by the forest ecosystem are important in the Basic Plan (Sedjo, 1989). If the forest role of mitigating the global warming is simulated on the Basic Plan, it becomes possible to assess the efficiency of forest policies to reduce the carbon stock in the atmosphere. This report focuses on the role of forests in reducing the global warming phenomenon as incorporated into the Basic Plan.

In Japan, there are 10.3 million hectares of man-made forests and 15.3 million hectares of natural forests. The growing stocks of man-made and natural forests are 160 million cubic meters and 154 million cubic meters, respectively. But, positive forestry management

1064-3389/97/$.50

activities are not carried out in natural forests. Discussions about the relationship between Japanese forest policies and carbon fixation are limited to plantation forests. The computer simulation is carried out to compare the effects of mitigating atmospheric carbon accumulation by using a model developed for the Forest Resource Basic Plan which provides the base for the Japanese forest assessments and policies.

2. FOREST ROLE FOR MITIGATING GLOBAL WARMING

There are several means by which the forest sector is thought to contribute to the reduction of global warming, and several models have been developed to evaluate these means (Row and Phelps, 1990; Turner et al., 1993). In this paper, the simulation model for the Japanese Basic Plan is utilized to evaluate some policies for mitigating global warming (Figure 1). Expansion of forest area is one of those means. Seventy percent of the country's topography is mountainous and there are few forests in the plains in Japan. Therefore, forests are the dominate land use in the mountains. The possibility that the forest will expand into the plains is extremely small for the foreseeable future. Accordingly, expansion of forest area is not argued here. The Japanese Forestry Agency is moving toward lengthening rotation ages of forest stands as a measure to mitigate global warming. The trend in recent years among private Japanese forest owners is also towards increasing their cutting rotation ages (Japanese Forest Agency, 1994). This tendency has been caused by rising costs of regeneration and tending stands (Johansson and Lofgren, 1985). Thus, the implementation of long rotation policy fits to the recent cutting behaviors of forest owners who have suffered from the high employee wages.

Decreases in the world's tropical forests has become a major problem because of the release of large amounts of carbon dioxide to the atmosphere (FAO 1993; Houghton et al., 1991). Since Japan is the greatest importer of tropical wood in the world (FAO, 1993), there have been some suggestions that Japan should soon stop importing tropical wood. It has been proposed to decrease the import of tropical wood over the next 10 years. However, a quantitative analysis has not been carried out because this matter is largely a political problem with a lot of uncertainty(Brooks, 1994). Furthermore, there is also an energy problem since wood, a heavy product, must be transported for long distances (Haynes 1994). However, this problem is not treated in this paper because it focuses on the reduction of global warming from the perspective of wood supply.

The role of forests for carbon sequestration by accumulation of biomass has been assessed in this paper. After harvesting trees, wood can still fix carbon in the forest sector because felled wood is used as wood products which hold carbon for a long time. Accordingly, extending the durability of wooden houses becomes one of the important means that can mitigate carbon buildup in the atmosphere.

2.1. Scenarios for Analysis

Three scenarios are set up to show future wood supply derived from domestic forest resources under different options. Case 1 is a policy that the wood supply will be carried out under stable economic conditions which is similar to the present situation. Japan imports 75 % of its total demand for wood; which totaled 81 million cubic meters from overseas in 1992. The amount of domestic supply is only 27 million cubic meters. Domestic plantation forest resources supply approximately 20 million cubic meters. The case 2 scenario almost doubles the rotation length. An average cutting age of Japanese plantations is 55 years old at present. According to the standard yield table for plantations, growth rates of plantations become smaller over 100 years old. Hence, it is not effective to expand the rotation length more than 100 years. Therefore the case 2 scenario is set to postpone a rotation from 55 to 100 years. Other various kinds of suppositions in Case 2 are the same as Case 1. Concretely, the average cutting year is prolonged 10 years for every five years. In Case 3, several positive policies are adopted to expand the domestic supply share in the Japanese timber market. These forest policies include positive investment for logging roads, introducing highly efficient forestry machines, subsidizing forest management and wood industry, tax deductions and so forth. Given this set of changes, the share of domestic production wood rises from 25 % at the present to a share of 35%. Table I shows that Japanese wooden houses have an average life of 36.8 years. Life spans of wooden houses are classified into two groups from Table I.

One group shows that houses are utilized for more than 50 years. Houses belonging to another group are utilized for approximately 25 years, on average, before houses are replaced. Physical durability of these houses are much longer than this figure, by at least 50 years (Arima, 1991). Reasons to rebuild and change houses before physical life expectancy occurs, in addition to the above factors, include situations where the family members or their life styles change. For simplicity, one scenario assumes extending durable years of current wooden houses from 36.8 years to 50 years within the next 20 years. The above three scenarios examined how building up growing stock of forests are different. A simulation model is used to estimate future growing stock of Japanese plantations under 3 different scenarios. Simulations also examine how growing stock of Japanese forests change, in three cases described above, when the housing life increases from 36.8 years to 50 years.

2.2. Constraints for the Basic Plan Model

The age distribution of Japanese plantation forests are skewed and 68% of plantations are concentrated between 16 to 40 years old. This means that the annual increment will vary over time. The model also needs to make clear the changing resource structures of forests according to age in order to project the changes of growing stock. The Japanese timber market depends on both domestic production timber and import timber to satisfy the wood demand in Japan. It is necessary for a model to be able to simulate the competing relationship with import wood and domestic wood. Therefore, it is not simply understand-

TABLE 1
Life Span of Wooden Houses

Year	9–9	10–14	15–19	20–24	25–29	30–34	34–39	40–44	45–49	50–	Ave.
Rate	3.5	5.4	15.6	17.9	12.0	9.5	5.1	5.9	2.0	21.7	36.8

Flow of Analysis

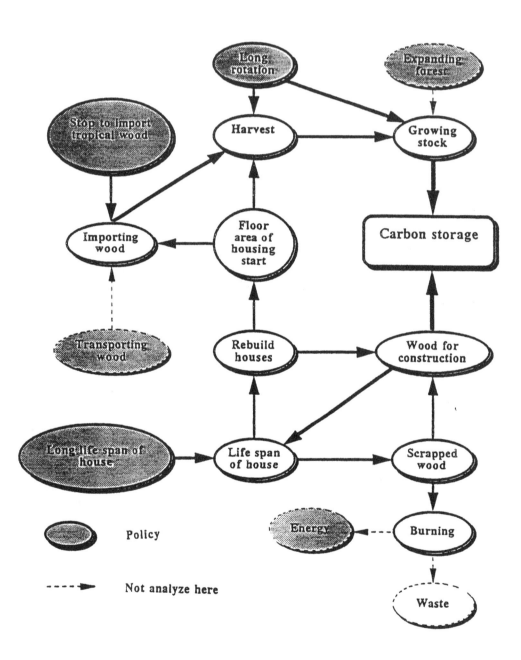

FIGURE 1. Flow of carbon in the Japanese forest sector.

able how to lengthen the house durability influences to the domestic forest resources and foreign forest resources. A model that can cover both the supply and demand sides of the forest sector becomes necessary to evaluate such influences. Since house durability is important, the wood supply - demand model used here has parameters related to wooden houses in the demand model. The parameters are the floor areas of buildings started and the rate of new wooden housing units started.

2.3. Structure of a Model

The following model was used for analysis developed in the above discussion. The model is made up of three submodules. These are supply, demand and roundwood market as shown in Figure 2. The supply model differentiate forests into two groups, one active and the other stagnant in management which are called economical timberland and non-economical timberland, respectively (Amano et al.1984). The boundary between the two timberland types shift with the fluctuation of timber price, logging cost, logging road construction and so forth. A simple equation to determine the area of timber land A_{timber} at a planning period t is derived by total plantation area A_p, timber price p, harvesting cost c, and new logging road construction mileage l.

$$A_{timber}(t) = f(A_p, p(t), c(t), l(t))$$

This boundary is considered the factor which determines the portion of economical timberland in the total forested area. Currently the economical timberland is the primary source of domestic timber supply. The amount of this supply is the sum of the products of harvesting probabilities (Suzuki, 1984) and the corresponding forest area. This probability was developed for predicting the timber supply using the Markov Process with age transition matrix P and forest area A_t at a planning period t.

$$A_t+1 = A_t\ P$$

where

$$A_t = (a(t)_0, a(t)_1, ..., a(t)_i,...,a(t)_n)$$

$$P = (p(j,k)), \quad j, k = 1,2,...,n$$

$a(t)_i$ is the forest area of i-th age class and $p(j,k)$ is the transition probability from j-th age class to k-th age class.

The forecasting parameters of harvesting probabilities are derived from a mean cutting age and its variance (Suzuki, 1984). Hence, it is easy to reflect the changes of cutting rotations length.

Demand for sawnwood, panelwood, and other uses including pulpwood were estimated by the demand submodule made using ordinary econometric models (Amano and Noda, 1987). The demand and the supply submodules are linked through the roundwood market submodule. Demand and supply equilibria are computed for 5-year periods over the

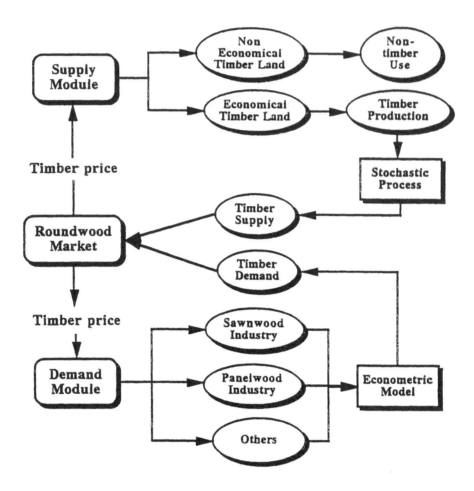

FIGURE 2. Structure of supply-demand model.

projection period. In the model, feedback iterations proceed in the three submodules until the demand and supply volumes converge upon an equilibrium.

3. RESULTS

Figure 3 is growing stock of forests in the future calculated under three scenarios. One period is equivalent to 5 years here. Figure 4 shows annual harvesting volume of each planning period calculated under scenarios. Differences of growing stock and harvest volume under each scenario is compared. In a case of stabilizing policy, growing stock of plantation forests increases 402 million cubic meters after 50 years more than a case of promoting domestic timber supply policy. Figure 4 shows how harvesting volume of the promoting domestic supply policy increases 59 % annually on average compared to the stabilizing policy.

When the environment surrounding forestry lasts for the next 50 years, as in the present situation, the volume of harvesting will increase. But, the growing stock continuously increases because the quantity of harvesting is less than the quantity of increment. Although the harvesting age of the long rotation scenario is double when compared with the stabilizing policy, harvesting volume is reduced by only 5.7 % at most. On the other hand, the growing stock of forests is increased by 8.1%. Harvesting volume increases 59.5% in the case of promoting domestic timber supply when compared with the stabilizing policy case, whereas growing stock of Case 3 decreases 12.5% comparing to the Case 1. Growing stock of Case 3 decreases more than 19.1 % compared to the long rotation scenario.

According to a well known method (Brown and Lugo, 1982), the above ground carbon storage of growing stock of plantations is calculated by whole biomass and wood density. The results in Cases 1, 2 and 3 are 1.03Gt, 1.12Gt and 0.90Gt, respectively 50 years later, where a wood density of plantation is 0.4. Assuming that 90% of harvesting volume from plantations is used for construction, and if these wood products are continuously used for 36.8 years on average, before the carbon stored in wood products is released into the atmosphere, the carbon content in wood products is calculated as follows (Arima, 1991):

Carbon content = Wood volume*density*0.5

where density of wood yielded from Japanese plantation is 0.4. Then 0.14Gt, 0.13Gt and 0.22Gt are stored in houses made by domestic timber in each case, respectively, 50 years later. If the total sum of growing stock of domestic forests and quantity of carbon fixation of wood produced from domestic timber is going to be maximized, then Case 2 is desirable. But this is not proper in the global scale because it is difficult to take account of the situation of foreign forest resources disturbed by Japanese wood importing.

In Figure 5, Case 1, Case 2, and Case 3 show the changes of growing stock caused by decrease of wood demand when the durability of houses is improved from 36.8 years to 50 years. Figure 6 shows harvesting volume for each case. Every scenario deteriorates harvesting volume according to decrease of demand. The influence causes increase of growing stock of 4.7%, 2.5% and 6.1% in Cases 1, 2 and 3, respectively.

4. CONCLUSION

The total carbon of growing stock of plantations and harvesting volume 50 years later shows the long rotation scenario is the best among the three scenarios. Extension of a rotation itself did not influence the increase of growing stock in a country with much import wood like Japan. It is said that 50 million cubic meters is the potential wood supply ability of a Japanese plantation forest. By the scenario's given, plantation forests supply wood from 20 million meters in case 2 to 38 million cubic meters in Case 3. This suggests that Japanese plantations have not established enough market comparing their domestic wood productive power. It seems that political efforts are not required to prolong rotation length, because forest owners cannot find out enough timber markets at the present timber price and will keep their forests without cutting incentives (Japanese Forest Planning Association, 1987). On this account, the extension of rotations is realized without a severe policy effort in Japan. The extension of the durable years of an average house will increase the

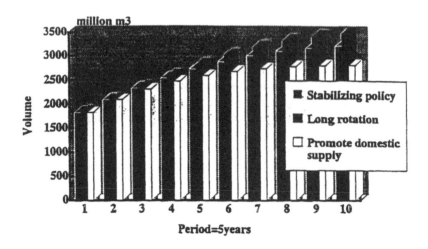

FIGURE 3. Three case projections of growing stock.

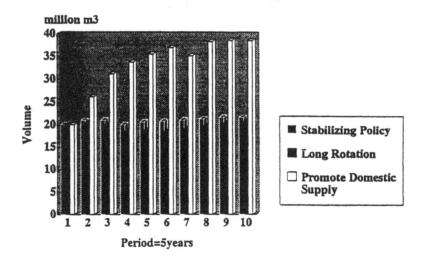

FIGURE 4. Three case projections of harvesting volume.

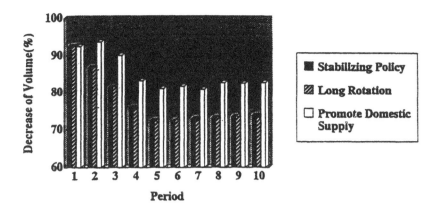

FIGURE 5. Decrease of harvest volume by expanding life span of houses comparing to original cases.

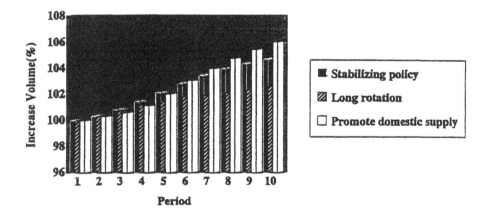

FIGURE 6. Increase of growing stock by expanding life span of houses comparing to original cases.

growing stock. Accordingly, this policy can become an effective means for letting growing stock increase.

ACKNOWLEDGEMENT

The author would like to thank Dr. Roger A. Sedjo of Resources for the Future and Professor Theodore Howard of University of New Hampshire for their help in reviewing this paper and their useful suggestions.

REFERENCES

Amano, M., I. Noda and M. Kumazaki (1984), *Timber Supply Projection Model for Private Forests in Japan.* Proceedings of IUFRO S4.04 Symposium on Forest Management Planning and Managerial Economics, Tokyo. 285-292.

Amano, M., I. Noda (1987), *Japanese Wood Supply in the Future and the Effects on Air Pollution, Forest Decline and Reproduction: Regional and Global Consequences,* IIASA, Austria, 515-522.

Arima, T. (1991), Life Cycle of Wooden Houses and Environment Protection, *Wood Industry* 46(12), 635-640 (in Japanese).

Brooks, D.B. (1994), Tropical Timber Market: Policy and Modeling, *Journal of Forest Economics* 126, 19-27.

Brown, S. and A.E. Lugo (1982), The Storage and Production of Organic Matter in Tropical Forests and their Role in the Global Carbon Cycle, *Biotropica*, 14,161-187.

Dixon, R. K., P.E. Schroeder and J.K. Winjum (1991), *Assessment of Promising Forest Management Practices and Technologies for Enhancing the Conservation and Sequestration of Atmospheric Carbon and Their Costs at the Site Level.* USEPA Environmental Research Laboratory. Corvallis, Oregon.

FAO (1993), *Forest Resources and Assessment 1990.* FAO Forestry Paper 112

FAO (1993), Forest Products Trade Flow Data 1991-1992, FAO.

Haynes, R.W. (1994), Modeling the United States Forest Sector, *Journal of Forest Economics* 126, 2-7.

Houghton, R.A., D.S. Lefkowitz and D.L. Skole (1991),Changes in the Landscape of Latin America between 1850 and 1985, *Forest Ecology and Management*, Vol. 38, 43-172.

Japanese Forest Agency (1994), Forestry White Paper 1992, Forest Agency, Japan.

Japanese Forest Planning Association (1987), *New Projection of Forestry in Japan.* Chikyusha, Tokyo (in Japanese).

Johansson, P.O. and K.G. Lofgren (1985), *The Economics of Forestry and Natural Resources.* Basil Blackwell Ltd, UK.

Row, C. and R.B. Phelps (1990), *Tracing the Flow of Carbon through the US Forest Products Sector.* Proceedings of the 19th World Congress of the International Union of Forest Research Organizations, Montral, Canada.

Sedjo, R.A. (1989), Forests: A Tool to Moderate Global Warming? *Environment*, 31(1): 14-20.

Suzuki, T. (1984), *The Gentan Probability, A Model for the Improvement of the Normal Forest Concept and for the Forest Planning.* Proceedings of IUFRO S4.04 Symposium on Forest Management Planning and Managerial Economics, Tokyo. 12-24.

Turner, D.P., J.F. Lee, Greg J. Koerper and J. R. Barker (1993), *The Forest Sector Carbon Budget of the United States: Carbon Pools and Flux under Alternative Policy Options*, US EPA.

Critical Reviews in Environmental Science and Technology, 27(Special): S123–S138 (1997)

ECONOMIC IMPACTS OF CLIMATIC CHANGE ON THE GLOBAL FOREST SECTOR: AN INTEGRATED ECOLOGICAL/ECONOMIC ASSESSMENT

JOHN PEREZ-GARCIA,[1] *LINDA A. JOYCE,*[2] *C.S. BINKLEY,*[3] *and A. D. MCGUIRE*[4]

[1]*Assistant Professor, Centre For International Trade in Forest Products, University of Washington, Seattle, WA 98195, U.S.A;* [2]*Project Leader, Rocky Mountain Forest and Range Experiment Station, USDA Forest Service, Fort Collins, CO, 80526-2098, U.S.A.;* [3]*Dean, Faculty of Forestry, University of British Columbia, Vancouver, B.C., Canada V6T 1Z4 [to whom correspondence should be addressed];* [4]*Research Associate, The Ecosystem Centre, Marine Biological Lab, Woods Hole, MA; U.S.A.*

ABSTRACT: Increased atmospheric concentrations of CO_2 and any associated climatic changes will affect many aspects of forests, including net primary productivity. Operating through normal timber-supply mechanisms, changes in forest growth will affect forest products markets—production, consumption, prices and trade—throughout the world. These impacts are simulated by linking the climatic change scenarios from the four commonly used general circulation models (GCMs) of climate with a model of global vegetation response (the Terrestrial Ecosystem Model) and a model of global forest products trade (the University of Washington Cintrafor Global Trade Model). The four GCMs produce surprisingly consistent economic results, with expanding net primary productivity of the world's forests generally hurting timber producing regions and benefiting consumers. The overall net economic impact is positive, with the net present value (computed with a 4% real discount rate) of the benefits to the forest sector ranging from $US1980 10.7 billion to 15.9, depending on the choice of GCM and economic scenario. Uncertainty in the economic models appears to be at least as great as the uncertainty in the GCMs.

KEY WORDS: global forest sector, ecological/economic assessment.

1. INTRODUCTION

Increased atmospheric concentrations of CO_2 (IPCC, 1990) may affect forest ecosystems both directly (since this gas is a fundamental plant nutrient) and indirectly through possible impacts on climate (Ceulemas and Mousseau, 1994; Idso and Idso, 1994; Norby, 1994; Wullschleger et al., 1994). These direct and indirect effects not only will alter the regeneration, phenology, growth, and mortality of the trees that dominate forests, but also the host of other organisms which inhabit these ecosystems. Humans, of course, depend on forests for a myriad of needs—products ranging from fuelwood to technically-advanced wood-based building materials and paper, water emanating from forested watersheds, and places of restful solitude. None of these relationships will escape changes in forests themselves.

1064-3389/97/$.50

This paper focuses narrowly on one aspect of climatic change and forests: the effects on the world's forest products economy. The analysis does not consider the use of wood for fuel, an activity that currently accounts for roughly half of all timber harvests. Our concern is specifically with how changes in forest growth induced by changes in climate will affect timber production, consumption, prices and trade over the next 50 years. Timber products comprise about $80 billion of the world's trade, or about 3% of the total, and dominate the economy in some countries (Laarman and Sedjo, 1992). To be sure, other, nontimber aspects of forests—recreation, water, wildlife, and so on—comprise significant forest values. For example, the USDA Forest Service (1990) estimated that, by 1995, nontimber "forest products" would comprise about 70% of the gross value produced by that country's national forests (32% in recreation, 23% in wildlife and fisheries, 14% in minerals, and 1% in other products). While some research has examined the potential impact of climatic change on the nontimber values of the forests, there has been virtually no analysis of the economic aspects of these impacts (Binkley and van Kooten, 1994). This paper deals only with industrial products manufactured from wood.

Integrated assessments of the impacts of global change on the forest sector require an understanding of the linkages among climate, forest ecology and economics (Binkley and van Kooten, 1994). Using one or more simulations of future climates (typically derived from general circulation models (GCMs) of the atmosphere), such assessments predict future forest conditions and estimate the economic implications of these ecological outcomes. Previous studies of the economic impacts of climate change on forests fall into one of two categories based on the kind of economic adjustments considered.

The first type evaluates changes in ecological conditions in the absence of price effects. Hodges et al., (1989) applied current market prices to value predicted climate-related changes in forest growth in the Southeastern US, and concluded that the impacts of a $2xCO_2$ warming would amount to only about 0.2% of the value of the region's forest products shipments. As part of Resources For the Future's MINK study, Bowes and Sedjo (1993) examined the impacts of a recurrence of the 1930's climate on the forests and forest sector of the United States Midwest. Without using a formal economic model they concluded that the economic impacts on the Midwest's forest sector were apt to be small, both because the simulated ecological impacts were small and because forest productivity and timber values were low.

These studies assume that overall changes in forest productivity are too small to affect production and consumption decisions, and, therefore, prices. Given the known responsiveness of plants to changes in atmospheric CO_2 concentrations, the possible impacts of forecasted changes in climate on plant growth, and the broad geographic scale over which global change is expected to occur, the output effects could be significant. With significant changes in the output of forest products will come changes in prices. The second type of assessment recognizes that price effects are likely to be consequential, and explicitly models market responses as well as ecological effects.

Binkley (1988) apparently performed the first such study by using a simple regression model of the impact of climate on forest growth to predict the effects of a CO_2 GISS climate scenario on the world's boreal forests. Simulations of a multi-region model of the world's forest sector (Kallio et al., 1987) found that the economic implications of these ecological changes differed among regions (the percentage change in timber revenues relative to a reference case ranged from +22.4% to -25.5%) but were positive overall (an increase in total economic surplus of from $US1980 25-150 billion, depending on the discount rate).

While the forest growth model used was exceedingly simple, and changes in forest growth outside the boreal zone were not considered, the general structure of Binkley's (1988) analysis is similar to that developed below.

Joyce et al., (in press) applied the same ecological model used in the present study to examine the economic impact of climatic change on forests in the United States. They linked the Terrestrial Ecosystem Model (TEM) estimates of changes in net primary productivity (NPP) under four GCM scenarios to a sophisticated forest-growth model used in the widely accepted Timber Assessment Market Model (Adams and Haynes, 1980) of the US forest sector. Their conclusions are similar to Binkley's (1988) (and, indeed, those of the present study), with higher forest growth rates producing larger timber supplies, lower prices and increased production and consumption in that country (Haynes et al., 1995). Some of the conclusions of the Haynes et al., (1995) study result from the fact that it treats the US in isolation from the rest of the world. For example, the modelled climatic change does not affect forest growth in Canada or the Canadian forest sector. The study concludes that increases in US domestic production will force down imports from Canada, despite the fact that the high latitude forests of Canada may respond more strongly to climatic change than those in the US.

Several studies have postulated changes in timber supply without any formal climatic or ecological modeling work, and then assessed the market implications of these changes. For example, Burton et al., (1994) examined six scenarios for changes in the growth of US forests on US forest products markets. Similarly, van Kooten and Arthur (1989) constructed a partial equilibrium model of North American wood products trade and estimated the effects of up to +/- 7.5% changes in timber supply in Canada and in the US. Because these studies did not link the economic analysis to a formal ecological analysis, they are technically not integrated assessments of the kind treated in this paper.

The present study goes beyond the previous research in several ways. In contrast to Binkley (1988), our analysis uses a sophisticated ecological model (Melillo et al., 1993) to estimate the response of forest vegetation to changes in climate. Unlike Joyce et al., (in press) who employed the same ecological model, our analysis covers the entire globe where that study focused on the US alone. We also examine the differences in economic outcomes from the different future climates simulated by the four GCMs in common use today.

Although the present analysis represents a clear advancement over previous work in this area, significant limitations remain. Specifically, we compare two steady states, one with current climate and a second with a modeled $2xCO_2$ climate, and do not examine the dynamics of how climate might change from the present to the future. Both the GCMs and TEM are equilibrium models. The GCM solutions represent equilibrium conditions resulting from modeled increases in radiative forcing associated with elevated CO_2 levels. The four GCMs differ in their treatment of oceans, and do not fully consider the thermal capacity of oceans which will induce lags in temperature response. TEM assumes that the vegetation is in equilibrium with the specified climate. Consequently we do not consider the impact of climatic change on the successional dynamics of forest or the dynamics of species migrations. Furthermore, we do not consider the impact of changes in the regimes of such disturbances as wildfires or outbreaks of insects or diseases, factors which could significantly affect the trajectory of forests in some locations (Kurz et al., 1995). While such disturbances can be managed on those lands which are readily accessible, a significant quantity of the world's forest products comes from remote areas under extensive development where management intervention is costly or impossible. Finally, using the terminology of Easterling et al., (1992) for agriculture, this assessment is a "dumb forester" model because we do not examine the many possible ways forest managers might adapt to future climatic conditions. Similarly, we do not explicitly consider how changes in forest growth or price-induced changes in the use of forest products will influence carbon sequestration and therefore future climate.

In this paper we first outline the integrated assessment methodology in three parts. (i) an overview of the climate change scenarios and how they were used to drive the ecological models, (ii) a summary of the ecological model describing how the climatic changes are translated into changes in net primary productivity (NPP), and (iii) a description of the econometric model of the world's forest sector with specific attention to how climate-driven ecological changes produce changes in the economic sector. The next section presents the simulation results for the four GCMs with two economic scenarios, one which assumes that currently undeveloped forest areas (e.g., Siberia) remain undeveloped, and a second which permits economic activity to expand into these areas. In the last section, we draw these results together, commenting on the total economic gain to the forest sector associated with climatic change, and the relative magnitude of uncertainty in the economic and biophysical models.

2. METHODOLOGY: AN INTEGRATED ECOLOGICAL/ECONOMIC ASSESSMENT

To assess the economic implications of climatic change on the world's forest sector requires analysis of three complex, interlocked systems: climate, the terrestrial biosphere, and global forest products markets. To do so we link extant, well-accepted models of each of these systems:

- Climate: the Goddard Institute of Space Studies (GISS) GCM (Hansen et al., 1983, 1984); the Oregon State University (OSU) GCM (Schlesinger and Zhao 1989): and two GCM simulations from the Geophysical Fluid Dynamics Laboratory (GFDL 1 and GFDL Q) (Manabe and Wetherald 1987, Wetherald and Manabe 1988);
- Terrestrial Biosphere: Terrestrial Ecosystem Model (TEM; Melillo et al., 1993); and
- Forest Products Markets: University of Washington Cintrafor Global Trade Model (CGTM; Cardellilchio et al., 1989).

Although each component is complex, the logic of the analysis is simple. Increased atmospheric concentrations of CO_2 and other greenhouse gases affect climate. The various GCMs forecast future temperature and precipitation regimes under the hypothesized future atmospheric concentrations of these gases. These changes in climate, as well as the changes in atmospheric CO_2 concentrations themselves, alter the growth of forests. TEM estimates the impacts of these modelled future climates on NPP. Changes in NPP translate directly into changes on forest growth which in turn affect forest inventories, especially over the long time periods considered in this analysis. Increases or decreases in forest inventory influence economic timber supply. The CGTM forecasts the impact of these shifts in timber supply on such key variables as forest products production, consumption, prices and trade. Our analysis depends on these three modeling systems, so it is important to review each.

2.1. The Climate Scenarios

The National Center for Atmospheric Research (Jenne 1992) provided output from the four GCM simulations, and these simulations were used to describe potential future climates under elevated atmospheric carbon dioxide levels. Each model was run under a baseline climate ($1xCO_2$) and under an elevated atmospheric carbon dioxide ($2xCO_2$) (Jenne 1992). Global temperature increases were 2.8°C for the OSU model, 3.0°C for the GFDL-1 model, 4.0°C for the GFDL-Q model, and 4.2°C for the GISS model. Global precipitation increases ranged from 7.8% in the OSU model to 11% in the GISS model. Differences in the global results across the models reflect the degree and extent of atmospheric physics modeled in each GCM (Jenne 1992). The output from these GCMs was interpolated to 0.5° longitude by 0.5° latitude resolution globally to provide the climate input needed for the ecological model (Melillo et al., 1993).

A contemporary climate was constructed using georeferenced datasets for air temperature, cloudiness, and precipitation converted to a 0.5° longitude by 0.5° latitude resolution globally (Raich et al., 1991). This contemporary climate provided a baseline for the ecological model. For each GCM, the relative changes in mean monthly temperature, monthly precipitation, and percent cloudiness between the $1xCO_2$ and the $2xCO_2$ scenarios were applied to the contemporary climate to produce four climatic change scenarios (Melillo et al., 1993). To examine the effects of potential changes in climate on forest productivity, the contemporary climate and these four altered climates were used to drive the ecological model.

2.2. The Terrestrial Ecosystem Model (TEM)

TEM is a process-based model which uses spatially referenced data on climate, soils, and vegetation to estimate NPP as affected by carbon and nitrogen cycling and environmental factors (Raich et al., 1991, McGuire et al., 1992, 1993, McGuire and Joyce, in press). Process-based models, such as TEM, carry fewer limitations than do regression-based approaches for predicting biogeochemical responses including net primary productivity to global change (McGuire et al., 1993). Other ecological models, such as BIOME (Prentice et al., 1992), and MAPPS (Nielson and Marks 1995), permit biomes to shift in response to changes in climate, but do not estimate productivity or changes in productivity in response to elevated atmospheric carbon dioxide and climate change. Nor do these models incorporate the interaction between nutrient cycling and climate change on productivity, effects that have been shown to be important (McGuire et al., 1992). TEM has been spatially extrapolated to the globe (Melillo et al., 1993). Other biogeochemical models such as BIOME-BGC (Running and Hunt, 1994) and CENTURY (Parton et al., 1987) are just now being spatially extrapolated to the globe. On a regional scale, these models have been compared in their response to elevated carbon dioxide and climate (VEMAP participants, in review).

TEM uses 18 vegetation types to describe the potential vegetation globally (Melillo et al., 1993). For this study, we considered only forest types: boreal forest, boreal forest wetlands, temperate coniferous forest, temperate mixed forest, temperate broadleaf evergreen forest, temperate forest wetlands, xeromorphic forest, tropical deciduous forest, tropical evergreen forest.

TEM computes monthly NPP as the difference between gross primary productivity and respiration, where respiration includes maintenance and construction respiration. The amount of available light, water and nitrogen, air temperature, and the atmospheric concentration of carbon dioxide all factor in the calculation of gross primary productivity (McGuire, 1992). For each vegetation type, intensive field data and literature values are used to calibrate the model. The globe was divided in to a 0.5° longitude by 0.5° latitude grid, and, based on a number of sources (Melillo et al., 1993), each grid cell was classified according to its dominant vegetation type. TEM's estimate of total global NPP, derived by summing NPP under the current climate over all grid cells, compares favourably with other global NPP values (Melillo et al., 1993).

For this study, contemporary global forest productivity was estimated using the contemporary climate to drive the environmental conditions (monthly temperature, monthly precipitation, percent cloudiness) in TEM. Forest productivity in a $2xCO_2$ world was estimated by simulating TEM under the four $2xCO_2$ climates described above. Under the $2xCO_2$ scenario, the atmospheric concentration of carbon dioxide in the ecological model was assumed to be 625 ppm (see Melillo et al., 1993).

Increases in global NPP estimated by TEM vary from 20.0% to 26.1% across the four GCMs. Within ecosystem types, the boreal and temperate forest types showed the greatest increases and tropical forests showed the largest decreases. The factors dominating the vegetation response differ across the globe. In the tropical and dry temperate ecosystems, elevated CO_2 is the primary determinant increasing NPP. In the northern and moist temperate ecosystems, elevated temperatures enhance nitrogen availability, essentially fertilizing the forests.

For the purpose of modeling timber supply, the CGTM classifies all vegetation as either softwood and hardwood. These broad classifications necessitate re-classifying the ecosystem types used in TEM into softwood and hardwood types. To do so, we overlaid the CGTM regional boundaries on the global vegetation map used in TEM and aggregated cells according to the two broad species types. To determine the changes in softwood forest productivity under climate change, within each CGTM region we computed, for softwood-dominated grid cells, the difference between the contemporary NPP results and the NPP result for each of the four $2xCO_2$ simulations. The calculations for hardwood productivity changes are similar.

These estimates describe the effects of climate change on forest productivity under the four $2xCO_2$ scenarios. Under the IPCC "Business-as-Usual" scenario (Jenkins and Derwent 1990), doubling of carbon dioxide (625 ppm) would occur by approximately 2065. To be consistent with this scenario we assumed the changes under the altered climate would be manifest by 2065 and exponentially interpolated between 1995 and 2065 to the annual values needed for the CGTM. Figures 1a and b show the resulting estimates of the percentage change in NPP in 2040 in each of the CGTM regions under each of the GCMs for softwoods (1a) and hardwoods (1b). The differences among GCMs tend to be small, except in SE Asia and Oceania for hardwoods, and the northern regions for softwoods.

2.3. The Cintrafor Global Trade Model (CGTM)

Forest products production, consumption, prices and trade are modeled with a computable spatial equilibrium model of the global forest economy. The CGTM (Cardellichio et al.,

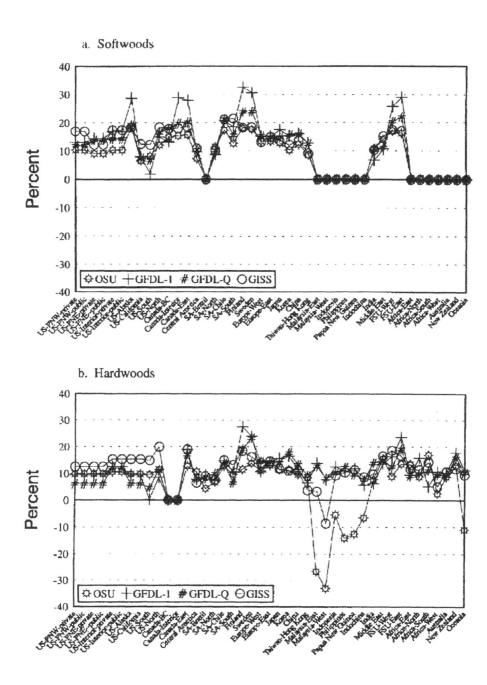

FIGURE 1. Productivity shifts over 50 years.

1989) is a modified version of the International Institute for Applied Systems Analysis' global trade model (Kallio et al., 1987). The CGTM computes market equilibria for regional forest product sectors considering constraints on processing capacity and available wood resources. CGTM divides the globe into 43 log-producing regions and 33 product-consuming regions. The CGTM is the most spatially detailed forest sector model currently available: it includes over 400 trade flows.

The model, a price-endogenous linear program, solves for equilibrium prices, production, consumption and trade levels for each region in each period. The demand for forest products in each region may be met by its own domestic production or by imports from other regions. A region will import some or all of its consumption needs if the price of the domestic product is higher than the cost associated with purchasing it outside of the region and transporting it home, including any tariffs. Increases in imports force down domestic prices until market prices are in equilibrium: the price in the importing region is equal to the price in the exporting region plus transportation costs. The competitive market structure imposed on the model assures that final demand is met employing the least-costly producer after accounting for transportation costs.

The CGTM computes a partial equilibrium solution for each year over the period 1990 to 2040. Recursive relations for timber growing stocks, processing costs, mill capacities and other factors are used to solve the market equilibrium problem for each subsequent time period based on the solution for the previous period. The dynamic structure of the model does not imply intertemporal market equilibrium because the CGTM does not compute the market solution for all periods simultaneously.

The CGTM is an integrated model since it describes all aspects of forest products production: forest growth, timber supply, processing capacity and final demand. It is a partial equilibrium model because it determines only wood use endogenously while currency exchange rates and the markets for other factor inputs—capital, labor and energy—are exogenous.

The effects of climatic change on forest productivity enter the CGTM through its forest growth and timber supply components. The timber supply equations specify the harvests in each region given information on log prices, forest inventory levels, and other information. Depending on the particular region, the CGTM uses one of four log supply functions. Appendix I summarizes the formulations used in each region.

The most common supply specification (used in most of the US regions, Finland, Sweden, Western Europe and Japan) estimates the quantity supplied as a positive function of both log prices and the level of growing stock inventory. Changes in growing stock inventory equal the difference between forest growth and harvests. Changes in NPP affect the timber growth term of this inventory accounting equation. The resulting increases or decreases in the timber inventory shift the supply curve out or in.

The second specification has an upward sloping supply curve but also places an upper limit to log supply. If a region is operating below the limit on harvests, timber supply will respond positively to increases in log prices. When harvest levels hit the limit, further increases in timber prices have no effect on harvest levels. The upper limit on harvests can be thought of as the maximum allowable annual cut (AAC). Timber supply in the Canadian regions, Chile and New Zealand is modeled in this fashion. Since the AACs are generally set with reference to long-run forest growth rates, changes in NPP are directly translated into changes in the AAC limit.

A third specification fixes log supply at a predetermined level for each period of the simulation. In any one period, log supply is wholly unresponsive to price changes. Examples of regions utilizing this strict AAC supply specification include China and the CGTM-defined Eastern Europe region where planning rather than prices determine harvest levels. Public timber supply regions in the US are also specified using projected harvest levels. The specified levels reflect a combination of constraints based on the physical productivity of the forest, constraints based on the infrastructure available to access the forest, and constraints based on policies which restrict industrial access to forest lands. As a result, we examine two cases, one where the change in NPP results in no change in the AAC level and a second where changes in NPP in the particular region results in proportional changes in the AAC constraint.

A fourth alternative for log assumes that log supply is perfectly elastic, so log supply is determined by product supply. An example is Brazil's hardwood forest sector where log supply is not a constraint on the production of hardwood products. In the case of Brazil, log supply behavior is implicit in product supply behavior.

3. RESULTS

The GCM and TEM results highly condition the economic results. Because NPP increases worldwide, timber becomes more abundant, prices fall and consumption increases. Regional differences in NPP changes translate into regional differences in economic impact, but the translation is not direct. Because infrastructure rather than forest growth rates constrain timber supply in some very large regions—e.g. Siberia—we present two economic scenarios. In the first (called the "Intensive Margin Scenario"), we constrain harvests to the reference case levels in all of the regions where the third type of supply curve— a fixed AAC—obtains, except for China. In effect, we permit no shift in the extensive margin of forest exploitation. China is an exception because it is currently a net importer of wood products. As a result, one would expect higher growth rates to translate directly into higher levels of domestic production.

In this scenario, increases in NPP force down timber prices even without increased supplies from the regions which are currently outside the extensive margin. As a result, there would be little economic incentive to push more deeply into these high cost, unexploited forests. Hence, this scenario might be a realistic expectation of market response despite the fact that the physical growth of some currently unexploited forests might increase.

The second set of simulations assumes that changes in growth rates would lead to changes in harvests in these remote regions. This scenario includes all of the supply responses from the Intensive Margin Scenario as well as shifts in timber supply which are proportionate to changes in NPP in all the other regions. Because this scenario implies outward shifts in the extensive margin for timber production, it is labelled "Extensive Margin Scenario" below.

Examining these two sets of scenarios provides a sense of the range of uncertainty in the economic model. Because we examine both of the economic scenarios with all four of the GCMs, we can also compare the sources of model-based uncertainty. It turns out that the uncertainty in the economic models is at least as great as the uncertainty in the climate models.

3.1. Intensive Margin Scenarios

Figure 2 presents the key results for these scenarios. For the TEM simulations of softwoods, NPP either increases or remains unchanged under all four of the GCM scenarios.

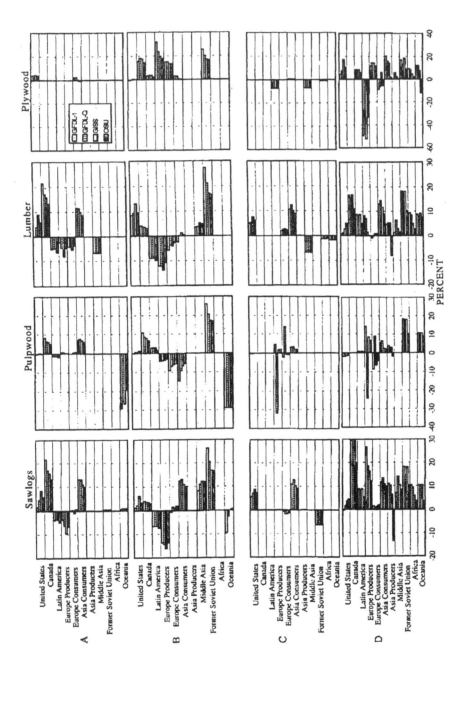

FIGURE 2. Changes in sawlog, pulpwood, lumber, and plywood production. Panel A. Softwoods, intensive margin. Panel B. Softwoods extensive margin. Panel C. Hardwoods, intensive margin. Panel D. Hardwoods, extensive margin. Bars ordered by GCM from top to bottom.

Forest sector activity increases as well, but generally somewhat less than the increase in NPP. Canada, China and other consumer countries in Asia capture nearly all of the domestic gains in forest productivity associated with the warmer climate scenarios, while other regions capture only a portion.

Average increase in NPP for Canada as a whole ranges from 15% (OSU) to 22% (GFDL 1). Additional softwood log production in Canada is slightly lower: 13% (OSU) and 22% (GFDL 1). In China, increases in forest productivity range from 12% (OSU) to 16% (GFDL 1 and GFDL Q). Increases in log production for consumer countries in Asia are slightly lower: 10% (OSU) and 13% (GFDL 1 and GFDL Q).

In other regions, however, the changes in log production are substantially lower than are the increases in NPP associated with the modelled climate changes. In the US, increases in log output range from 1.5% (GFDL 1) to 8% (GISS), considerably less than the increases in NPP (13%, GFDL 1; 17%, GISS). In European producer countries—Finland and Sweden—log output *declines* from between 4% and 10% (depending on the GCM), even though the climatic change scenario suggested significant increases in forest growth—18% (OSU) to 32% (GFDL 1). A similar response occurs in Latin America. In consumer countries of Europe—Western and Eastern Europe—increases in forest growth associated with warmer climate average 14% across climate scenarios, but there is no appreciable change in log output.

Higher production of pulpwood and residual chips from lumber and plywood manufacturing in Canada and China displace pulpwood production in the Oceanic region. The result will hold as long as the cost of hauling residual chips derived from lumber manufacturing from Canada to Japan, for example, is less than harvesting and transporting roundwood chips from New Zealand and Australia.

Finally, production of structural panels in the US increases dramatically under all GCM-modelled climates. The US has a significant cost advantage in this class of products, and, in the reference case, accounts for nearly 75% of the global production. In the US, increases in timber supply translate into increases in panel production because other regions—especially Canada—enjoy cost advantages in lumber production. The gains in production, however, are significantly less than the estimated increases in forest growth.

Although the NPP impacts for hardwood forest types are significantly less than they are for softwoods, a somewhat similar story for the hardwood sector emerges, albeit with different players (Figure 2b). China, facing constraints in hardwood timber supply as well as with softwoods, captures as increased sawlog production nearly all of the increased forest growth associated with climatic change. Increases in NPP range from 10% (OSU) to 13% (GFDL Q) where increases in log output range from 9% (OSU) to 13% (GFDL Q). The US also captures in domestic consumption most of the gain in NPP in the form of expanded sawlog output. Increases in forest growth in the US South and North regions—the major hardwood producers for the US—range from 4% (GFDL 1) to 17% (GISS) and log outputs increases from 6% (GFDL 1) to 9% (GISS).

Higher log output in the US and China reduces log production in the former Soviet Union and European consumer countries regardless of the positive impact climatic change has on forest growth in these regions. In Europe, log production falls, but increases in log imports from the US allows lumber manufacturing to expand. Higher log output in the US and China, combined with lower log exports from the former Soviet Union, also redirects trade flows of lumber, and these new trade patterns reduce lumber manufacturing activities in the Middle Asia region, Africa and Oceania.

3.2. Extensive Margin Scenario

Whether the former Soviet Union will develop its forest resources is one of the enigmas of modern forest economics. The region represents one fifth of global softwood production and has two-thirds of the global softwood inventory. Under extensive management, the increases in NPP associated with climate change would not necessarily translate into greater timber supply as investments needed to reach new timber areas would be required. However, if timber prices were to increase over time, investments to expand the area under forest management would become economic. Climate-induced productivity gains will, all else equal, improve the economics associated with extensive forestry such that harvests occurs in new areas. The Extensive Margin Scenario responds to these uncertainties.

Results from the analysis indicate that the former Soviet Union can have a substantial impact on the outcome. Most notably, production from the former Soviet Union substitutes for Canadian production. If harvest levels in the former Soviet Union expand in response to climatic change, increased Canadian log production is reduced dramatically. Output gains in the US also suffer, although the impact is much smaller. Similarly, the higher output in the former Soviet Union decreases log production in Europe.

3.3. Economic Welfare Effects

Figure 3 shows the changes (in $US1980/yr for the year 2040, measured from the reference case conditions) in surplus measures for timber owners, mill owners and consumers for the most important softwood markets. The figures present average surplus change for the four climate change scenarios. Figure 3a presents the intensive margin results and Figure 3b presents the extensive margin results.

Recall that harvest levels generally rise as a consequence of climatic change, and prices fall. As a result of the demand conditions, the decline in timber prices is generally greater than the increase in harvest levels, and timber owners suffer. Figure 3a shows that, in all regions, producer's surplus for timber owners declines. These results differ from earlier estimates by Binkley (1988) where climatic change produces increased timber income from all northern latitude forest areas, with the exception of Sweden. The difference occurs because the present study permits forest growth and harvest levels to expand where ever NPP increases, while the previous study considered boreal forests alone. Figure 3b shows that losses to timber owners are greater when forests at the extensive margin expand production.

The regional distribution of economic welfare depends on the structure of forest sector production and consumption in each region. The larger the proportion of timber producers in a region, the greater the potential for regional economic losses. The larger the proportion of forest products consumers in a region, the greater the potential for regional economic gains. Overall, the US benefits from climatic change under the scenarios presented here. The large gains to US consumers and a smaller gain to US mill owners more than offset the losses to timber owners. The economic welfare for the US as a whole amounts to slightly more than $US1980 1 billion annually by 2040.

Consumers of softwood products in Europe and Japan—two other large consuming economies—also gain economic welfare. However, timber owners in Europe, primarily the

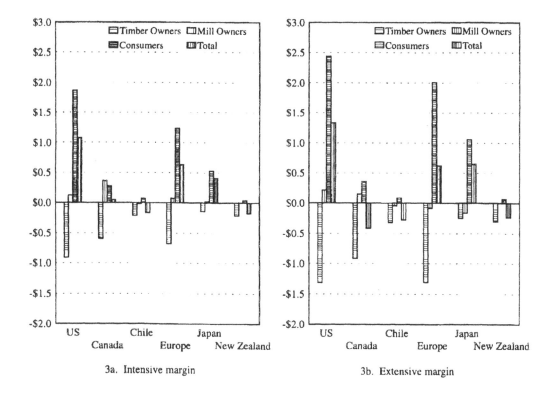

FIGURE 3. Changes in economic surpluses, averaged over four GCMs.

Scandinavian regions, suffer losses. The losses are particularly large if the extensive margin shifts in response to higher NPP.

As a result of these market effects, regions which are major net timber producers do not necessarily benefit from climatic change-induced increases in NPP. For example, the total economic gains in Canada is positive but small. Both Chile and New Zealand suffer significant economic losses. Increased production of logs at the extensive margin exacerbate these losses to timber producers (compare Figure 3a and 3b). The lack of any substantial amount of processing capacity in New Zealand and Chile and small domestic consumption of softwood products results in a greater loss in economic welfare for these regions. These results would differ if energy policies strongly favoured renewable sources so that the demand for the additional timber expanded.

4. CONCLUSIONS

This study links state-of-the-art models of global climate, forest productivity and the forest sector to examine the economic effects of climatic change. As a result of climatic change, global net primary productivity will apparently increase (a finding reported previously by Melillo et al., 1993). These increases in forest growth should unambiguously benefit consumers of forest products. The impact on timber producers is mixed, but generally negative as prices fall more than output expands. These conclusions are robust against the different GCMs and different specifications of the possible global supply response.

Losses to timber producers arise not because forest decline in the face of climatic change, but, ironically, because growth rates increase. The resulting increase in output is not sufficient to offset price declines, so, taken as a whole, timber producers suffer economically.

This study omits a myriad of important ecological and economic questions surrounding climatic change and forests. Foremost among these are (i) possible changes in catastrophic events such as fire and outbreaks of insects and diseases which in many parts of the world already consume more timber than humans do, (ii) the details of the dynamics of ecosystem change which could produce rather different outcomes, particularly in the short run and at different regional scales, (iii) the capacity of humans to adapt to the modelled changes in climate, and (iv) impacts of climatic change on the nontimber values of the forests which, important in their own right, will also influence the availability of land for timber production.

A complex regional distribution of losses and gains emerges from the analysis, depending primarily (but not exclusively) on the region's position as either a net consumer or producer. Under the scenarios analyzed here, the big consuming economies—the United States, Japan, and Europe—all benefit from climatic change. Two of the most dynamic timber supplying regions—New Zealand and Chile—lose. These results can probably be generalized to other regions which have made major investments in forest plantations for export markets. The present value of the total increase in economic surplus related to the forest sector ranges from $US1980 10.7 to 15.9 billion using a 4% real discount rate (and $0.5 billion to 1.8 billion using a 10% discount), positive but lower than the range previously reported by Binkley (1988). Hence, including more of the possible economic adjustments reduces the measured impact of climatic change.

Differences among the GCMs produce relatively minor differences in the forest products markets. Trade tends to dampen the effects of these differences. Difference in economic assumptions about timber supply response in the comparatively undeveloped regions produce variations in economic outcome of comparable magnitude to the differences among GCMs. From the perspective of the forest sector, it appears to be at least as important to have a more precise understanding of economic and institutional responses to climatic change as it is to improve our understanding of the biophysical responses.

Different vegetation response models may produce estimated economic responses which differ significantly from those reported here, and these different models might also alter the relative sensitivity of the economic results to the different GCMs and economic assumptions. Completing the full factorial experiment of climate, vegetation response, and economic models would illuminate these remaining uncertainties.

LITERATURE CITED

Adams, D.M. and R.W. Haynes, (1980), The 1980 timber assessment market model: structure, projections and policy simulations. *Forest Science Monograph* 22.

Binkley, C.S. (1988), A case study of the effects of CO_2-induced climatic warming on forest growth and the forest sector: B. economic effects on the world's forest sector, in M. L. Parry, T.R.

Carter and N.T. Konijn, eds. *The impact of climatic variations on agriculture.* (Kluwer Academic: Dordrecht).

Binkley, C.S. and G. C. van Kooten, (1994), Integrating climatic change and forests: economic and ecologic assessments. *Climatic Change* **28**: 91-110.

Bowes, M.D. and R.A. Sedjo, (1993), Impacts and responses to climatic change in forests of the MINK regions. *Climatic Change* **24**: 63-82.

Burton, D.M, B.A. McCarl, D.M. Adams, R. Alig, J.M. Calloway and S.M. Winnett, (1994), An exploratory study of the economic impacts of climate change on southern forests: preliminary results. xerox.

Cardellichio, P.A., Y.C. Youn, D. Adams, R.W. Joo and J. Chmelik, (1989), A preliminary analysis of timber and timber products production, consumption, trade and prices in the Pacific Rim until 2000. Cintrafor *Working Paper* **22**, University of Washington, College of Forest Resources, Seattle.

Ceulemans, R., and M. Mousseau, (1994), Effects of elevated atmospheric CO_2 on woody plants. *New Phytology* **127**:425-46.

Easterling, W.E., M.S. McKenney, N.J. Rosenberg, and K.M. Lemon, (1992), Simulations of crop responses to climatic change: effects of present technology and no adjustments (the "dumb farmer" scenario). *Agricultural and Forest Meterology* **59**: 53-73.

Gates, W.L., J.F.B. Mitchell, G.J. Boer, et al., (1992), Climate modelling, climate prediction, and model validation. in Houghton, (J.T. et al., eds.) *Climate change 1992: the supplementary report to the IPCC scientific assessment.* Cambridge, UK: Cambridge University Press: 97-134.

Hansen, J., A. Lacis, D. Rind, et al., (1984), Climate sensitivity: Analysis of feedback mechanisms. in (J.E. Hansen, T. Takahashi, eds.) Climate processes and climate sensitivity. *American Geophysical Union*: 130-163.

Hansen, J., G. Russel, D. Rind, et al., (1983), Efficient three-dimensional global models for climate studies: models I and II. *Monthly Weather Review* **111**: 609-662.

Haynes, R.W., D.M. Adams, and J.R. Mills, (1995), The 1993 RPA timber assessment update. USDA Forest Service *General Technical Report* RM-GTR-259.

Hodges, D.G., J.L. Regens and F.W. Cubbage, (1989), Evaluating the potential impacts of global climatic change on forestry in the southern United States. *Resource Management Options* **6**: 235-251.

Idso, K.E., and S.B. Idso, (1994), Plant responses to atmospheric CO_2 enrichment in the face of environmental constraints: a review of the past 10 years research. *Agricultural and Forest Meteorology.* **69**: 153-203.

IPCC. 1990. *Climate Change. The IPCC Scientific Assessment.* (J.T. Houghton, G.J. Jenkins, J.J.†Ephraums, eds.) Cambridge University Press, Cambridge. 362 p.

Jenkins, G.J. and R.G. Derwent, (1990), Climatic consequences of emissions. in (J.T.†Houghton, G.J. Jenkins, and J.J. Ephraums, eds.) *Climate Change. The IPCC Scientific Assessment.* World Meteorological Organization United Nations Environment Programme, Cambridge University Press, Cambridge. 364 pages.

Jenne, R.L. (1992), Climate model description and impact on terrestrial climate. in (Majumdar, et al., eds.) *Global climate change: implications, challenges, and mitigation measures.* Easton, PA: Pennsylvania Academy of Sciences: 145-164.

Joyce, L.A., J. Mills, L. Heath, A.D. McGuire, R.W. Haynes, and R.A. Birdsey, in press. Forest sector impacts from changes in forest productivity under climate change. *Global Ecology and Biography Letters.*

Kallio, M., D. P. Dykstra, and C. S. Binkley, (1987), *The global forest sector: an analytical perspective.* (John Wiley and Sons: London).

Kurz, W.A., M.J. Apps, S.J. Beukema, and T.Lekstrum, (1995), 20th century carbon budget of Canadian Forests. *Tellus* **47B**:170-177.

Laarman, J.G. and R.A. Sedjo, (1992), *Global forests: issues for six billion people.* (McGraw Hill: New York).

Manabe, S. and R.T. Wetherald, (1987), Large scale changes in soil wetness induced by an increase in carbon dioxide. *Journal of Atmospheric Science.* **44**: 1211-1235.

McGuire, A.D., L.A. Joyce, D.W. Kicklighter, J.M. Melillo, G. Esser, and C.J. Vorosmarty, (1993), Productivity response of climax temperate forests to elevated temperature and carbon dioxide: A North American comparison between two global models. *Climatic Change* **24**:287-310.

McGuire, A.D., J.M. Melillo, L.A. Joyce, D.W. Kicklighter, A.L. Grace, B. Moore III, and C.J. Vorosmarty, (1992), Interactions between carbon and nitrogen dynamics in estimating net primary productivity for potential vegetation in North America. *Global Biogeochemical Cycles.* **6**: 101-124.

Melillo, J.M., A.D. McGuire, D.W. Kicklighter, B. Moore III, C.J. Vorosmarty, and A.L. Schloss, (1993), Global climate change and terrestrial net primary production. **Nature** 363:234-240.

Nielson, Ronald P., and D. Marks, (1995), A global perspective of regional vegetation and hydrologic sensitivities from climatic change. *Journal of Vegetation Science* 5:715-730.

Norby, R.J. (1994), Issues and perspectives for investigating root responses to elevated atmospheric carbon dioxide. *Plant and Soil* **165**: 9-20.

Parton, W.J., Schimel, D.S., Cole, C.V., and D.S. Ojima, (1987), Analysis of factors controlling soil organic matter levels in Great Plains Grasslands. *Soil Sci. Soc. Am. Proc.* 561:1173-1179.

Prentice, I.C., W. Cramer, S.P. Harrison, R. Leemans, R.A. Monserud, and A.M. Solomon, (1992), A global biome model based on plant physiology and dominance, soil properties and climate. *Journal of Biogeography* 19:117-134.

Raich, J.W., E.B. Rastetter, J.M. Melillo, D.W. Kicklighter, P.A. Steudler, B.J. Peterson, A.L. Grace, B. Moore II, and C.J. Vorosmarty, (1991), Potential net primary productivity in South America application of a global model. *Ecological Applications* **1**: 399-429.

Running, S.W., and E.R. Hunt, (1994), Generalization of a forest ecosystem process model for other biomes, Biome-BGC, and an application for globa-scale models. Pages 141-158 in J.R. Ehleringer and C. Field editors. *Scaling processes between leaf and landscape levels.* (Academic Press, San Diego).

Schlesinger, M.E. and Z. Zhao, (1989), Seasonal climatic changes induced by doubles CO_2 as simulated by the OSU atmospheric GCM/mixed-layer ocean model. *Journal of Climate* **2**: 459-495.

van Kooten, C.G. and L.M. Arthur, (1989), Assessing the economic benefits of climate change on Canada's boreal forest. *Canadian Journal of Forestry Research* **19**: 463-470.

VEMAP participants. in review. Vegetation/Ecosystem modeling and analysis project (VEMAP): Comparing biogeography and biogeochemistry models in a continental-scale study of terrestrial ecosystem response to climate change and CO_2 doubling. *Global Biogeochemical Cycles.*

USDA Forest Service, (1990), The Forest Service program for forest and range resources: Recommended 1990 RPA program. (Forest Service: Washington, DC).

Wetherald, R.T. and S. Manabe, (1988), Cloud feedback processes in a general circulation model. *Journal of Atmospheric Science.* **45**: 1397-1415.

Wullschleger, S.D., L. H., Ziska, and J.A. Bunce, (1994), Respiratory responses of higher plants to atmospheric CO_2 enrichment. *Physiology Plantae* **90**: 221-29.

Critical Reviews in Environmental Science and Technology, 27(Special): S139–S149 (1997)

SILVICULTURAL OPTIONS TO CONSERVE AND SEQUESTER CARBON IN FOREST SYSTEMS: PRELIMINARY ECONOMIC ASSESSMENT

ROBERT K. DIXON

Director, U.S. Support for Country Studies to Address Climate Change, PO-63, 1000 Independence Ave., SW, Washington, DC, 20585 USA

ABSTRACT: Forest establishment and management can be employed to conserve and sequester carbon (C) in the terrestrial biosphere. A 40 nation assessment of silvicultural practices and techniques revealed that forest drainage, thinning, fertilization, weeding and modified harvesting can be employed to sequester or conserve 1-64 Mg C ha^{-1}. Sequestration or conservation of C in forest systems due to the application of silvicultural practices can be achieved for $2-56 Mg C. A small proportion of the world's forests are actually managed, and, theoretically, silvicultural practices could conserve or sequester 0.1-0.3 Pg C annually. Silvicultural practices can contribute to future C sequestration of forests, but the biologic and economic opportunities and constraints merit careful consideration before large-scale implementation in the field.

KEY WORDS: silviculture, carbon, climate change, greenhouse gases, mitigation.

1. INTRODUCTION

The accumulation of greenhouse gases in the Earth's atmosphere has stimulated interest in land-use sector mitigation options (Wisniewski and Sampson, 1993). Recent assessments have identified forest establishment options to conserve and sequester carbon (C) in forest and agroforest systems (Dixon et al., 1994; Winjum et al., 1993a,b; Swisher, 1991). Biologic, financial and logistical options to sequester CO_2 in tropical, temperate and boreal forest systems have been identified in key nations (Winjum et al., 1993a). Over the past few years, forest-sector C offset pilot projects have been established to field test viability of C conservation and sequestration options (Dixon et al., 1993b).

Historically, silvicultural practices (e.g., forest drainage, thinning, fertilization, weeding, and modified harvesting) are directed at increasing the quality and quantity of the forest resource. These practices may also promote forest CO_2 sequestration and C conservation (e.g., Plantinga and Birdsey, 1993; Birdsey et al., 1993). Although considerable information is available on the cost of forest establishment and juvenile plantation management (e.g., Krankina and Dixon, 1994; Sedjo, 1983), there has been relatively little comparative economic analysis of sequestering CO_2 by applying silvicultural practices in forest systems of key nations (Sedjo et al., 1995). Assessments by Harmon et al. (1990) and Kershaw et al. (1993) examine the effects of harvesting old growth Douglas-fir stands on C storage, but neither provides economic insight into various management possibilities.

1064-3389/97/$.50

Similarly, the economic dimensions of the forest management activities were reviewed by Sampson et al. (1993), but no analysis was presented.

At high- (boreal) and mid- (temperate) latitudes, there are few studies that consider the macro- or microeconomics of CO_2 sequestration attributable to silvicultural practices. Moulton and Richards (1990) developed a project cost curve which characterizes C sequestration on forest lands, where, in addition to artificial regeneration, they considered both active and passive forest management. The economic "efficiency" of various strategies or approaches to sequester CO_2 in Norway forest systems was examined by Solberg and Hoen (1995). The approach and units in the study make direct comparison with the literature somewhat difficult, but the quantity of CO_2 sequestered for most silvicultural strategies examined was economically comparable to other options. In a related analysis, Hoen and Solbert (1994) asserted that while stand thinning is not a cost effective means of sequestering CO_2, forest fertilization stimulates marginal C sequestration at a cost of about $71 Mg C.

Within low latitudes (tropical), the biologic C storage potential and project costs of silvicultural options to sequester C have only recently been examined. Early analysis has focused on selective logging practices that reduced damage to residual trees and decreased soil disturbance which reduced C emissions. Putz and Pinard (1993) estimate that modest modifications in harvesting techniques in the forests of Sabah, Malaysia, a technique called reduced impact logging, can conserve up to 46 Mg C ha at a cost of less than $3 Mg C. However, the applicability of this analysis to temperate forests may be limited because, while the tropical forest harvest regime involves selective logging, temperate and boreal forest systems are harvested by selective, shelterwood or clear cuts.

Forest fertilization stimulates biomass production on some sites, but economic analysis of the C sequestration potential has not been thoroughly examined. Turner et al. (1993) analyzed several forest management options and determined that the least promising U.S. silvicultural practices, from a C conservation perspective, included thinning, fertilization and other stand improvement treatments. In contrast, Huettl and Zoettl (1992) and Nilsson (1993) suggest forest fertilization methods may be a promising C storage stimulation option in central and northern Europe. Recycling of municipal waste, as a fertilizer to stimulate forest productivity is also a promising possibility (Nilsson, this volume).

Many silvicultural practices are designed to improve some aspect of the wood and/or tree form and may do little to increase C sequestration (Sharma, 1992). Thus, silvicultural techniques that allow greater portions of the forest to be harvested and converted to long-lived wood products may conserve a greater volume of C in the long term (Row and Phelps, 1992). Thinning of pulpwood plantations may not increase the total volume of the stand, but this technique may generate a greater total amount of sequestered C in solid wood products and a smaller portion of residual short-lived paper products and wood by-products. However, the comparative C implications and project costs of this hypothesis have not been carefully examined.

Given the scope of earlier reports on silvicultural options to conserve or sequester C in forest systems, this analysis has three primary objectives: (1) review incremental C storage potential of applying intermediate silvicultural practices in boreal, temperate and tropical forests; (2) assess project financial costs of applying silvicultural practices at the site level in selected key nations; and, (3) identify biologically and economically promising silvicultural practices that can be applied on technically suitable land to conserve or sequester forest C using current infrastructure and technology.

2. MATERIALS AND METHODS

2.1. Data Collection

The assessment was based, in part, on the assembly of a global database of biologic and economic information on promising silvicultural options to conserve and sequester forest C (Winjum et al., 1993a,b). Information and analysis of silvicultural practices and economic factors associated with their implementation in approximately 40 nations representing boreal, temperate, and tropical biomes on six continents were collected using previously described methods (Dixon et al., 1993a; Schroeder, 1991; Winjum et al., 1993a,b). Regional and national data were collated and analyzed in four major categories: (1) impact of silvicultural practices on C flux; (2) biomass accretion (CO_2 sequestration) or C conservation resulting from silvicultural practices; (3) economic project costs for application of silvicultural practice(s); and, (4) area of land potentially suitable for application of silvicultural practices. The preliminary data base structure and organization has been described previously (Winjum et al., 1993a,b). Specific assumptions and procedures associated with data presented in this paper are described in the next three subsections: Silvicultural Practices, Economic Factors, and Land Technically Suitable.

2.2. Silvicultural Practices

A 40 nation survey of silvicultural practices and their impact on CO_2 sequestration or conservation was completed. The practices were grouped into major types: site drainage, fertilization, thinning, weeding and harvesting. Data collected included type of practice (e.g., fertilization), forest crops, site conditions, and potential for CO_2 sequestration or conservation. Given the myriad of silvicultural practices, only a representative group of options are presented in this report.

Vegetation biomass production (C sequestration) and C conservation data were collected for silvicultural practices currently being implemented on an operational basis in the nations surveyed (Dixon et al., 1993b). The primary criteria for selection of promising practices was the potential for C sequestration or conservation (Winjum et al., 1993a,b). Silvicultural practices were emphasized in this assessment. Other forest management practices and technologies (e.g., forestation) were the subject of earlier reports (Dixon et al., 1994; Dixon et al., 1993a).

The relevant parameter in terms of C accretion calculation is the average amount of C on site over an indefinite number of rotations (Schroeder, 1991). If it is assumed that the ecosystem is sustainable and there is no yield reduction in later rotations, the result is the same as the average amount of C on site over one full rotation. Because any number of biologic, climatic, or socioeconomic events could contribute to some level of yield reduction that cannot be predicted, this approach may represent an upper limit of C storage at the site level (Dixon et al., 1994). Carbon accretion and storage in this analysis was calculated by summing up the C content of biomass for every year in the rotation and dividing by the rotation length. In this report, temporary C storage is considered in the time frame of years to decades, and long-term implies C storage from decades to centuries (Dixon et al., 1994).

The assumptions of this analysis simplify historical C budget calculations in two important ways (Harmon et al., 1990). First, the best evaluation of the C benefits of a silvicultural practice is the difference between the C balance over the same number of rotations of that practice compared to the practices currently utilized. Second, forest regrowth processes and the introduction of a new land use(s) increase the C balance on a given site, in contrast to the assumption of instantaneous release of C to the atmosphere (Brown et al., 1993). A static C budget, based on the assumptions stated above, was employed in the analysis to project the net C conserved and/or sequestered by implementing silvicultural practices on technically suitable land (Kauppi et al., 1992).

2.3. Economic Factors and Silvicultural Practices

The relative project financial costs of promising silvicultural practices employed in this assessment were estimates of direct costs, at the site level, for labor, materials, transportation, and the initial infrastructure (for up to two years) to employ the options (Dixon et al., 1993 a,b; Winjum and Lewis, 1993; Rubin et al., 1992). Scaling of costs was not considered because previous analyses suggest this approach may be invalid (Row, 1978). Indirect costs (e.g., project administration, road construction) and secondary economic impacts (e.g., effects of increasing supply of a commodity on market price) were not considered in this analysis but have been considered in parallel assessments (Winjum and Lewis, 1993). The cost and/or rent of land was not included because data are highly variable (Dixon et al., 1993 a,b; Trexler and Haugen, 1994). Land cost and/or rent varies widely around the world and values are difficult to establish where land is held in common by communities, or where virtually all land is government-owned and no market values exist (Nordhaus, 1993).

Economic data are reported in 1994 US dollars. Costs for any reference year were adjusted, based on the inflation and exchange rates for individual nations according to the International Financial Statistics (IFS) tables published by the International Monetary Fund (1994). A nation's inflation rate for the reference year, as measured by the Consumer Price Index, was extracted from the IFS tables. The reference year cost was then converted to a 1994 value and converted back to US dollars at the 1994 exchange rate (Dixon et al., 1993; Winjum and Lewis, 1993).

Because forest systems are periodically harvested (i.e., intermediate or final harvest) and regenerated, the costs of applying silvicultural practices are sometimes recurring (Trexler and Haugen, 1994). In estimating direct costs, it is important to account for these additional investments that will occur at more or less periodic intervals in the future. The net interest rate used was 5% because this value is representative of long-term global rates for forest commodities (Winjum and Lewis, 1993). Project costs in U.S. $ per Mg C were calculated as the present value of all establishment costs over a 50-year period divided by mean C storage (Dixon et al., 1993a). Financial costs computed in this manner do not account for land rent nor any financial benefits resulting from the initial investment (Nordhaus, 1993; Manne and Richels, 1990). If full direct and/or indirect benefits assessment data were available and compared with establishment and maintenance costs over the 50-year time horizon, many practices may have positive benefit-to-cost ratios

(Winjum and Lewis, 1993). These economic assumptions were verified by evaluating forest sector C offset projects in six nations in 1991-1993 (Dixon et al., 1993b).

2.4. Land Area Technically Suitable

Estimates of suitable land area potentially available for selected silvicultural practices are required in order to estimate total amount of CO_2 removal and C storage (Turner et al., 1993; Winjum et al., 1993b). Estimates of land available on which to implement silvicultural practices are based on assessments by Dixon et al. (1994) and Trexler and Haugen (1994). These assessments consider only technically suitable lands which are currently under-utilized. Within low latitudes, the assessment relies primarily on Advanced Very High Resolution Radiometer (AVHRR) analyses of land-use patterns. For the temperate zones, national inventories of land-use practice and patterns were consulted (e.g., Kauppi et al., 1993; Birdsey et al., 1993). In Russia and other high-latitude regions, national inventories of land-use practices were employed (Krankina and Dixon, 1994). The C accretion and storage values were also calculated on a land unit area basis (e.g., Mg C ha^{-1}).

3. RESULTS

A myriad of silvicultural practices to sequester CO_2 or conserve C were identified in the 40 nations surveyed. For ease of comparability, these silvicultural practices and the range of C sequestered or stored per hectare (ha) were grouped by latitude (high-, mid- and low-) in Table 1. At high latitudes, site drainage, weeding, thinning, and fertilization sequestered or conserved 1-16 Mg C ha^{-1} forest systems. Thinning and weeding of mid-latitude forests can conserve or sequester 8-65 Mg C ha^{-1}, with greatest efficiency achieved in *Pinus, Pseudotsugu* and *Populus* stands. Compared to high- and mid-latitude stands, thinning and fertilization of forests in low-latitude sequestered or conserved more C per hectare. In general, fast-growing tropical conifers and broad-leaf species responded to silvicultural treatments which improved resource availability, with the exception of weeding.

Initial project costs ($ ha^{-1}) and costs per unit of C ($ Mg C) of implementing silvicultural practices to conserve and sequester C in key nations are presented in Table 2. Forest drainage, thinning and fertilization were the most efficient C conservation and sequestration techniques at mid-and high-latitudes. Project costs were highly variable in all nations ranging from $18-573 ha^{-1} with labor cost being the dominant economic term. Application of silvicultural practices (e.g., thinning, selective logging and fertilization) in Malaysia, Mexico, New Zealand, and Russia can sequester or conserve C at less than $10 Mg C. In Germany, C conservation or sequestration via silvicultural practices was over $40 Mg C, the highest among the nations surveyed.

A large area of technically suitable forest land is available to apply silvicultural practices to stimulate C storage in key nations (Table 3). Currently, global forest area is estimated to be over 4 billion ha. Russia, containing 20% of the world's forest reserves, has over 100 million ha of mature forests which could sequester or conserve more C if drained or thinned. At mid-latitudes, some forests of U.S. and China could sequester more C if

Table 1. Range and Median Values of Incremental Carbon (C) Sequestration and Storage Values by Implementing Silvicultural Practices in Boreal, Temperate, and Tropical Forest Ecosystem

Latitude	Practice	MgC ha^{-1}
High- (60-90o)	Fertilization	1 - 16
	Thinning	3 - 7
	Weeding	1 - 3
	Drainage	3 - 11
	Median Value	5
Mid- (30-60o)	Fertilization	2 - 28
	Thinning	15 - 65
	Weeding	8 - 34
	Median Value	22
Low- (0-30o)	Fertilization	26 - 71
	Thinning	18 - 64
	Median Value	41

intermediate silvicultural practices are implemented. A small proportion of low-latitude forests are currently subject to silviculture practices, but the potential for their application is large. Worldwide, 112×10^6 ha forest land are in plantations, implying limited potential for application of silvicultural practices.

4. DISCUSSION

The analysis of biologic and economic data from approximately 40 nations, representing diverse climatic and edaphic conditions, reveals a variety of silvicultural practices that can be applied to conserve or sequester C. Some silvicultural practices, such as stand thinning and fertilization, can be employed to stimulate C sequestration and conservation in high-, mid- and low-latitude nations. The median conservation and sequestration values were 5, 22 and 41 Mg C for all practices evaluated in high-, mid- and low-latitude forests. In economic terms, the mean initial project cost was $13 Mg C, which compares favorably when considered to reduce greenhouse gas emissions in energy or land-use sectors (Rubin et al., 1992).

Large areas of technically suitable forest land are currently available to apply silvicultural techniques to conserve and sequester C. Globally, only 10% of the world's forests are intensively managed and the opportunity to employ intermediate silvicultural practices is significant at high-, mid- and low-latitudes (Sharma, 1992). If appropriate silvicultural practices were implemented on 600×10^6 ha of technically suitable land in all 40 nations surveyed, it would be possible to conserve 0.1 - 0.3 Pg C annually over a 50-year period.

Table 2. Median Initial Project Costs ($ ha^{-1} and $MgC) of Implementing Silvicultural Practices to Conserve and Sequester Carbon (C) in Forest Systems of Key Nations

Nation	Practice	$ ha^{-1}	$MgC
Australia	Fertilization	362	12
	Thinning	573	14
Canada	Fertilization	188	19
	Thinning	350	30
China	Thinning	195	14
Germany	Fertilization	378	43
	Thinning	295	56
Malaysia	Thinning	193	2
	Sel. harvesting	127	3
Mexico	Thinning	137	3
New Zealand	Fertilization	80	4
	Thinning	299	7
Russia	Fertilization	25	9
	Thinning	37	11
	Drainage	18	8
USA	Thinning	93	3
	Fertilization	59	4
All[1]	All[1]	186	13

[1] Mean value.

Compared to the global potential of forestation, stimulation of C conservation or sequestration by employing silvicultural practices may be modest (Dixon et al., 1994). Ancillary C conservation benefits attributable to thinning of stands, such as a reduction in catastrophic wildfire and associated large greenhouse gas emission, are significant but were not considered in these global calculations (Binkley and Van Kooten, 1991).

Relative to forest establishment practices, application of silvicultural techniques to conserve or sequester C may not be as efficient, biologically or economically (Dixon et al., 1994). A large body of scientific and economic literature (Sedjo et al., 1995) reveals that forestation can sequester or conserve C at less than $5 Mg C, compared to $13 Mg C for silvicultural practices. A full benefit-cost analysis of forestation practices indicates that on many sites worldwide the C sequestration costs are negative (e.g., benefits outweigh costs).

Table 3. Forest Land Area (×106) of Key Nations that is Potentially Available for Application of Ailvicultural Practices (from Dixon et al., 1994)

Latitudinal belt	Current forests	Protected forests	Plantations
High -			
Russia	884	178	43
Canada	436	9	3
Alaska	52	2	1
	1372	189	47
Mid -			
Continental U.S.	241	14	2
Europe	283	40	1
China	118	Trace	31
Australia	396	18	1
	1038	72	35
Low -			
Asia	310	49	22
Africa	527	113	2
America	918	105	6
	1755	267	30
Total	4165	528	112

Silvicultural techniques may not always stimulate significant C sequestration, and project costs of application are variable.

Application of silvicultural practices at high-, mid-, or low-latitude can be labor intensive, with some practices (e.g., fertilization) requiring high-cost, energy-intensive inputs (Hoen and Solberg, 1994). Low-cost fertilizers, such as recycled municipal waste, may help reduce some of these costs (Nilsson, this volume). Mechanization of thinning operations at some mid- and high-latitude sites improves economic efficiency, but this intrusive, energy-intensive technique disturbs the site, stimulating greenhouse gas emissions (Sampson et al., 1993). Some practices, such as site drainage, can stimulate long-term (beyond 50 yr rotation considered in this analysis) increases in site productivity and C sequestration (Krankina and Dixon, 1994). The complex biologic and economic interactions of applying silvicultural practices to conserve and sequester C merit further careful consideration before application in the field (Sedjo et al., 1995).

Silvicultural techniques are generally applied by foresters to stimulate both the quantity and quality of biomass, with an emphasis on bole wood quality (Sharma, 1992). Sequestration of biomass in the tree bole increases the proportion of C which flows into long-term wood products (Row and Phelps, 1992). At a regional or national scale, storage of C in wood products can be significant. However, at a global scale, C stored in permanent wood products over the past 50 years is less than 30 Pg C, a small fraction of C in the terrestrial biosphere (Dixon et al., 1994). Although C flow into wood products has been the subject of prior study, this topic requires further investigation before nations implement forest sector greenhouse gas mitigation options.

Generally, the biomass production of fast-growing, shade-intolerant tree species are significantly stimulated by intermediate silvicultural techniques (Schroeder, 1991; Sedjo, 1983). However, wide-scale implementation of intermediate silvicultural practices to conserve or sequester C may not be feasible or desirable, due to limiting ecologic factors or species composition of forest stands. Decisions to apply silvicultural practices are often driven by a complex myriad of biologic, economic and logistical variables, leading to an improved flow of high-quality wood and non-wood products. The marginal benefits of sequestering or conserving C attributable to silvicultural practices must be carefully considered relative to other forest management goals (Solberg and Hoen, 1995).

In conclusion, application of intermediate silvicultural practices can contribute to C sequestration and conservation in forest systems. However, the biologic and economic opportunities and constraints associated with these practices merit careful consideration before large-scale implementation (King, 1993). Of the forest-sector C offset pilot projects recently implemented, only a small number are based on the application of silvicultural practices (Dixon et al., 1993). In some regions of the world, such as the Nordic countries, application of silvicultural practices to sequester or conserve CO_2 may be efficient (Solberg and Hoen, 1995).

ACKNOWLEDGEMENTS

Discussions with J.K. Winjum and P.E. Schroeder contributed to the development of this paper. This document has not been subject to U.S. EPA administrative review and should not be construed as Agency policy.

REFERENCES

Binkley, C.S. and G.C. van Kooten (1991) Integrating Climate Change and Forests: Economic and Ecological Assessments. *Climatic Change* 28, 91-110.

Birdsey, R.A., A.J. Plantinga and L.S. Heath (1993) Past and prospective carbon storage in United States forests. *Forest Ecology and Management* 58, 33-40.

Brown, S., C.A. Hall, W. Knabe, J. Raich, M.C. Trexler and P. Woomer (1993) Tropical forests: their past, present and future role in terrestrial carbon budget *Water, Air and Soil Pollution* 70, 71-94.

Dixon, R.K., K.J. Andrasko, F.A. Sussman, M.C. Trexler and T.S. Vinson (1993a) Forest sector carbon offset projects: Near-term opportunities to mitigate greenhouse gas emissions *Water, Air and Soil Pollution* 7, 561-577.

Dixon, R.K., J.K. Winjum and P.E. Schroeder (1993b) Conservation and sequestration of carbon: the potential of forest and agroforest management practices *Global Environmental Change* 2, 159-173.

Dixon, R.K., S. Brown, R.A. Houghton, A.M. Solomon, M.C. Trexler and J. Wisniewski (1994) Carbon pools and flux of global forest systems *Science* 263, 185-190.

Harmon, M.E., W.K. Farrell and J.F. Franklin (1990) Effects on carbon storage of conversion of old-growth forest to young forests *Science* 247, 699-702.

Hoen, H.F. and B. Solberg (1994) Potential and economic efficiency of carbon sequestration in forest biomass through silvicultural management *Forest Science* 40, 429-451.

Huettl, R.F. and H.W. Zoettl (1992) Forest fertilization: Its potential to increase the CO_2 storage capacity and to alleviate the decline of global forests *Water, Air and Soil Pollution* 64, 229-250.

IMF (1994) International Financial Statistical Yearbook. International Monetary Fund, Washington, DC, USA.

Kauppi, P.E., K. Mielikainen and K. Kuusela (1992) Biomass and carbon budget European forests, 1971-1990 *Science* 256, 70-78.

Kershaw Jr., J.A., C.D. Oliver and T.M. Hinckley (1993) Effect of harvest of old growth Douglas-fir stands and subsequent management on carbon dioxide levels in the atmosphere. *Journal of Sustainable Forestry* 1, 61-77.

King, G.A. (1993) Conceptual approaches for incorporating climate change into the development of forest management options for sequestering carbon *Climate Research* 3, 61-78.

Krankina, O.N. and R. K. Dixon (1994) Forest management options to conserve and sequester terrestrial carbon in the Russia Federation. *World Resource Review* 6, 88-10.

Manne, A.S. and R.G. Richels (1990) CO_2 emission limits: An economic cost analysis for the U.S.A. The Energy Journal 11, 51-74.

Moulton, R. and K. Richards (1990) Costs of Sequestering Carbon Through Tree Planting and Forest Management in the United States. U.S. Department of Agriculture, Forest Service, General Technical Report WO-58, Washington, DC, USA.

Nilsson, L.O. (1995). Can recycling of waste help us to sequester carbon in forestry: experimental results and economic visions? **this volume**.

Nilsson, L.O. (1993) Carbon sequestration in Norway spruce in South Sweden as influenced by air pollution *Water, Air and Soil Pollution* 70, 177-186.

Nordhaus, W.D. (1993) Reflections on the economics of climate change *Journal of Economic Perspectives* 7, 11-25.

Plantinga, A.J. and R.A. Birdsey (1993) Carbon fluxes resulting from U.S. private timberland management *Climatic Change* 23, 37-53.

Putz, F.E. and M.A. Pinard (1993) Reducing the impacts of logging as a carbon-offset method. Conservation Biology 7, 755-759.

Row, C. 1978. Economics of track size in growing timber. *Journal of Forestry* 76, 576-582.

Row, C. and R. Phelps (1992) Carbon cycle impacts of improving forest products utilization and recycling (A. Quereshi, ed.) Proc. of North American Conference on Forestry Responses and Climate Change. Climate Institute, Washington, DC. p.208-219.

Rubin, E.S., R.M. Cooper, R.A. Frosch, T.M. Lee, G. Marland, A.H. Rosenfeld, and D.D. Stine (1992) Realistic mitigation options for global warming. *Science* 257, 148-266.

Sampson, R. N., L.L. Wright, J.K. Winjum, J.D. Kinsman, J. Benneman, E. Kürsten and J.M.O. Scurlock (1993) Biomass management and energy. *Water, Air and Soil Pollution* 70, 139-159.

Schroeder, P.E. (1991) Can intensive management increase carbon storage in forests. *Environmental Management* 15, 475-481.

Sedjo, R.A. (1983) The Comparative Economics of Plantation Forestry. Resources for the Future. Washington, D.C.

Sedjo, R.A., J. Wisniewski, A.V. Sample and J.D. Kinsman (1995) The economics of managing carbon via forestry: assessment of existing studies. *Environmental and Resource Economics*. in press.

Sharma, N.P., ed. (1992) Managing the World's Forests. Kendall Hunt Publishing, Dubuque, IA, USA, 605p.

Solberg, B. and H.F. Hoen (1995) Economic aspects of carbon sequestration - some findings from Norway (M. Apps, ed.) The Role of Forest Ecosystems and Forest Resource Management in the Global Carbon Cycle. Springer Verlag, New York, NY, USA.

Swisher, J. (1991) Cost and performance of CO_2 storage in forestry projects. *Biomass and Energy* 6, 317-323.

Trexler, M.C. and C.M. Haugen (1994) Keeping it Green: Evaluating Tropical Forestry Strategies to Show Global Warming. World Resources Institute, Washington, DC, USA, 52 p.

Turner, D.P., J.J. Lee, G.J. Koerper and J.R. Barker, eds. (1993) The Forest Sector Carbon Budget of the United States: Carbon Pools and Flux Under Alternative Policy Options. U.S. EPA, Corvallis, OR, USA, EPA/600/3-93/093, 202p.

Winjum, J.K. and D.K. Lewis (1993) Forest management and the economics of carbon storage: the non-financial component. *Climate Research* 3, 111-119.

Winjum, J.K., R.K. Dixon and P.E. Schroeder (1993a) Forest management and carbon storage: an analysis of 12 key forest nations. *Water, Air and Soil Pollution* 70, 239-257.

Winjum, J.K., R.A. Meganck and R.K. Dixon (1993b) Expanding global forest management: an easy-first proposal. *Journal of Forestry* 91, 38-42.

Wisniewski, J. and R.N. Sampson, eds. (1993) Terrestrial Biosphere Carbon Fluxes: Quantification of Sinks and Sources of CO_2, Kluwer Academic Publishers, Dordrecht, The Netherlands, 692 p.

Critical Reviews in Environmental Science and Technology, 27(Special): S151–S162 (1997)

CO_2-TAXING, TIMBER ROTATIONS, AND MARKET IMPLICATIONS

HANS FREDRIK HOEN[1] and BIRGER SOLBERG[2]

[1]Norwegian Forest Research Institute, Postbox 5044, N-1432 ÅS, Norway; [2]European Forest Institute, Torikatu 34, 80100 JOENSUU, Finland.

ABSTRACT: The build-up of greenhouse-gases (GHG's) in the atmosphere can be seen as a depletion of a non-renewable resource, namely the absorptive capacity of the atmosphere. A tax scheme on fossil fuels could provide incentives so that the substitution from fossil fuels to a non-GHG backstop energy technology is optimally timed. We argue that a consistent tax scheme should include subsidies and taxes as biomass is produced and decayed in order to let the economy utilize the possibility to, at least temporarily, increase the amount of carbon stored in forest biomass and thus delay the time at which a higher cost backstop energy technology must be introduced. A study of how such a subsidy/tax scheme would impact forest rotation ages is provided. The consequences of a carbon tax on the markets for wood based products and the total build-up of CO_2 in the atmosphere are discussed, and, finally, some opinions are presented on what types of quantitative models would be preferable to analyse these consequences.

KEY WORDS: carbon sequestration, optimal rotation, climate change, market impacts, models

1. INTRODUCTION

Biomass production on forest areas represents a main active sink in the global carbon cycle. The maximum level or rate of biomass production is limited by area and available technology (e.g. tree species and different silvicutural management options like fertilization). Globally we are far away from utilizing the maximum potential forest growth. Afforestation and intensified management can increase the rate of carbon assimilation from the atmosphere (see e.g., Hoen & Solberg 1994; Solberg & Hoen 1995). A second option related to forest biomass is to undertake actions that lower the decay of the current stock of wood biomass, be it in the forest or in various forest products. Recycling of paper, wood/lumber used in constructions or improved care of wood products may be actions that have potential of increasing the lifetime of carbon in forest products.

A tax on fossil fuel to lower the emission of CO_2 has been introduced in several countries (e.g. Finland, the Netherlands, Norway, Sweden). Such a tax gives a corresponding value of fixation of atmospheric carbon.

Let us assume that all activities causing emissions of CO_2 into the atmosphere are taxed according to a general CO_2 price reflecting the social marginal cost of

1064-3389/97/$.50

increasing the accumulation of CO_2 in the atmosphere. We further assume that a corresponding net payment, a subsidy, is payed by the government for a documented net accumulation of CO_2 during a specified time period. In forestry this could be estimated on an areal basis from forest inventory and growth data. An even-aged timber rotation will typically be represented with a biomass growth period, with net CO_2 assimilation, preceeding a biomass decay period, with net CO_2 emission. In the presence of a CO_2 tax/subsidy regime this sequence of CO_2 assimilation/ emission will generate a cash-flow consisting of a number of years with positive net payments[1] (income), from regeneration to clearfelling, followed by a number of years with negative net payments (costs), from clearfelling until the biomass has decayed. This differs somewhat from a traditional timber production cash-flow where net negative payments (costs) may incur at different points of time during the rotation period, depending on the investment intensity, and the positive payments is obtained when the timber stand is harvested. Figure 1 illustrates the typical flow of CO_2 assimilation and emission related to one timber rotation, based on specific assumptions regarding enduse and corresponding decay ratios.

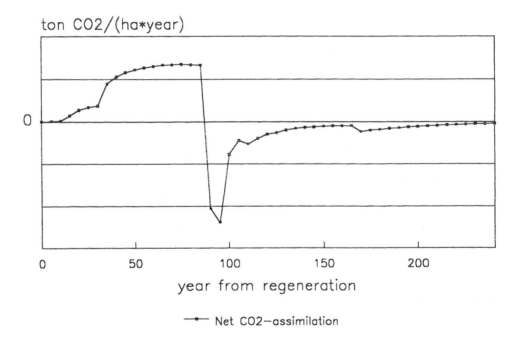

FIGURE 1. CO_2 fixation from, and emission to, the atmosphere related to one timber rotation.

---

[1]   We assume costless regeneration of a stand.

Several authors have studied how a subsidy/tax regime related to CO_2-assimilation from, and emission to, the atmosphere may impact the economically optimal management of a forest stand (e.g. Englin & Callaway 1993, Hoen 1994, Price 1994). The purpose of this article is, first, to analyse the question of economic optimal rotation at forest stand level. Here, we will comment particularly on some of the results presented in Englin & Callaway (1993). We proceed with a qualitative discussion of how such a subsidy/tax regime may impact the markets of forest based products and their substitutes, as well as the likely net consequences on the build-up of atmospheric CO_2. Finally, we discuss what types of quantitative models would be preferable to analyse these consequences more in detail.

2. ON THE OPTIMAL FOREST ROTATION IN THE PRESENCE OF A POSITIVE CO_2-PRICE

2.1. CO_2 as Single Output

Given a positive shadow price for CO_2 assimilation, we can calculate the value of forest land in terms of carbon sequestration by forest biomass production. We assume that subsidies are paid at the time of carbon sequestration and that taxes are imposed at the time of decay. For practical reasons, sequestration and decay may be estimated on periodical basis, e.g. annually. Following the Faustmann tradition, see e.g., Hartman (1976), Strang (1983) or Johansson & Löfgren (1985), the land value, V(T), can be stated as:

$$V(T) = \left[\int_0^T f'(n) \cdot p_1(n) \cdot e^{-rn} \mathrm{d}n + \int_0^{T+d} g'(m) \cdot (-p_1(m)) \cdot e^{-rm} \mathrm{d}m \right] \cdot (1 - e^{-rT})^{-1} \quad (1)$$

where f'(n) gives the annual assimilation of CO_2 in a stand n years after regeneration[2], $p_1(\cdot)$ is the marginal value of atmospheric CO_2 reduction at the corresponding point of time in prices on time 0 ($\cdot$ is n or m), r is the real rate of discount (RRD), g'(m) is the emission of CO_2 from the biomass m years after regeneration and d is the decay lag, i.e., the time it takes from harvesting until all the biomass is decayed and the CO_2 is emitted back to the atmosphere[3]. The decay lag will be a function of the timber quality when harvested. When timber is used for purposes like bioenergy or pulp and paper production this implies short decay lags (possibly from 0 to 5 years), while end uses based on saw timber indicates substantially longer decay lags (possibly from 20 up to 200 years). If we know the flow of

---

[2]   f(n) defines the total quantity of CO_2 fixed in the stand at time n.
[3]   The decay lag can alternatively represent the time it takes until a fraction of the biomass is left, e.g., when 90% of the biomass is decayed.

assimilation and emission of CO_2 related to one timber rotation, the RRD and the vector of prices $\mathbf{p}_1$, Equation (1) can be used to calculate the net present value of this flow (NPV_{CO_2}), or V(T) for a chosen T.

As forest biomass is grown during the rotation, the benefit in terms of CO_2 sequestering is obtained. Simultaneously, a potential source of costs is established related to decay and emission of CO_2 from the same biomass which will take place after clearfelling. Analytically, this situation can be compared with tax subsidies that are established due to favourable tax deduction rules for certain investments.

Let us, for simplicity, assume a constant CO_2 price over time. We further assume that any natural mortality during the rotation decay immediately, thus f'(n) represents net assimilation over the rotation. At time T, f(T) gives the total amount of CO_2 in the growing stock. Decay starts when the stand is clearfelled and the function q(m) gives the proportion of f(T) that decays m years after clearfelling. The integral of q(m) over the decay time d, is by definition set equal to 1. Then taking the derivative of V(T) with respect to T, setting this equal to zero and rearranging gives the following first order optimum condition:

$$f'\left(T^O\right)\cdot\left(1-\int_0^d q(m)\cdot e^{-rm}\mathrm{dm}\right)+r\cdot f\left(T^O\right)\cdot\left(\int_0^d q(m)\cdot e^{-rm}\mathrm{dm}\right)=r\cdot V\left(T^O\right) \quad (2)$$

The first expression on the lefthand side of Equation (2) gives the net discounted value of a marginal increase in CO_2 assimilation, i.e. the value of the marginal biomass grown at time T^O **minus** the value of the discounted flow of the future decay of the marginal biomass grown at time T^O. The second expression gives the value (the interest) of delaying the decay and emission of CO_2 from the standing biomass at time T^O. The sum of these should be equal to the interest on the land value at the optimal rotation age.

When r is strictly positive (r > 0) it will always hold that

$$0<\left(1-\int_0^d q(m)\cdot e^{-rm}\mathrm{dm}\right)<1,$$

$$\text{since} \quad \int_0^d q(m)\mathrm{dm}\equiv 1$$

Similarily,

$$r\cdot f\left(T^O\right)\cdot\int_0^d q(m)\cdot e^{-rm}\mathrm{dm}>0$$

when r > 0. Thus the left hand side of Equation (2) always remain positive as long as f'(T) ≥ 0. For a virgin forest ecosystem in balance it is reasonable to assume that f'(T*) ≈ 0. The optimality criterion then reduces to a comparison between the benefit obtained by postponing the emissions related to decay of the standing biomass and the interest on the land value. If the discounted value of decay and emission related to clear felling of the initial stand is larger than the discounted value of the biomass production in a regenerated stand, the best strategy will be to leave the stand. Clear felling of a virgin forest will be optimal only if the net present value of the decay rate is lower than the net present value of assimilation in a new rotation.

The fact that the decay lag d is depending on the timber quality or actually the end uses of the timber when harvested indicates that there might be several local optima. This means that the decay lag (d) and the decay proportion (q(m)) are functions of T. This should be kept in mind when analyzing the issue numerically. See Hoen (1994) for a treatment of this problem.

2.2. CO_2 and Timber in Double Output Production

The value of the forest biomass production in terms of timber can easily be incorporated in the model. Let p_2 be the net price per m^3 for timber and k a coefficient defining the timber volume in m^3 related to one unit of CO_2 given by f(n). Equation (1) can be rewritten:

$$V(T) = \left[\int_0^T f'(n) \cdot p_1 \cdot e^{-rn} dn - \left(f(T) \cdot p_1 \cdot \int_0^d q(m) \cdot e^{-rm} dm \right) \cdot e^{-rT} \right.$$
$$\left. + \left(f(T) \cdot k \cdot p_2 \cdot e^{-rT} \right) \right] \cdot \left(1 - e^{-rT} \right)^{-1}$$

(3)

Following the same procedure as above the first order condition can be written:

$$V'(T) = 0$$
$$\Updownarrow T \neq 0$$

$$f'(T^O) \cdot p_1 \cdot \left(1 - \int_0^d q(m) \cdot e^{-rm} dm \right) + r \cdot f(T^O) \cdot p_1 \cdot \int_0^d q(m) \cdot e^{-rm} dm$$

(4)

$$+ r \cdot V(T^O) + r \cdot f(T^O) \cdot k \cdot p_2$$

The left hand side of Equation (4) express the marginal revenue of an increase in the rotation age i.e., the net discounted value of a marginal increase in CO_2

assimilation (biomass growth) at time T^O **plus** the value (the interest) of delaying the decay and emission of CO_2 from the standing biomass **plus** the value of the change in timber volume. The right hand side of Equation (4) expresses the marginal costs of an increase in the rotation age i.e., the interest on the land value **plus** the interest on the value of the growing stock. As in the pure timber case, the interest on the value of the growing stock of timber will contribute significantly to the costs of extending the rotation age. If we compare Equation (4) with Equation (2), we see that the timber value of the marginal biomass growth is added to the revenue and the interest on the timber value of the growing stock is added to the costs, for a marginal increase in rotation age.

Englin & Callaway (1993) present a slightly different formulation of the land expectation value when timber and carbon sequestration/emission are priced outputs of forest production. Their formula (2) includes, in addition to the terms valuing the timber harvest and the flow of carbon out of and back to the atmosphere, a term $P_cF_te^{-rt}$ where P_c is the price of sequestered carbon f, F_t is the total (cumulative) amount of carbon sequestration at time t and r is the discount rate[4]. This looks suspiciously like double-counting, at least when the same price (P_c) is used to express the value of the carbon flow and the value of the carbon stock at time t. As we see it, one should either evaluate the carbon benefits and costs related to a timber rotation with a subsidy per unit of carbon in the stock for a given time period or preferrably one should pay a subsidy at the time of build up and impose a tax at the time of decay. The latter approach has intuitive appeal since a fossil fuel tax follows the same logic of taxing the flow as it marginally contributes to the depletion of the absorptive GHG-capacity of the atmosphere. Assuming that one unit of CO_2 from decay of (forest) biomass has the same effect on the atmosphere as one unit of CO_2 from fossil fuel combustion, the same tax level should be applied. Finally, if the subsidy/tax regime is connected to the stock it would not be proper to use a once and for all payment at the time of harvest, but rather to pay a smaller amount on a per period basis. This would create (correct) incentives for promoting actions that extend/delay the decay of the carbon already sequestered. A once and for all payment, similar to how timber is valued at harvesting, would generate counterproductive incentives as it at the margin would represent a substantial cost to increase the rotation age and postpone the payment related to the carbon sequestered.

2.3. Numerical Results

Table 1 gives results from numerical simulations where both timber and CO_2 benefits and costs are valued for a timber rotation of spruce. The results correspond

---

4   The land expectation value as defined in Equation (2) in Englin & Callaway (1993:1993).

TABLE 1
Optimal Rotation Age, in Years, for Norway Spruce, Site Index G17, for Different RRD's and Four Different Shadow Prices on CO_2. 6.3 NOK ≈ 1 US $, June 1995

CO_2 shadow price	RRD = 2% p.a.	RRD = 3% p.a.	RRD = 4% p.a.
0 NOK/ton CO_2	80	70	65
125 NOK/ton CO_2	105	100	90
250 NOK/ton CO_2	155	145	135
600 NOK/ton CO_2	>400	>400	>400

with Price (1994) and van Kooten, Binkley and Delcourt (1995). The optimal rotation age increases with increasing CO_2 price. The results further indicate that under "moderate" CO_2-prices the optimal rotation period decreases when the real rate of discount increases. Englin & Callaway (1993) come to a different conclusion regarding the impact of different levels of the CO_2-price. Their optimal rotation ages show minor differences compared with a pure timbervalued rotation, when the CO_2-price is included, and increased by a factor of 10 and 20. They further report that an increase in the discount rate will increase the rotation age when carbon sequestration is given a positive price and timber prices are set equal to zero. Whether this is related to an incorrect formulation of the problem as mentioned above or to different growth and decay properties of the species analysed (Douglas fir) is not clear to us. Englin & Callaway have applied a constant rate of decay for forest biomass irrespective of the rotation age. The decay estimate is based on current, aggregate, end use patterns (Row & Phelps 1992). This quality aspect is central both related to the valuation of a timber harvest and for the valuation of a sequence of decay of carbon when forest biomass is removed from the forest and utilised elsewhere in the economy (e.g., as pulp, paper, lumber or bioenergy).

The results indicate that if the CO_2 fixation and emission related to timber production are assumed to have the same impact on atmospheric CO_2 concentrations as emissions from burning of gasoline (or other fossil fuels) this might have a large impact on the optimal rotation age in forestry. The mortality function, or better, the increasing probability of a "collapse" in timber stands as they grow old, will be an important parameter in the analysis of the optimal rotation age.

The tentative simulations presented above show that the rotation age is significantly influenced by a CO_2-price. A level equal to the current Norwegian CO_2 tax on gasoline (250 NOK/ton CO_2), roughly doubles the rotation age. The simulations reported here are suffering from poor treatment of the mortality in old growth stands. A constant mortality rate equal to 0.7% of the tree number pro annum is applied over the whole rotation (this is the mortality function which has the highest statistical significance in Norway - but it is based on data where the majority of stands are younger than 100 years). Thus, the results of the simulations should only be regarded as indicative, especially for the higher CO_2-prices.

Given that optimal rotation ages increase substantially when a tax is introduced, timber supply might decrease, at least temporarily. This could lead to increases in the price of timber, which would moderate the effect. The long term impact can be the opposite. When a stand has grown really old and the probability of a "collapse" has become significant, it might be a better alternative for the forest owner to sell the timber at a low price and therby transfer the potential costs inherent in the biomass (related to tax payments as the biomass decays) to the buyer. The willingness to pay for timber among buyers will reflect the tax that have to be paid when the biomass starts to decay. With a positive long term RRD, e.g., in the range of 2-7% p.a. or higher, the net present value of the future decay will be significantly different between sawn timber and pulpwood. For example, if we assume a linear decay profile (equal amounts each year) over 5 and 50 years for respectively pulpwood and lumber and a RRD of 4%, the net present value of pulpwood decay would represent 89% of the same total emitted immidiately. For lumber the corresponding figure is 43%.

3. IMPACTS ON MARKETS AND CO_2 NET EMISSION

The natural sciences tell that the main problem is the cumulative build-up of GHG's in the atmosphere, thus CO_2 and other GHG's have the feature of stock pollutants. As time passes and fossil fuels are consumed this contributes to the build up of GHG's in the atmosphere, and the absorptive capacity of the atmosphere is depleted - see e.g. Anderson (1991) for a clear presentation. A main problem is related to the uncertainty regarding the absorptive capacity of the atmosphere, as we do not know the safe maximum limit (interval) of GHG accumulations. A carbon tax on fossil fuels will increase fossil fuel prices and shift the fossil fuel supply curve upwards. One effect will be improved competitiveness of energy substitutes. If the tax rate is increased over time, the user price of fossil fuels will ultimately reach the level of a non-fossil backstop technology, e.g. solar, nuclear or biomass energy. The latter alternative would represent a renewable energy source where CO_2 sequestration and emission can balance in a steady state equilibrium. A second effect will be a general substitution away from energy-intensive products towards less energy-intensive products.

If a GHG-tax is imposed on fossil fuels to reflect the social cost of depleting the non-renewable absorptive capacity of the atmosphere, carbon fixing processes such a forest biomass production should represent a benefit. Production of forest biomass temporarily restores some of the absorptive capacity of the atmosphere. Thus, the time at which the higher cost backstop energy bearers have to be used is delayed (Anderson 1991). Since there is a substantial lag between sequestration and decay of carbon in forest biomass, increased forest biomass production adds significant flexibility in the short to medium term.

A consistent tax/subsidy policy would have to tax all activities that lead to emission of GHG's to the atmosphere and to subsidize activities that involve assimilation of GHG's from the atmosphere. Thus, fossil fuel consumption should be taxed, (forest) biomass production should be subsidized and (forest) biomass decay should be taxed. This would create incentives to reduce fossil energy consumption, increase forest biomass production and undertake actions that delay decay of forest biomass. Whether it is practically feasible to implement such a tax/subsidy scheme is difficult to say.

It is necessary to consider rather carefully how various types of taxes on fossil fuel consumption and subsidies on carbon sinks influence the build-up of CO_2 in the atmosphere. The simplest solution will be to put a tax on fossil fuel. Its main effect would be to change the production (through changes in relative prices on input factors) from production with a high direct and indirect use of fossil fuel products, to production with a relatively lower use of fossil fuel products.

This will increase the production of most forestry based products (including wood based energy). In the short run this could lead to higher total net emissions of CO_2 from forest biomass and in the total economy. The uptake of atmospheric carbon in the forest will most likely be less in the initial stage before, the forest areas being harvested, because of the increased demand for wood, start accumulating enough CO_2 to offset the increased emission of CO_2. In the longer run, there will be a balance here, but on a higher level of consumption of wood products. This will, however, depend upon how the GHG taxes are channelled back to the economy -e.g., as general income tax reductions, increased public consumption, or higher savings.

The higher timber prices caused by the increase in wood demand will also lead to higher investments in forest management, including afforestation and regeneration, and thus increasing the absorption of atmospheric CO_2. This effect will, however, in most cases be less than what is optimal from a national or international point of view if no special measures are taken. The main reasons for this is that the forest owners will not be sure of getting any benefits of their extra long term investments in increased forest stocks because of three types of risks. The first risk factor is that today it is not certain that increased emission of GHG's causes climate change. The second risk factor is that even if increased emissions of GHG's cause climate change, it is uncertain how damaging these changes are and how much resources should be spent, at the margin, to mitigate them. The third risk factor, which the forest owners will be facing, is that even if the damages by climate change will be high if no preventing steps are taken, technology changes (e.g. fusion energy) may make higher taxes on fossil fuel unneccessary 40-100 years ahead when the forest owner is harvesting todays' seedlings. These risks are mainly caused by the fact that the forest owner is carrying all forestry investment costs himself, wheras the benefits of decreased concentration of atmospheric CO_2 is a true public good, globally seen.

The risk situation is assymmetric, and the society should contribute to get a more optimal resource allocation by decreasing the risk of the forest owners' investments in building up the carbon stock. This could be done by, in addition to a tax on fossil fuel, introducing a subsidy for absorption of carbon. This subsidy should be linked as closely as possible to the fixation of carbon in the forest biomass over time, and it could be estimated rather accurately based on the available yield-tables. The size of the tax should equal the emission tax on CO_2 (e.g. in $ per ton CO_2 per year), but subtracted by the cost of emission of CO_2 from the future decay of the biomass fixed. An alternative to this net subsidy is to introduce the gross fossile fuel tax, similiar on all fixations and emissions of atmsopheric carbon in forest, and charge the end-user (paper, sawnwood, and bioenergy users) for the emission. But that is probably more difficult to measure and control. Another alternative could be to be to pay subsidies according to estimated net growth and impose a tax to account for decay when the biomass is removed (harvested). This (harvest) tax should reflect the assumed decay profile of the forest biomass and should be varied according to end use. To reflect current decay profiles, sawn timber would be given a lower tax than pulpwood, while wood for bioenergy would be given a higher tax than pulpwood.

In this way, one may say that the risks of the investments in increasing the forest biomass stock for GHG control are taken over by the society or public at large, who also get the benefits of the investments. Relative to a base scenario with no carbon subsidies/taxes, this will most likely imply that the growing stock of forest biomass increases, the use of wood fibre (for forest industry and bioenergy) increases, and the use of fossil fuel decreases in all sectors, in particular for the sectors which today depend heavily on fossil fuel and which have product substitutes which are not so fossil-fuel dependent. The overall effect will be that the build-up of CO_2 in the atmosphere is reduced.

4. MODELS

The main control or decision problem for forestry related to climate change is what type of investments (including time and intensity of harvesting) should be done over time. To analyse the above discussed problems more in detail, one need different types of quantitative models. At stand level it may be advantageous to have process models - i.e., models which directly could incorporate effects of possible climate changes. Such models exist, but it seems that they, at present, at best can be used for calibrating the existing traditional yield models on climate change effects (through e.g., relative changes in the latter). Nevertheless, this kind of modeling will be important for predicting factors like yield natural regeneration rates, mortality at various ages, etc. as a function of climate changes.

These stand (or ecosystem) level models are the basic block for forestry models, which is one key element for market analysis. However, also the demand side must be included, and this is possible in several ways. The simplest will be to use expected exogenous prices on the various forest products (including environmental goods and services). Different climate change scenarios will give different prices over time, and they become exogenous model inputs. A typical example of this kind of modelling is Hoen & Solberg (1994).

The next level will be to include the forest industries and the other demanders of forest products in one model - i.e., to develop forest sector models. Many such models exist, but few have been used for consistent analyses of the net flux of carbon. Their structure is however very well suited for such analysis, as the use of fossil fuel in the production as well as the transport is easy to include, and also the timber harvest and forest growth can be taken care of. A drawback is that the forest growth is rather static in these models, and this has to be addressed (e.g., by having a more detailed forestry model as a supplement). This type of modeling could be done at a national level, but to capture the dynamics on the market fully, a global model approach would be more appropiate. Again, the global aspects could be taken care of exogenously by specifying a scenario of prices corresponding to the climate change scenario assumed. A typical example of this kind of modelling is Perez-Garcia, Joyce, Maquire and Binkley (this volume).

None of these approaches incorporate the impacts of substitution between forest products and products like steel, aluminium, plastics, fossile fuel energy, etc., and the direct and indirect effects of the total economy. To do that, general equilibrium models will be necessary. The strength of these models is that they will give the total CO_2 impacts of introducing carbon taxes, and also the important impact related to how the tax is used (cf. above). Such models exist in most industrialized countries, but their treatment of forestry and forest industry is rather crude, although they could in some cases give meaningful analysis for certain types of forest sector questions (see e.g. Solberg 1986). Again, regarding analysis related to GHG emissions, one will have to supplement with more detailed forest sector and forestry models. To get a consistent picture one may have to use global general equilibrium models, or regional general equilibrium models (e.g., the GREEN model for OECD) in iterative procedures. The main output from these models for forest sector models will be the size of the carbon taxes necessary to keep the concentration of atmospheric CO_2 to a desired level.

Based on this it is our opinion that consistent analysis of forestry and global climate change issues may have to use a hierarchical set of models -from stand level process models up to general equilibrium macroeconomic models. Each model type in this hierarchy will need exogenous inputs from the model being the next up in the hierarchy. The need for models will of course depend upon the problem analysed, and not all problems need the whole set of models. Only experience can tell what is the appropriate level of model detailness.

A special feature is of course related to the aspects of risks involved in the global change issue. Several possibilities exist to incorporate risk explicitly in such models. Again, only trial and experience will show what is useful. It is a great challenge for research to look into these issues.

$$V_t = \frac{P_g G_t e^{-rt} + P_c F_t e^{-rt} + \int_0^t e^{-rt} P_c f_\tau d_\tau - \int_t^\infty e^{-rt} P_c}{1 - e^{-rt}}$$

where,

V_t = the present value of forest land,
P_g = the price of timber,
G_t = the quantity of timber at time t,
r = the discount rate,
P_c = the price of sequestered carbon f,
F_t = the total amount of carbon sequestration at t,
f_t = the rate of carbon sequestration at t,
D_t = the rate of release of carbon back into the atmosphere at time t.

REFERENCES

Anderson, D. (1991), 'The forest industry and the greenhouse effect', Report for the Scottish Forestry Trust and the Forestry Commission, Edinburgh. 27 p.

Englin, J. and Callaway, J.M. (1993), 'Global climate change and optimal forest management', *Natural Resource Modelling,* **7(3),** 191-202.

van Kooten, G.C., C.S. Binkley and G. Delcourt (1995), 'Effect on carbon taxes on the optimal forest rotation age and the supply of carbon services', *American Journal of Agricultural Economics* (in press).

Hartman, R. (1976), 'The harvesting decision when a standing forest has a value'. *Economic Inquiry,* **14,** 52-58.

Hoen, H.F. (1994), 'The Faustmann rotation in the presence of a positive CO_2-price', pp. 278-287 in Helles, F. and Lindahl, M. (eds.). Proceedings of the Biennial Meeting of the Scandinavian Societym of Forest Economics, Gilleleije, Denmark, November 1993. *Scandinavian Forest Economics,* No. 35, 1994.

Hoen, H.F. and Solberg, B. (1994), 'Potential and economic efficiency of carbon sequestration in forest biomass through silvicultural management'. *Forest Science,* **40,** 429–451.

Johansson, P.O. and Löfgren, K.G. (1985), *The economics of forestry and natural resources.* Basil Blackwell. 279 pp. + bibliography and index.

Perez-Garcia, J., L. Joyce, D.A. Maguire, and C.S. Binkley (1995), 'Economic impacts of climate change on the world's forest sector', *Environmental and Resource Economics,* this volume.

Price, C. (1994), 'Time profiles and pricing of CO_2 externalities', pp. 320-336 in Helles, F. and Lindahl, M. (eds.). Proceedings of the Biennial Meeting of the Scandinavian Society of Forest Economics. Gilleleije, Denmark, November 1993. *Scandinavian Forest Economics,* No. 35, 1994.

Row, C. and R. Phelps (1992), 'Determining the flows and disposition of carbon in timber harvest and wood in use', unpublished manus, Institute of Forest Analysis, Baltimore, MD.

Solberg, B. (1986), 'Macro-economic models for long-term forest sector analysis in Norway', *TIMS Studies in the Management Sciences,* **21,** 113-121.

Solberg, B. and Hoen, H.F. (1995), 'Economic aspects of carbon sequestrian — some findings from Norway (in press).

Strang, W.J. (1983), 'On the optimal forest harvesting decision', *Economic Inquiry,* **21,** 576-583.

Critical Reviews in Environmental Science and Technology, 27(Special): S163–S176 (1997)

COMPENSATING FOR OPPORTUNITY COSTS IN FOREST-BASED GLOBAL CLIMATE CHANGE MITIGATION

ALAN GRAINGER

School of Geography, University of Leeds, Leeds LS2 9JT, UK.

SUMMARY: Compensation to developing countries for mitigating global climate change by conserving forest or improving forest management should include the opportunity cost of foregone national development and not just the cost of carbon stored.Estimation of the opportunity cost is discussed and alternative compensation schemes compared.

KEY WORDS: Climate change mitigation, tropical forestry, sustainable development, compensation schemes.

1. INTRODUCTION

The prospect of global climate change has forced the international community to consider taking concerted action to mitigate it, but no future international mitigation programme can succeed if it fails to treat developed and developing countries equitably. The global external costs of climate change will fall on the whole of humanity, but since historically the distribution of carbon emissions from deforestation and fossil fuels has been spatially heterogeneous, some nations have contributed more to these costs than others. Similar disparities in economic development affect the ability of nations to pay for mitigation activities.

Historical spatial preferences in forest exploitation have left remaining forest concentrated in low and high latitudes. This clearly influences the ability of countries to participate in forest-based mitigation programmes, particularly those intended to stabilise the atmospheric concentration of carbon dioxide by conserving forests or making existing management more sustainable. Many of the countries that can make the greatest contribution have not yet have exploited their forests very much. Most will also be developing countries, so that when developed countries try to internalise on a global basis the global external costs of their past development, they could reduce the scope for developing countries to pursue their own development. The first part of this paper uses a sustainable development approach to show that the true cost of forest conservation and management is not merely determined by the value of forests as carbon stores, but also by the substantial opportunity cost of foregone development.

Good design will help to ensure that an international mitigation programme is equitable, but it will also be necessary to devise a compensation scheme (Hicks, 1939; Kaldor, 1939) to balance the consequences of the spatial heterogeneity of historical carbon emissions and development. The Framework Convention on Climate Change gives scope for

1064-3389/97/$.50

developed countries to make transfer payments to developing countries to help meet the cost of their mitigation activities. One approach is to use the revenue from taxes on carbon emissions to subsidise forest-based strategies (Barrett, 1991), but such taxes will be based only on current carbon emissions from fossil fuels. The second part of the paper discusses possible alternative arrangements, one of which bases each country+s share of the global cost of forest-based mitigation on its historical record of forest exploitation. The paper closes with a discussion of design challenges for mitigation programmes. For convenience, the analysis of mitigation here is restricted to carbon dioxide and excludes other greenhouse gases.

2. THE ECONOMIC VIABILITY OF FOREST-BASED MITIGATION PROJECTS

Whether or not a national forest-based mitigation programme needs subsidies depends upon the economic viability of the strategies used. There are three main types of forest-based strategies:

1. Afforestation: increasing the size of the carbon store and the annual carbon sequestration rate by expanding the area of forests and plantations.
2. Conservation: maintaining the size of the carbon store by protecting existing forests from deforestation and timber exploitation.
3. Sustainable Management: increasing the sustainability of existing forest management in order to stabilise the size of forest carbon stores.

2.1. Assessing Economic Viability

The economic viability of a forest-based mitigation project is assessed by using cost-benefit analysis to compare its net returns with those that would be obtained from the next best land use (Pearce and Nash, 1981). Economic viability is indicated by:

$$R = (B_m - C_m) - C_o > 0 \qquad (1)$$

where R represents the net benefits, B_m the benefits of mitigation, C_m the cost of mitigation, and C_o the opportunity cost of not using the land for the alternative land use.

Some afforestation projects are commercially viable, in which case R > 0 even if Bm only includes the value of marketable goods produced. However, in many instances, R > 0 only if the values of all net environmental benefits are included, by estimating Bm in total economic value terms so it embraces direct and indirect use values, option values and existence values (Barde and Pearce, 1991). If B_m is divided into B_{mp}, which refers to direct use values (e.g., receipts from timber sales), and B_{mc} which refers to local indirect use values (e.g., stabilisation of river flows from catchments) and existence values, the viability rule becomes:

$$R = (B_{mp} + B_{mc} - C_m) - C_0 > 0 \qquad (2)$$

2.2. The Economic Viability of Different Strategies

Establishing artificial plantations on degraded lands will yield marketable goods (B_{mp}) and local environmental benefits (B_{mc}), but if degradation is severe then planting and management will be expensive, so C_m will be high. A lot of degraded land is still farmed, so the opportunity cost C_0 will reflect the cost of switching to the new use. If the land is privately owned, C_0 will determine its actual availability for mitigation (Grainger, 1995a). If not, it will reflect public costs, e.g., those incurred by government in resettling current users, like nomadic farmers.

The main aim of conservation and management projects is to maintain carbon stores and stabilise the concentration of carbon dioxide in the atmosphere, but some sequestration will occur as protected degraded forests regenerate. B_{mc} will be high in conservation schemes, but without earnings from tourism or gene bank exploitation fees B_{mp} will be zero. C_m will include the costs of protecting forest both physically and socially, i.e., reducing the pressures on forest by helping local people to farm more productively and sustainably. The opportunity cost C_0 will be high as no income can be gained by logging forest or clearing it for agriculture.

Managing existing forests more sustainably will have major environmental benefits, but there will be some environmental impacts so Bme will be lower than in the conservation strategy. The protection costs in C_m will be similar to those for conservation but higher than normal for forest management. The opportunity cost will reflect income lost by not exploiting the forest in the usual way and the costs of expanding management capacity rapidly.

3. THE OPPORTUNITY COST OF FOREST PROTECTION AND MANAGEMENT

Over the last thirty years environmental economists have paid a great deal of attention to estimating the environmental opportunity costs of development projects, and much less to the opportunity costs of environmentally beneficial projects. There have been various economic studies of the potential for forest-based climate change mitigation (e.g., Andrasko et al., 1991; Dixon et al., 1993) but they have focused more on management costs than social costs. This section examines the opportunity costs of conservation and sustainable management strategies.

3.1. Forest Resource Depletion and Sustainable Development

Some reduction in the stocks of forests and other natural resources is necessary as a country develops. Using the terminology of sustainable development, a reduction in Natural Capital (NC) is needed to achieve an expansion in Human Capital (HC), in the form of agriculture, settlements, industries, buildings, services etc.

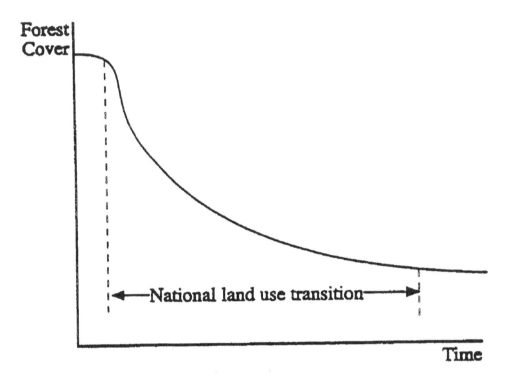

FIGURE 1. The National land use transition.

Sustainable development has been defined as that leading to non-declining human welfare over time (Pearce, 1991), and to achieve it requires imposing constraints on the flows of Natural and Human Capital. Pearce and Turner (1990) defined sustainability very strictly in terms of a constant stock of Natural Capital, so that any fall in the stock of non-renewable resources must be balanced by a rise in the stock of renewable resources. Expanding forest cover to offset fossil fuel burning is an interesting example of this.

However, Pearce (1994) later defined two conditions for achieving sustainability. In the strong condition, even if it is not possible to keep Natural Capital constant, there must be no decline in +critical+ Natural Capital: that which is essential for human well being or survival, e.g., areas of forests with high biodiversity. In the less restrictive weak condition, the present generation passes on to future generations a constant stock of total capital, i.e., the rise in Human Capital balances the fall in Natural Capital:

$$\partial NC = \partial HC \qquad (3)$$

Natural Capital has two main components: Resource Capital and Environmental Quality. Converting it into Human Capital is usually inefficient, i.e.:

$$\partial NC = \partial HC + e \qquad (4)$$

because waste (represented by 'e') is generated when natural resources are extracted, processed and consumed. This waste is often emitted into open access "sinks", i.e., air and water, resulting in a decline in Environmental Quality when the assimilative capacity of the sink is exceeded (Daly, 1990). Some wastes result in local external costs, but others, e.g., carbon dioxide, lead to global externalities.

Assuming weak sustainability, Grainger (1995b) proposed measuring the sustainability of national development using a Sustainable Development Index (S), estimated as a function of the relative annual changes in NC and HC (which are usually negative and positive, respectively):

$$S = f\ (\partial HC + \partial NC) \tag{5}$$

Since development is a long-term trend, sustainability must be evaluated over a long period, e.g., by comparing each country's time series for S with a normative national trend. The closer the two trends are, the more sustainable is national development. The degree of sustainability rises over time, and full sustainability may only be reached when a country has reached an advanced stage of development.

To estimate a normative trend for S requires establishing normative trends for the stocks of forests and all other natural resources, and for environmental quality too. The normative trend in forest area could be based on the *national land use transition* - the period of significant forest decline that usually occurs as a country develops, and one of a number of generic long-term trends in national land use that are most evident in countries which were largely forested before human settlement expanded (Figure 1) (Grainger, 1995c). Development would be sustainable if the decline in national forest cover (the percentage of national territory under forest) followed a normative transition curve, characterised by a moderate slope, and by deforestation ending with a substantial amount of forest remaining in the country, as happened in the USA.

3.2. The Opportunity Cost of Forest Conservation

Two forms of income are obtained from forest exploitation. The first comes from the sale of wood extracted from the forests, and the second from agriculture or other uses to which forest land is put after logging and deforestation. The opportunity cost of forest conservation must include not only this income, but also the vital contribution which agricultural and urban development would make to national development. The income lost by limiting the expansion of agriculture and settlements could be estimated by assuming that the country develops sustainably so that national forest cover declines as in the normative national land use transition curve. The forest area to be set aside for the international mitigation programme (A_s) could be compared with the area remaining at the end of the national land use transition (A_t). If $A_s > A_t$, various scenarios could be constructed for trends in farm (and where appropriate urban) income that might be obtained from the extra cleared land, and from selling wood from the cleared forest. This income could then be used to estimate the larger developmental benefits.

3.3. The Opportunity Cost of Sustainable Management

If sustainable forest management were to be imposed on a country quickly this would incur two types of opportunity cost: one concerns the fall in the net income (and developmental benefits) obtained in the early stages of forest exploitation, and the other refers to the extra costs of increasing management capacity more rapidly than would otherwise be the case.

3.3.1. The Loss of Income from Logging

In most countries the early stages of forest exploitation resemble the mining of a non-renewable resource, rather than the sustainable management of a renewable resource. Overexploitation is common, and environmental impacts are ignored, as forests are seen as a limitless bounty of nature. Not until the initial wave of exploitation is over is the importance of good forest management generally appreciated.

Initial overdepletion, and limited investment in management and environmental impact alleviation, mean that the net income received in the early stages of forest exploitation is much greater than if management were fully sustainable This foregone income can be included in the opportunity cost of introducing sustainable management (and that of conservation too) by comparing the net benefits obtained from sustainable management with those from the more customary, over-exploitative form of management.

3.3.2. The Costs of Institutional Development

Concern about deforestation has led to calls for tropical rain forest to be managed for timber production in a more sustainable way than in the past. Only a small proportion of tropical forest is managed sustainably, but it is known that the main constraints on sustainability are institutional, not technical (Poore, 1989). This author has argued that it is wrong to expect that a rapid rise in the degree of sustainability can be achieved simply by establishing new international obligations, e.g., through the International Tropical Timber Agreement or a new Forest Convention, because sustainability takes time to evolve in all countries, whether tropical or temperate. Usually in developing countries only a limited number of forestry department personnel are stationed in forested areas to monitor forest management, check that it complies with official norms, and if it does not, force loggers to change their practices. The pace of change depends on how fast this institutional capacity develops and when economic and political pressures to protect forest become irresistible (Grainger, 1995d). Even if a government were to adopt sustainable forest management as a policy goal, there would still be a long delay before this was achieved, because of the time needed to strengthen institutions.

Sustainable management of forests and farmland is not the same as sustainable economic development, but Grainger (1995b) has proposed that national institutional capacity for resource and environmental management accumulates over time as a country develops and can be regarded as a specific form of Human Capital, called Resource Sustainability Capital (RSC). The gradual rise in the quality of management as a resource is exploited illustrates the argument of Klaasen and Opschoor (1991) that economic development is an evolutionary process in which there is a co-evolutionary feedback between the economy and the environment.

Developed countries that have exploited much of their forests and built up their RSC will be far more able to protect or replenish them to mitigate global climate change than countries whose forests have only been exploited partially or not at all. But in countries that are still to fully exploit their forests it will be expensive to speed up the rate of increase of RSC so that large areas of forest can be effectively protected from illegal clearance or managed sustainably. This cost should also be included in the opportunity cost of sustainable management.

It might be argued that if a large part of a country is still forested there is no reason why this should change, and that it will not cost anything to retain the status quo. However, development is bound to happen at some point in the future, and if national forests are not protected in an appropriate way, it will lead to deforestation just as it has in other countries.

3.4. Estimating the Opportunity Cost

When estimating the economic viability of mitigation projects the opportunity cost of conservation and sustainable management should therefore include some or all of the following: (1) income lost from logging: (2) the contribution this makes to national economic development; (3) the specific contribution to the rise in Resource Sustainability Capital; (4) the opportunity cost of not overlogging in the early stages of forest exploitation; (5) income and developmental benefits lost when forests are not replaced by farming and settlements. Actually estimating these values will not be easy, but this is a problem faced in all new fields.

4. COMPENSATION SCHEMES

Compensation will be at the heart of any global climate change mitigation programme. The main aims of the programme will be to physically compensate for historic emissions of carbon dioxide and other greenhouse gases by sequestering as much carbon as possible from the atmosphere, and to reduce future emissions from fossil fuels and deforestation. Forest-based mitigation will compensate for historic spatial preferences in deforestation by expanding forest in areas where forest cover is low and maintaining it where forest cover is still high.

Some form of financial compensation scheme will also be necessary so that the overall costs of the programme can be shared in an equitable way between developed and developing countries. For while the programme will convey benefits to the whole world, historically greenhouse gas emissions have not been shared equally between different nations, and disparities in development status mean that some countries are more able to pay the costs of mitigation than others. The notion of compensation is by no means universally accepted among the governments of the developed countries who will be the main donors, but developing countries are unlikely to take much action without compensation. It has been suggested that conditional transfers from developed countries are

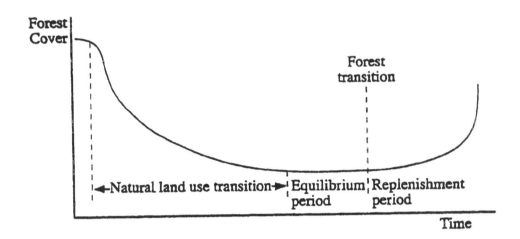

FIGURE 2. Unified model of national land use transition and forest replenishment period.

actually the most economically efficient way to improve the protection of tropical rain forests generally (Thiele and Wiebelt, 1993).

A key question is how the overall costs of a mitigation programme should be shared between countries. This will eventually be decided by international negotiations, but it would be desirable to devise some form of rational framework for calculating relative national shares, so they do not have to be determined in an entirely ad hoc manner. Three main types of compensation scheme to use in funding an international forest-based mitigation programme are considered here: (1) single source schemes, based on emissions from fossil fuels or deforestation; (2) dual source schemes; (3) bilateral compensation arrangements.

4.1. Single Source Schemes

A global compensation scheme will transfer money from a central fund to balance the deficits on national forest-based mitigation programmes. A possible criterion for compensation might be that total net returns are negative in monetary terms but positive in total economic value terms (Equations (1) and (2)). Including in B_{me} the net benefit to a country from global climate change after mitigation would allow more projects to be subsidised, but would be difficult to implement because of uncertainty about the timing and magnitude of global climate change.

Contributions to the fund could come from national carbon taxes (Hoel, 1993), but this would mean that national contributions would be based on just one of the two main sources of carbon dioxide - fossil fuels - and only on current emissions. Experience in Europe also suggests that a universal carbon tax will not be politically acceptable (Anon, 1995).

An alternative approach would be to rank each country according to its historic carbon emissions from the other main source - forests. The ranking system would be used to divide countries into net donors to the compensation scheme and net beneficiaries. Essentially, present national forest cover would be compared with the original forest cover before human settlement, the difference calculated, and the net amount of carbon transferred into the atmosphere determined. Original forest cover could be estimated approximately using standard maps of climax forest cover, or more reliably using recent maps produced by rule-based computer biome models (Neilson et al., 1990).

The procedure would have to be modified, however, in order to deal equitably with developed and developing countries. For example, in a number of developed countries forest cover declined markedly in the past but then increased owing to natural regeneration or intentional afforestation. This is shown in Figure 2 by a forest replenishment period following the national land use transition (Grainger, 1995c), the two being separated by a turning point called the forest transition (Mather, 1992). Two countries could therefore have the same forest area and have made the same net contribution to the external costs of global climate change, but have gained very different developmental benefits from forest exploitation. One could be in the early stages of its national land use transition, i.e., on the left arm of the curve, while the other lost a lot of forest in the past but is now well advanced in its forest replenishment period, i.e., on the right arm. Major deforestation and logging are yet to occur in the first country, but the second country has had a long history of forest exploitation and greatly benefited from this.

This suggests that the ranking system should also take into account aggregate income received from forest exploitation and the resulting increase in human capital, rather than just the net decline in forest area or forest carbon.Thus, the same factors which would determine the opportunity cost of conserving forest or improving forest management in developing countries could be used to ensure that the overall costs of mitigation are shared in an equitable way.

The advantages of this scheme are that compensation for forest-based mitigation is based on a country's full historic record of forest exploitation, rather than current forest activities alone. It takes account of the role of forests in national development, although it will not reward developing countries that have exhausted their forest resources but still remain in the low income category. One objection to the scheme might be that the present generation cannot be blamed for the "carbon sins" of past generations — who did not realise they were acting wrongly — and so governments will not accept responsibility for past actions. Of course, they may be unwilling to make transfer payments anyway (Andersson, 1991). Another is that poor data on trends in forest cover could lead to unfair assessments of liability for compensation.

4.2. Dual Source Schemes

The schemes proposed in section 4.1 suffer from the limitation that national contributions were based on carbon emissions from forest exploitation or from fossil fuels, but not both. It would be possible to devise a scheme in which countries were ranked according to their historic carbon emissions from both sources, but problems would arise since for most countries the former would refer to the depletion of one of its domestic resources, while the latter would refer to consumption of resources that might be domestic or imported.

4.3. Bilateral Compensation Arrangements

An entirely different approach would be to rely on bilateral rather than global arrangements. Thus, a developed country could pay a developing country to implement a forest-based mitigation programme to offset the developed country+s obligations to reduce net carbon emissions. These obligations might be specified in a future protocol to the Framework Convention on Climate Change, or imposed as part of a system of tradeable emissions permits (Markandya, 1990). A centrally administered international compensation scheme would not be needed, and projects could be based on the experience of electricity generating utilities from developed countries which have voluntarily funded forestry schemes in developing countries to offset their carbon emissions (Trexler et al., 1990).

5. DESIGN CHALLENGES FOR MITIGATION PROGRAMMES

5.1. Stabilizing Timber Prices After Mitigation

One of the perils of a large-scale forest-based climate change mitigation programme is that it might lead to such an over-production of wood that world wood markets will be destabilised and prices collapse, causing existing forests to be taken out of production and even cleared (Grainger, 1990, 1995a). The problem could be prevented or at least alleviated by a judicious balancing of the three mitigation strategies above. Expanding the area of plantations or forests in general will lead to a downward price trend as supply expands, though this will depend upon the balance between industrial wood and fuelwood plantations, and the mixture of rotations chosen. There is great potential for fuelwood to substitute for fossil fuels. On the other hand, expanding the areas of conserved and sustainably managed forests will decrease the availability of wood and so raise timber prices. Any future mitigation programme would need to carefully calculate the right mix of strategies so that overall wood supply keeps pace with demand and prices remain at least relatively level.

5.2. Adopting an Integrated Approach to Mitigation

Another challenge when designing an international mitigation programme will be deciding how to integrate action on forests and fossil fuels. So far there has been little integrated analysis of the economics of reducing emissions from these two main sources (Sathaye and Christensen, 1994). A good starting point might be to look at how societal inertia will delay action in the two sectors and influence the relative priorities accorded to them. Grubb (1993) has stated that neglect of societal inertia is a major limitation of current energy policy models.

The national land use transition (Figure 1) shows that some deforestation is inevitable as a country develops, so it will be difficult to halt tropical deforestation quickly. Grainger (1995a) suggested that a country's position on the national land use transition — forest replenishment period curve (Figure 2) will influence its ability to undertake forest-based mitigation activities.

Attempts to reduce emissions from fossil fuels will also encounter delays. It is easy to construct possible future global scenarios for reducing overall carbon emissions from energy production to zero (Boyle, 1994), but more difficult to do it in a way that takes account of societal inertia. For if the whole world is to make the transition to a fossil fuel free economy, then all countries will have to pass through a national carbon transition (Figure 3), in which they shift from a carbon-intensive economy to one that is independent of fossil fuels.The transition curve shows the rise in energy use and changes in dominant energy sources as a country develops. In the early stages of development fuelwood is the main energy source - as it still is in many developing countries - and then fossil fuel consumption rises. For the transition to end, renewable and nuclear energy must replace fossil fuels as the sole energy source.

Because of societal inertia some countries will complete their carbon transitions faster than others, just as with national demographic transitions (Notestein, 1945). However, unlike the national land use transition and the demographic transition, the national carbon transition is not an inevitable consequence of national development, except in technologi-cally optimistic scenarios. So, under other circumstances, the most economically efficient mix of energy sources in a mature developed economy in the next century would still have a place for fossil fuels. Only 3% of UK electricity is now generated from renewable sources, although this is expected to rise to 6% by the year 2000 (Simonian, 1994). However, fossil fuel consumption should continue to rise in developing countries. Its share of all energy consumption in West and Central Africa, for example, has been forecast to go from 34% in 1988 to 56% in 2025, while the share of fuelwood and other renewables falls from 66% to 45% (Adegbulugbe and Oladosu, 1994). Consequently, completion of the global carbon transition will probably only be possible if there is market intervention in response to global externalities. Even so, to finish the global carbon transition within 50 years will require major changes in the energy economies of developed countries, and some developing countries might have to bypass the fossil fuel phase and switch straight from fuelwood to renewable and nuclear energy.

When trying to formulate an integrated mitigation programme it would help to com-pare possible trends in the mitigating effect of forest strategies with trends in carbon emissions from fossil fuels in the framework of the national carbon transition. An attractive feature of forest-based mitigation is that it can sequester a large amount of carbon over a much shorter time scale than energy consumption patterns are likely to change. This could particularly help developing countries which otherwise would finish their national carbon transitions last of all. Many developing countries still depend heavily on fuelwood for energy, and fuelwood produced by a forest-based mitigation programme could delay the rise in fossil fuel consumption and +buy time+ while technological and social change takes place.

6. CONCLUSIONS

This paper has argued that compensation paid to developing countries for participating in global climate change mitigation should not just be based on the estimated value of forests as carbon stores but allow for the huge opportunity costs faced when they set aside forests or manage them more sustainably. Developed countries heavily exploited their forests in

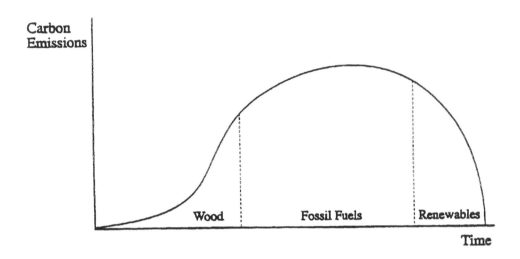

FIGURE 3. The idealized national carbon transition, divided into three phases by dominant energy source.

the past, incurring large global external costs while benefiting enormously from the resulting income and economic development. So they have a greater need to pay and a greater ability to pay than the developing countries. None of the international compensation schemes proposed so far is perfect. Carbon taxes, as proposed, would be levied only on current carbon emissions from fossil fuels, and dividing national shares of the total cost of mitigation according to the benefits obtained from historic forest exploitation would be equally partial, although it would represent national development more fairly and focus on domestic resources in a way that a carbon tax approach would not. However, it is important that an acceptable scheme is devised that takes account of historical spatial preferences in resource use and development, and can provide the foundation for an equitable international mitigation programme.

REFERENCES

Adegbulugbe, A.O. and G.A. Oladosu (1994), Energy and CO2 emissions in West and Central Africa, Energy Policy 22, 499-588.
Andersson, T. (1991), Government failure the cause of global environmental mismanagement, Ecological Economics 4, 215-236.
Andrasko, K., K. Heaton and S. Winnett (1991), Evaluating the costs and efficiency of options to manage global forests, in C.Sargent (ed.). Proceedings of Technical Workshop to Explore Options for Global Forest Management, Bangkok, 24-30 April 1990, pp 216-233, Office of the National Environment Board, Government of Thailand/US Environmental Protection Agency, Bangkok.
Anon (1995), Rules for EU tax on energy, Financial Times, 11th May, 2.
Barde, J.P. and D.W. Pearce (1991), Valuing the Environment, Earthscan Publications, London.

Barrett, S. (1991), Global warming: the economics of a carbon tax, in D.W. Pearce (ed.), Blueprint 2, Greening the World Economy, pp 31-52, Earthscan Publications, London.

Boyle, S. (1994), A global fossil fuel free energy scenario: towards climate stabilization, Energy Policy 22, 106-109.

Daly, H.E. (1990), Toward some operational principles of sustainable development, Ecological Economics 2, 1-6.

Dixon, R.K., J.K. Winjum and P.E. Schroeder (1993), Conservation and sequestration of carbon: the potential of forest and agroforest management practices, Global Environmental Change 3, 160-173.

Grainger, A. (1990), Modelling the impact of alternative afforestation strategies to reduce carbon emissions, in Proceedings of the IPCC Conference on Tropical Forestry Response Options to Global Climate Change , Sïo Paulo, 9-12 January 1990, pp 93-104, US Environmental Protection Agency, Washington DC.

Grainger, A. (1995a), Integrating the socio-economic and physical dimensions of global climate change mitigation, in M. Apps (ed.),The Role of Forest Ecosystems and Forest Resource Management in the Global Carbon Cycle, Springer Verlag, Berlin.

Grainger, A. (1995b), The role of biophysical indicators in evaluating national progress in sustainable development, in D. Tunstall (ed.) Proceedings of a Workshop on Global Environmental Indicators, World Resources Institute, Washington DC, 7-8 December 1992. World Resources Institute, Washington.

Grainger, A. (1995c), The forest transition: an alternative approach, Area 27 (in press).

Grainger, A. (1995d), Changes in land use and forest management in Southeast Asia - an evolutionary perspective, in .Sandbukt (ed.), Management of Tropical Forests: Towards an Integrated Perspective , pp 3-17, Centre for Development and the Environment, University of Oslo.

Grubb, M. (1993), Policy modelling for climate change. The missing models, Energy Policy 21, 203-208.

Hicks, J.R. (1939), The foundations of welfare economics, Economic Journal 49, 696-712.

Hoel, M. (1993), Harmonization of carbon taxes in international climate agreements, Environmental and Resource Economics 3, 221-231

Iverson, L.R., S. Brown, A. Grainger, A. Prasad and D. Liu (1993), Carbon sequestration in South/ Southeast Asia: an assessment of technically suitable forest lands using geographic information systems analysis, Climate Research 3, 23-38.

Kaldor, N. (1939), Welfare comparisons of economics and interpersonal comparisons of utility, Economic Journal 49, 549-552.

Klaasen, G.A.J. and J.B. Opschoor (1991), Economics of sustainability or the sustainability of economics: different paradigms, Ecological Economics 4, 93-115.

Markandya, A. (1991), Global warming: the economics of tradeable permits, in D.W Pearce (ed.), Blueprint 2, Greening the World Economy, pp 53-62, Earthscan Publications, London.

Mather, A.S. (1992), The forest transition, Area 24, 367-79.

Neilson, R.P., G.A. King and G. Koerper (1990), Towards a rule-based biome model, US Environmental Protection Agency, Corvallis, Oregon.

Notestein, F. (1945), Population - the long view, in T.W. Schultz (ed.), Food for the World, University of Chicago Press, Chicago.

Pearce, D.W. (1991), Introduction, in D.W Pearce (ed.), Blueprint 2, Greening the World Economy, pp 1-10, Earthscan Publications, London.

Pearce, D.W. (1994), Blueprint 3, Earthscan Publications, London.

Pearce, D.W. and C.A. Nash (1981), The Social Appraisal of Projects. A Text in Cost-Benefit Analysis, Macmillan, London.

Pearce, D.W. and R.K. Turner (1990), Economics of Natural Resources and the Environment, Harvester Wheatsheaf, London.

Poore, M.E.D. (ed.) (1989), No Timber Without Trees, Earthscan Publications, London.

Sathaye J. and J. Christensen (1994), Cost of stabilizing GHG emissions. Focus on developing countries. Energy Policy 22, 891-893.

Simonian, H. (1994), Windmills and waste continuing to produce much heat, Financial Times, 23 December, 10.

Thiele, R. and M. Wiebelt (1993), National and international policies for tropical rain forest conservation - a quantitative analysis for Cameroon, Environmental and Resource Economics 3, 501-531.

Trexler, M.C., P.E. Faeth and J.M. Kramer (1990), Forestry as a response to global warming: an analysis of the Guatemala agroforestry and carbon sequestration project, in Proceedings of the Intergovernmental Panel on Climate Change (IPCC) Conference on Tropical Forestry Response Options to Global Climate Change, Sïo Paulo, 9-12 January 1990, pp 506-509, US Environmental Protection Agency, Washington DC.

Critical Reviews in Environmental Science and Technology, 27(Special): S177–S184 (1997)

OPTIMAL SUBSIDIES FOR CARBON: COST-EFFECTIVENESS AND DISTRIBUTIONAL CONSIDERATIONS

EDUARDO LEY and ROGER A. SEDJO

Resources for the Future, Washington DC 20036, USA

ABSTRACT: This paper develops a simple static model to derive the optimal subsidy structure to achieve cost-effectiveness and, in a more general framework, the paper develops a welfare function approach to explore the determinants of the optimal level of carbon sequestration and its social value.

KEY WORDS: public goods, subsidies, discount rate, distribution, welfare.

1. INTRODUCTION

On the assumption that climate change is a problem, and that it is largely driven by the accumulation of greenhouse gases (GHG) in the atmosphere — mainly CO_2 —, then carbon sequestering by trees can be seen as a pure public good.[1]

In this paper we explicitly recognize that the size of the optimal carbon subsidy depends upon the responsiveness of supply to the subsidy. Also, in the second part of the paper, we specifically treat changing 'climate' itself as having a positive or negative social value — *i.e.,* we have climate directly entering into individuals' utility functions. Thus, some of the distribution aspects of climate change are explicitly addressed. Finally, the paper develops a welfare function approach to formalize and explore the determinants of the optimal level of carbon sequestration and its social value.

Carbon sequestration (carbon, henceforth.) is produced jointly with other nonmarket or market goods. In the case of a park, the carbon is produced jointly with the recreational services that the trees in the park provide. In the case of a tree plantation, carbon is produced jointly with timber. Since there is no market for carbon, it is not taken into account in the individual producers' private profitability calculations. As a result, on a first approximation, too little carbon sequestration may be supplied.

In the next section we develop a simple static model which we use to obtain the optimal subsidy structure given a fixed amount to be spent in subsidizing carbon sequestration. In section 3 we go one step beyond and we look at the optimal level of carbon sequestering from a global-welfare perspective.

1064-3389/97/$.50

2. COST-EFFECTIVE CARBON POLICY

For the moment, we shall focus on plantations and develop a simple static model abstracting from any intertemporal issues. If p is the market price of one cubic meter of timber which is storing θ tons of carbon whose *social* value is p_c, and $MC_i(\cdot)$ is i's marginal cost schedule, the individual i is going to plant $\tilde{x}_i$ according to

$$p = MC_i\left(\tilde{x}_i\right) \tag{1}$$

while the optimal amount from a social perspective, x_i^*, is going to be given by

$$p + \theta p_c = MC_i(x_i^*)$$

since $MC'(\cdot) > 0$ then[2]

$$X^* \equiv \sum_i x_i^* > \sum_i \tilde{x}_i$$

(The implicit optimal amount of carbon would be given by $C^* = \theta X^*$.)

The total socially optimal amount of timber-carbon, X^*, could be achieved if each individual producer received a payment — *i.e.*, a subsidy — of θp_c for each unit of the jointly supplied products.

2.1. Differential Subsidies

When the value of p_c is uncertain, a practical approach consists in choosing the aggregate expenditure, S, to be dedicated to carbon sequestration, and then spending it in the most efficient way from a carbon perspective. In this case, in general, as we show below, different producers should receive different subsidies. In practice, however, normally the difference in subsidies could best be based on geographical differences. If a subsidy of s_i per m^3 of timber provided by individual i Is established, then equation (1) becomes

$$p + s_i = MC_i\left(x_i(s_i)\right) \tag{2}$$

The planner's problem is given by

$$\max_{s_i} \quad \theta \sum_i x_i(s_i)$$
$$\text{subject to} \quad \sum_i s_i x_i(s_i) = S$$

The first-order conditions are, for all i,

$$\theta x_i'\left(s_i^*\right) = \lambda * \left[x_i'\left(s_i^*\right) + x_i\left(s_i^*\right)\right] \tag{3}$$

the left-hand side shows the value of the incremental addition of carbon and the right-hand side the incremental cost of raising the subsidy to producer i. This last part is composed of two sources; the subsidy on the additional amount supplied and the additional subsidy on all the units supplied.

Equation (3) implies that for any two producers, i and j, the following must hold:

$$s_i^*\left(1 + \frac{1}{\eta_i\left(x_i\left(s_i^*\right)\right)}\right) = s_j^*\left(1 + \frac{1}{\eta_j\left(x_j\left(s_j^*\right)\right)}\right) \tag{4}$$

where η_i is the elasticity of i's supply to the subsidy.

$$\eta_i \equiv x_i'\left(s_i\right)\frac{s_i}{x_i\left(s_i\right)}$$

Equation (4) is just a version of the well-known Ramsey's inverse-elasticity rule. Here, the more elastic the supply with respect to the subsidy, the higher the subsidy rate should be.

Equation (4) calls for different subsidy rates for different establishments. It also expresses the idea that subsidies should only be paid when they make a difference at the margin. Ideally, one would like to be able to tell the agent's 'type' when she applies for a subsidy. Perhaps, the best that can be done is to know the agent's geographical location. Geographical areas can serve as crude proxies of the agents' types and the subsidy rates can be based on the establishment's location. This notion is shown in Figure 1.

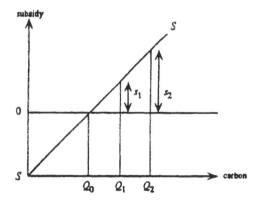

FIGURE 1. Supply of carbon.

The schedule SS in Figure 1 shows the supply of carbon for different subsidy rates. Note that the supply when there is no subsidy is Q_0 — from the policymaker's perspective, Q_0 is free. With perfect discrimination, optimal subsidies would involve different subsidies for each unit produced — *i.e.*, each unit would receive the vertical distance from 0 to SS. Suppose that the policy target amount of carbon is Q_2; then a subsidy of s_2 per unit is required at the margin. When a uniform subsidy of s_2 is being paid to *all* suppliers, the amount $Q_0 s_2 + 1/2(Q_2 - Q_0)s_2$ is being wasted from the policy's perspective since it doesn't affect marginal decisions. Suppose now that we can associate suppliers with portions of the supply schedule. Assume that suppliers to the left of Q_0 are mostly from geographical region A, suppliers in the Q_0–Q_1 range can be associated with region B, and that suppliers to the right of Q_1, are mostly from region C. Then a subsidy scheme that pays no subsidy to plantation establishments in region A, pays a subsidy of s_1 to plantations in region B, and pays a subsidy of s_2 to plantations in region C, achieves the goal with a much smaller waste — namely, $1/2(Q_0 - Q_1)s_1 + 1/2(Q_2 - Q_1)s_2$.

Identifying the agent's type when perverse incentives prevent truthful revelation is often a problem in many economic situations. Here, however, many of the most important attributes determining the producer's elasticity, η_j, — *e.g.*, weather, soil characteristics, proximity to markets, transportation infrastructure — are closely linked to location which is readily observable. Therefore, differential subsidies based on geographic areas are feasable solution. In principle, one could establish an optimal subsidy scheme based on 'iso-elasticity' topological maps.

The problem with this approach, however, is that there is no reason to expect that the actual subsidy is going to be close to the Pigouvian subsidy, p_c. When funds are limited and there is reason to believe that $s < p_c$, then a cost-effective subsidy program will generally be welfare-enhancing. In the next section we turn our attention to p_c.

3. THE SOCIALLY OPTIMAL AMOUNT OF CARBON

There is a great uncertainty associated with the probable effects of the accumulation of greenhouse gases in the atmosphere. As a result, the *value* of one ton of carbon to society at time t, is not known. A substantial amount of research has focused on the *cost* of emission reduction (or carbon sequestration) — see, *e.g.*, the surveys by Darmstadter (1991), or Weyant (1993). There have also been some attempts to estimate the shadow price of CO_2 — most notably Nordhaus (1982, 1991, 1993); see also Uzawa (1992) who pays attention equity considerations and country differences in the stages of econormc development.

From a global-policy perspective, the basic questions are:

[Q_0] How does climate change ultimately translate into welfare change?

[Q_1] How does accumulation of GHG (interacting with other gases like aerosols) translate in climate change?

[Q_2] How is climate change likely to affect production activities?

While $[Q_1]$ and $[Q_2]$ are positive scientific questions,[3] $[Q_0]$ has a moral or ethical content since, in a world of heterogeneous agents, the individual welfare changes must be weighted in order to be aggregated or otherwise summarized. Often, economists will get the answer to $[Q_0]$ directly from $[Q_2]$ under the argument (explicit or implicit) that a policy which results in efficiency gains is always preferable since the bigger pie can always be redistributed latter. This is the approach, *e.g.*, in Nordhaus (1982) when a single-representative individual is used.

However, climatic change will affect different people (countries) in different ways. For instance, a farmer in Siberia might be better off if warming occurs while a farmer in Mexico might be worse off. Or an avid skier might suffer while somebody with a taste for a warmer weather might enjoy the climate change. (Catastrophic changes could result in dramatic and generalized loses.) Dealing with the climate change problem, it is especially important to explicitly deal with the distributional issues. Climate change, if it ever happens, is probably going to result in large redistributions across the planet (and across generations).[4]

3.1. The Welfare Function Approach

One way to deal with $[Q_0]$ is to assume that we have a general welfare function, $W(\cdot)$ (see Bergson (1938)). The *planner* has M to be spent either in direct transfers to individuals (indexed by h, for 'household'), Δm_h, or in a public good G which somehow represents the 'atmosphere quality' — see, *e.g.*, Chichilnisky and Heal (1994). We can think of G as an index which tells us how close the atmospheric concentrations of GHG are to, say, century's.[5]

For simplicity, assume that G costs c per unit. Let $v_h(m, p(G); G)$ be individual's h indirect utility function (conditional on the level of G), where m_h is h's income, $p(G)$ is a vector of prices which possibly depend on G, and G is the public good, again somehow representing the atmosphere's quality.

We allow the quality of the atmosphere, G, to affect each individual's welfare (*i*) directly since different people will have different tastes for climate,[6] and (*ii*) indirectly via prices. We shall return to this poiat later.

The planner's objective will be to maximize

$$
\max_{\Delta m, G} \quad W\!\left(v_1\!\left(m_1 + \Delta m_1, \mathrm{p}(G); G\right), \ldots, v_H\!\left(m_H + \Delta m_H, \mathrm{p}(G); G\right)\right)
$$
$$
\text{subject to} \quad \sum_h q_h \Delta m_h + cG = M \tag{5}
$$

The price q_h tells us the cost of transferring funds to individual h. Note that the individual subscript h can also be interpreted as a time subscript. In that case if $h = T$ means that the transfer is to be made to somebody T periods away, then $q_T = (1 + r)^{-T}$ where r is the interest rate. In other cases, $q_h < 1$ could capture leaks in the buckets when transferring funds.

The first-order conditions with respect to Δm_h imply

$$\frac{\partial W}{\partial v_h} \alpha_h = \lambda q_h, \qquad \forall h \tag{6}$$

where α_h is the marginal utility of income of individual h and λ is the Lagrange multiplier associated with the planner's constraint — $i.e.$, the shadow cost of public funds. Let us introduce a measure of dealing with the marginal social benefit of transferring income to individual h: $\theta_h \equiv (\partial W/\partial v_h)\alpha_h$. Then (6) implies

$$\frac{\theta_h}{\theta_t} = \frac{q_h}{q_t}, \qquad \forall h, l \tag{7}$$

which is the usual equalization of marginal effects across alternative uses.

Differentiating with respect to G, using Roy's Law and rearranging, we get

$$\sum_h \frac{\partial W}{\partial v_h} \alpha_h \left\{ MRS_h + \sum_i x_h^i \frac{\partial p^i}{\partial G} \right\} = \lambda c \tag{8}$$

where p^i is the i-th commodity's price, and x_h^i is consumer h's demand for good i; and

$$MRS_h \equiv \frac{\dfrac{\partial v_h}{\partial G}}{\alpha_h}$$

Note that MRS_h is also h's the inverse demand for the public good G.

Define h's net benefit of a marginal change in G, β_h as:

$$\beta_h \equiv \sum_i MRS_h + \sum_i x_h^i \frac{\partial p^i}{\partial G}$$

At the individual level, climate change (G) affects the individual (i) directly because 'climate' enters her utility function, and (ii) indirectly through the effects of 'climate' in prices. For instance, a person with a taste for warmer weather could be better off through the first effect if global warming occur and worse off through the second if the prices of the commodities that she consumes went up. Note that when $x_h^i = 0$ so that individual h is not consuming commdity i, h is not affected by changes in p^i.

Let $\bar{\theta} \equiv 1/H \sum_h \theta_h$, and noting that $\theta_h = (\theta_h - \bar{\theta}) + \bar{\theta}$, we can manipulate (8) to get:[7]

$$\bar{\theta} \left\{ \sum_h \beta_h + H\mathrm{Cov}\left(\frac{\theta_h - \bar{\theta}}{\theta}, \beta_h \right) \right\} = \lambda c \tag{9}$$

This condition is a generalized version of the famous Samuelson condition for optimal public good provision. When there are no distribution considerations, all the θ's are the same and every individual generates the same social welfare when receiving a transfer. Then there is no variability of the θ's within the population. This implies that the covariance term in (9) would disappear and we would obtain the usual Samuelsonian condition which mandates the equalization of the sum of the MRS_h's to the marginal rate of transformation.

When distributional considerations do matter — i.e., the θ_h's differ across h — then the sum of the net benefits, β_h's, is corrected by a term which reflects distributional concerns. If the β_h's are positively correlated with the θ_h's so that groups with higher weight in the welfare function derive higher benefits from the public good, then the covariance term in (9) is positive and a larger amount of G will be provided. Another way of expressing this is to take the covariance term to the left-hand side, subtracting from the cost or marginal rate of transformation. Then, something which is good for welfare-favored groups becomes 'cheaper'.

Equation (9) not only provides guidance as to the optimal level of G but also gives us the social value of G. This is the amount of the left-hand side, which is then equated to the marginal social cost.[8] The social value of sequestering a ton of carbon is given by the weighted sum of the individual valuations. These, in turn, are the sum of a direct effect and a secondary effect via prices.

4. SUMMARY AND CONCLUSIONS

We have explored some policy issues related to carbon sequestering, or more generally climate change. We argue that differential subsidies are both feasible and desirable in the context of a cost-effective policy. In a more general context, we explore the determinants of the social value of carbon sequestration. We show how the individual welfare changes can be appropriately weighted and aggregated in the context of a welfare function. We also derive the optimality conditions which determine the optimal level of 'carbon effort' and its social value.

The contributions of this paper are three. First, it specifically recognizes that changing climate may, in itself, have a positive or negative value and this it explicitly recognizes the distribution aspects of climate change. Second, it explicitly recognizes that the size of the optimal carbon subsidy depends upon the responsiveness (elasticity) of supply to the subsidy. Finally, the paper develops a welfare-function approach to formalize and explore the determinants of the optimal level of carbon sequestration and its social value.

ACKNOWLEDGMENT

We thank the World Bank's GEF for funding our study on "Argentina: Carbon and Forests" that raised some of the issues which we explore here, and to the Vetlesen Fund for partially financing this research. We also benefitted from many helpful discussions with Jose Luis Darraidou, Luis Constantino, Charles Feinstein, Hugo Iza, Bob Kirmse, Mirta Rosa Larrieu, Ricardo Larrobla, Mike McGarry, Jens Rosebrock, and comments from two anonymous referees.

NOTES

1. A good is *non-rival* when each unit can be consumed by all agents. It is *non-excludable* when each unit has to be consumed by all agents. A good that has both characteristics is a *pure* public good.
2. While it is possible to have a flat marginal cost schedule for a single plantation, over some range, from a country's perspective the marginal cost is better approximated by a strictly increasing curve.
3. See, *e.g.,* Reilly and Richards (1993), and Richards (1995) for a framework to deal with them.
4. It is worth noting that the cost of bringing down the levels of GHG to, say, 1990 levels is quite large relative to foreign aid figures (Schelling (1992)).
5. In this interpretation, we assume in what follows that $G \leq 100\%$ to avoid dealing with the calculus of corner solutions.
6. As noted by Schelling (1992), most people retire to and spend their vacations in warmer places.
7. Below $Cov(\cdot,\cdot)$ represents the usual covariance functional applied to the individuals which enter into the welfare function. It must be noted that there is nothing stochastic here, it is just a convenient way to simplify and interpret the equation.

REFERENCES

Bergson, A. (1938), 'A Reformulation of Certain Aspects of Welfare Economics', *Quaterly Journal of Econ omics,* **521,** 310–334.

Chichilnisky, G. and G. Heal (1994), 'Who should abate carbon emissions? An international viewpoint', *Economics Letters,* **44,** (4), 443–450.

Darmstadter, J. (1991), 'The Economic Cost of CO_2 Mitigation: A Review of Estimates for Different World Regions', Discussion Paper ENR91-06, Resources for the Future, Washington DC.

Nordhaus, W. (1982), 'How fast should we graze the Global Commons?', *American Economic Review,* **72,** (2), 242–246.

Nordhaus, W. (1991), 'To Slow or Not to Slow: The Economics of the Greenhouse Effect', *Economic Journal,* **101,** 920–937.

Nordhaus, W.D. (1993), 'Reflections on the Economics of Climate Change', *Journal of Economic Perspectives,* **7,** (4), 11–26.

Reilly, J. and K. Richards (1993), 'Climate Change Damage and the Trace Gas Index Issue', *Environmental and Resource Economics,* **3,** 41–61.

Richards, K. (1995), 'The Time Value of Carbon in Bottom-Up Studies', *Environmental and Resource Economics,* this issue.

Schelling, T.C. (1992), 'Some economics of global warming', *American Economic Review,* **82,** (1), 1–14.

Uzawa, H. (1992), 'Imputed Prices of Greenhouse Gases and Land Forests', Beijer Discussion Paper Series No. 9, Stockholm.

Weyant, J.P. (1993), 'Costs of Reducing Global Carbon Emissions', *Journal of Economic Perspectives,* **7,** (4), 27–46.

Critical Reviews in Environmental Science and Technology, 27(Special): S185–S192 (1997)

CARBON SEQUESTRATION AND TREE PLANTATIONS: A CASE STUDY IN ARGENTINA

EDUARDO LEY and ROGER A. SEDJO

Resources for the Future, Washington, DC 20036, USA

ABSTRACT: This study undertakes two tasks. First, it estimates the carbon sequestering potential for two alternative tree plantations programs in two regions of Argentina, which have markedly different carbon sequestering potentials. While it estimates the physical volume of the carbon sequestered, unlike a host of other studies it does not try to estimate the costs of sequestration. Second, the study then develops a method for evaluating the comparative benefits of the carbon sequestered in the two cases based upon the price (value) of the external benefits, the growth of this price over time, and the discount rate. Estimates of the comparative benefits or value provide guidance as to the optimum comparative size of subsidies for the two regions.

KEY WORDS: carbon sinks, discount rate, carbon storage value.

1. INTRODUCTION

Argentina is currently developing its program for carbon sequestration as called by the Framework Convention on Climate change. Such a program will involve a variety of measures to reduce carbon releases and/or increase carbon sequestration. Included in the mix of carbon sequestration measures will be forestry activity including forest preservation and natural forest management, as well as plantation establishment.

It has been recognized for some time that tree plantations have the potential to sequester large amounts of carbon (Sedjo and Solomon, 1989). Furthermore, a recent study indicates that the costs of carbon sequestration using forestry projects might be quite modest compared to other approaches (Sedjo et al., 1995).

This study involves an assessment of anticipated effects of the use of tree plantations to sequester atmospheric carbon in Argentina. It examines the carbon sequestration implications of two plantation establishment approaches. The first involves the establishment in Mesopotamia of an even-aged regulated forest with a rotation of 25 years. The other the envisages the establishment of plantations in Patagonia with a 35 year rotation. The methodology involves keeping tract of carbon in forest ecosystems as well as in long-lived wood products. The approach suggests that the absence of a value for sequestered carbon will result in too little planting from a social perspective. The per ha time profiles of the captive carbon are developed for each case. The results are presented in Figure 1.

While the study estimates the physical volume of carbon sequestered, unlike a host of other studies it does not try to estimate the costs of sequestration. Rather, study develops

1064-3389/97/$.50

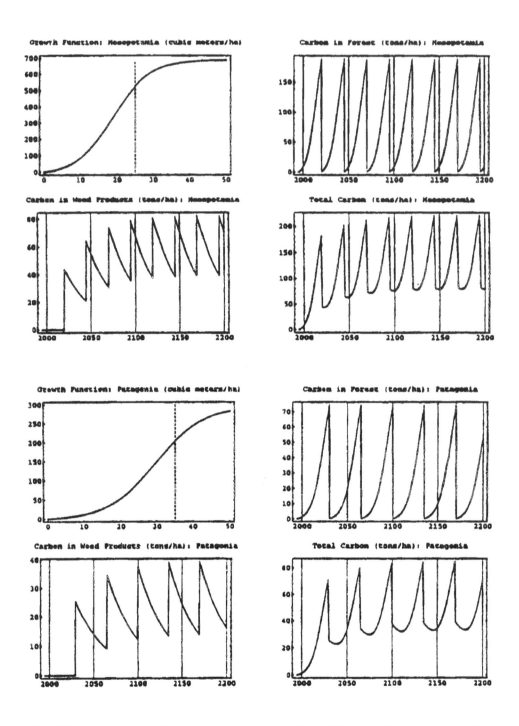

FIGURE 1. Examples of forest carbon sinks over time. (Years on the horizontal axis.)

an innovative method for evaluating the comparative benefits of the carbon sequestered in the two cases based upon the 'price' of the benefits, the growth of this price over time and the discount rate. Estimates of the comparative benefits or values provide guidance for policy makers as to the optimum comparative size of subsidies in the two regions.

2. INTERTEMPORAL CARBON MODEL

Let $c_i(t)$ the total stock of carbon associated with the biomass in one particular ha of site i at time t,

$$c_i(t) = \underbrace{\alpha_i z_i(t)}_{\text{forest}} + \underbrace{\int_{s \leq t} \sum_j h_{i,j}(s)\phi_j(t-s)\,ds}_{\text{wood products}} \tag{1}$$

where $z_i(t)$ is the stock of *commercial* wood in the forest,[1] α_i is a coefficient that converts commercial wood into total associated-biomass carbon, $h_{i,j}(t)$ is the amount of wood harvested for use j (waste will be included as a particular type of use). We capture the carbon decay phenomenon by the function $\phi_j(\cdot)$ which is a carbon 'survival' function for use j, thus $\phi_j(0)$ is a coefficient that converts j-timber into carbon and $\phi_j(\Delta t)$ gives the carbon which remains sequestered Δt years after the harvest. The stock of wood in this ha of forest, over the rotation, evolves according to:

$$\frac{dz_i(t)}{dt} = \theta_i(t - t_0) - \sum_j h_{i,j}(t), \quad t \in (t_0, T)$$

where $\phi_j(\cdot)$ is a growth function, t_0 is the initial planting time, and T is the rotation length.
The individual producer i's net profits (per ha) over the rotation are:

$$\Pi_i = \int_{t_0}^{t_0+T} \sum_j \left(p_{i,j}(s)h_{i,j}(s) - k_i(s)\right)e^{-rs}\,ds \tag{2}$$

where r is the interest rate and $p_{i,j}(t)$ is the price that i obtains for j-type wood at time t,[2] and $k_i(t)$ represents the costs incurred. Note that T, the rotation length, is a planter's control variable.

---

[1] We use 'commercial wood' as the *numeraire* because it is the unit of most of the forestry research.

[2] Allowing the $p_{i,j}(t)$'s to vary by producer they can reflect, among other things, location differences.

2.1. Social Profits

Provided that carbon sequestering has a positive social value, the absence of forest carbon sequestration considerations will have, at least, two important affects in the private choices involved. First, too little planting will be realized. Second, the harvest times and the total rotation length, T, will be chosen from a timber perspective[3] and might be sub-optimal from the social point of view when carbon is properly accounted for. At time T of the harvest, the *social* profits are given by

$$\Pi_i^s = \Pi_i + \int_{t_0}^{\infty} p_c(s) \left(\alpha_i z_i(t) + \int_{t_0}^{s} \sum_j h_{i,j}(t) \phi_j(s-t) dt \right) e^{-ps} ds$$

$$= \underbrace{\int_{t_0}^{t_0+T} \sum_j \left(p_{i,j}(s) h_{i,j}(s) - k_i(s) \right) e^{-rs} ds}_{\Pi_i = \text{commercial wood}}$$

$$+ \underbrace{\int_{t_0}^{\infty} p_c(s) \left(\alpha_i z_i(t) + \int_{t_0}^{s} \sum_j h_{i,j}(t) \phi_j(s-t) dt \right) e^{-ps} ds}_{\Pi_i^c = \text{carbon}}$$

(3)

when $p_c(t)$ is the carbon's price at t, and ρ is an appropriate discount rate — see, e.g., Arrow et al. (1994) and Rosebrock (1993) for the difficulties associated with this concept in this context. The inclusion of carbon considerations will, in general push to the future the optimal harvest time — see, e.g., van Kooten et al. (1995). (Note that the integrals in (3) are forward-looking while the integral in (1) is backward-looking.)

In principle, carbon-sequestering policies can involve subsidy schemes directed to two distinct objectives: (i) Increase plantation activities, and (ii) Increase rotation times. In the remaining discussion, we shall focus only on (i).

3. PROTOTYPICAL MODELS FOR ARGENTINA

In this section we develop two models used for Argentina. Lacking specific empirical growth models we use a logistic function fitted to anticipated forest growth. The carbon coefficient is handled in the common way for an analysis of this type — see, e.g., Marland (1988), Crosson et al. (1994).

Here we present prototypical models for two Argentinian regions: Mesopotamia (subtropical) and Patagonia (temperate). We use a biomass coefficient of 1.40 and a carbon coefficient, of 0.26 (Crosson et al., 1994) so that $\alpha_i \equiv \alpha \equiv 0.26 \times 1.4$. Only two types of

---

[3]   Essentially, the wood would be harvested when the growth rate of the commercial wood equals the interest rate.

harvest types will be considered here: waste and commercial use — i.e., $h_{i,1}$ will represent waste and short-lived wood products, and $h_{i,2}$ long-lived wood products.

The growth functions, $\theta_i(\cdot)$, used for the commercial wood are constructed to be consistent with the average annual growth figures in SAGyP (1994).[4] Over the rotation period, they accumulate the same amount of commercial wood as the models in SAGyP (1994). We use a logistic growth curve which is defined by

$$\theta_i\left(t - t_0\right) \equiv \frac{dz_i(t)}{dt} = \frac{\gamma_i z_i(t)\left(\alpha_i - z_i(t)\right)}{\alpha_i}$$

where $z_i(t)$ is the volume of commercial wood at site (region) i, at time t; which implies

$$z_i(t) = \frac{\alpha_i}{1 + \beta_i e^{\gamma_i(t_0 - t)}} + C$$

Note that $z_i(t_0) = 0 \Rightarrow C = -\alpha_i/(1+\beta_i)$. The graph of $z_i(t)$ is S-shaped symmetric about its inflexion point — see Figure 1. Table 1 shows the values of the different parameters for the two regions considered in the prototypical models that follow. The column AAG represents the annual average growth over the rotation.

The carbon sequestered in wood products is assumed to be declining exponentially, so that the survival function is given by

$$\phi_2\left(t - s\right) \equiv 0.26e^{-0.030(t-s)}$$

After 50 years, about 22% of the carbon remains sequestered, less than 5% remains after 100 years and only a bit over 1% stays sequestered after 150 years.[5] We assume that the carbon associated with waste (which includes dead trees and limbs on forest floor) and other short-lived wood products is released immediately, i.e., $\phi_1(t - s) = 0$.

TABLE 1
Logistic Growth Function Parameters of Two Prototypical Models of Forest Carbon Sequestration

i	Region	α_i	β_i	γ_i	AAG	Rotation
1	Mesopotamia	706.21	45	0.20	20 m^3/ha/yr	25 years
2	Patagonia	301.50	85	0.15	13 m^3/ha/yr	35 years

[4] SAGyP stands for *Secretaría de Agricultura, Ganadería y Pesca* of the Republic of Argentina.
[5] Row and Phelps (1995) have developed a sophisticated modal (HARVCARB) which tracks wood carbon flows and storage after the timber harvest for different types of trees and regions. They use different (exponentiai) decay functions for each of 12 major end-use classes. Given the uncertainty of the destiny of the wood from the Argentinean regions here considered at the time horizons which we are looking at, we chose for a simple aggregate, decay function.

3.1. Valuing Carbon Flows

Figure 1 shows the status of the potential carbon sinks per ha in two different regions of Argentina: Mesopotamia and Patagonia. Masopotamia with better growth conditions produces substantially more sequestering sooner in time. We assume that a larger proportion of the commercial wood goes into long-lived wood products in Patagonia than in Mesopotamia ($\frac{1}{2}$ versus $\frac{1}{3}$) but this cannot compensate for the natural conditions favoring Mesopotamia. We ask here where (if anyplace) tree-planting should be subsidized for the carbon-sequestering services that it renders.

We therefore turn now to compute the 'carbon' part, Π_i^c, in Equation (3). In order to do that, we assume that the social value of one-year carbon sequestering evolves according to

$$p_c(t) = p_0 e^{\pi t}$$

We can then compute the ratio of carbon benefits from one ha in Mesopotamia to the carbon benefits of one ha in Patagonia: Π_1^c/Π_2^c. This ratio is a function of the difference between the growth rate for carbon prices and the discount rate. Table 2 shows both the absolute values and the ratio for different values of $(\pi - \rho)$. (Note that for the integral in Π_i^c to converge we need $(\pi - \rho) < 0$.)

Let us focus on the last column in Table 2. When $(\pi - \rho) = -0.01$, the value of the carbon-sequestering services of 1 ha in Mesopotamia is 3.20 times its value in Patagonia. When there is more impatience, e.g., $(\pi - \rho) = -0.10$, the ratio goes up, to 4.86. Suppose, at $(\pi - \rho) = -0.10$, that the *effective subsidy per incremental ha* was 5 times lower in Patagonia. Then, it would pay to choose Patagonia over Mesopotamia despite Mesopotamia's higher yield.

TABLE 2
Economic Valuation of the Forest
Carbon-Sequestering Services of 1 ha in
Two Proptotypical Models

	Mesopotamia	Patagonia	
$(\pi - \rho)$	Π_1^c/p_0	Π_2^c/p_0	Π_1^c/Π_2^c
−0.01	94750.00	29605.30	3.20
−0.02	24039.80	7376.22	3.26
−0.03	10195.60	3010.73	3.39
−0.04	5433.32	1528.75	3.55
−0.05	3287.46	876.98	3.75
−0.06	2157.07	544.53	3.96
−0.07	1497.92	357.91	4.19
−0.08	1085.06	245.85	4.41
−0.09	812.36	175.09	4.64
−0.10	624.65	128.58	4.86

SAGyP (1994, 1995) estimated the required (one lump sum) subsidies to make tree-planting activities financially attractive for Mesopotamia and Patagonia: US$ 438/ha versus US$ 778/ha. We stress here the notion of effective subsidies per incremental ha planted. Because of political and informational constraints, uniform subsidies must be offered within each region. As a result, the subsidy must be paid to *all* the ha's planted regardless of whether they are incremental or not — see Ley and Sedjo (this volume) for a more detailed discussion. Suppose, for the sake of argument, that in Mesopotamia the subsidy doubles the annual planting — i.e., half of the subsidized planting would have occurred in the absence of any subsidy — while in Patagonia no planting would occur without the subsidy. Then, the effective subsidy per incremental ha would be US$ 876 (=438 × 2) in Mesopotamia while it would remain at US$ 778 in Patagonia.

What about the absolute value of Π_i^c? Table 2 shows Π_i^c/p_0. What is a reasonable value for p_0? Suppose that the optimal carbon tax today was \$$\tau$/ton, then: $p_0 = \tau(\rho - \pi)$; since a carbon tax preventing the release of 1 ton is equivalent to the eternal sequestration of 1 ton. As an example, if $(\pi - \rho) = -0.05$, and $-\tau = \$5$/ton (Nordhaus, 1993); we get $p_0 = 0.25$ so that the value of the carbon services would be worth US$ 821.75/ha in Mesopotamia (=0.25 × 3287.46) and US$ 219.25 (= 0.25 × 876.98) in Patagonia.

We should note that the application of the above is made difficult by the problems associated with the estimate of the "effective subsidy." In our example, for a given effective subsidy, Mesopotamia will always be chosen. However, it is also clear that the effective subsidy will vary by region with location, value of commercial wood, access to the market, and so forth. Thus, this approach provides a basis for subsidizing plantations in regions that might not be favored by commercial considerations.

4. CONCLUDING REMARKS

This study undertakes two tasks. First, it estimates the carbon sequestering potential for tow alternative tree plantations programs in two regions of Argentina, which have markedly distinct carbon sequestering potentials. Second, the study develops an innovative method for evaluating the relative benefits of the carbon sequestered in the two cases based upon assumptions about the evolution of the shadow price of carbon and the discount rate. Estimates of the comparative benefits or value provide guidance as to the optimum comparative size of subsidies for the two regions. Alternative assumptions could be used in a similar framework to explore other scenarios.

The approach demonstrates that the more efficient plantation subsidy may go to the region that is disadvantaged commercially, since the subsidy will not be allocated to activities that would be undertaken anyway without the subsidy.

ACKNOWLEDGMENT

We thank the World Bank's GEF for funding our study on "Argentina: Carbon and Forests" that raised some of the issues which we explore here, and to the Vetlesen Fund for partially financing this research. We gratefully acknowledge the assistance from the Secretaría de Agricultura, Ganadería y Pesca (SAGyP), República de Argentina; especially Jose Luis

Darraidou, Hugo Iza, Mirta Rosa Larrieu, and Ricardo Larrobla. We also benefitted from many helpful discussions with Luis Constantino, Charles Feinstein, Bob Kirmse, Mike McGarry, Jens Rosebrock, and comments from Joel Darmstadter and three anonymous referees.

REFERENCES

Arrow, K., W.R. Cline, K.-G. Mäler, M. Munasinghe, and J.E. Stiglitz (1994), 'Intertemporal and Equity Discounting', Chapter 3 in the forthcoming IPCC report on Climate Change.

Crosson, P., J. Adamoli, K. Frederick, and R. Sedjo (1994), 'Potential Environmental and Other External Consequences of the Program to Increase the Area of Plantation Forests in Argentina by 400,000 Hectares', Final Report to the Secretaría de Agricultura, Ganadería y Pesca, Gobierno de Argentina.

van Kooten, G. Casey, G. Cornelis, Clark Binkley, and Gregg Delcourt (1995), 'Effect of Carbon Taxes and Subsidies on Optimal Forest Rotation Age and Supply of Carbon Services', *American Journal of Agricultural Economics,* 77, (August), forthcoming.

Ley, E. and R.A. Sedjo (this volume), 'Differential Subsidies and Equity Considerations in Carbon Policy', *Environmental and Resource Economics,* this Volume.

Marland G. (1988), 'The Prospect of Solving the CO_2 Problem Through Global Reforestation', DOE/ NBB-0082, U.S. Department of Energy, February.

Nordhaus, W.D. (1993), 'Reflections on the Economics of Climate Change', Journal of Economic Perspectives, 7, (4), 11-26.

Rosebrock, J. (1993), 'Time-Weighting Emission Reductions for Global Warming Projects — A Comparison of Shadow Price and Emission Discounting Approaches', mimeo, Harvard University, JFK School of Government.

Row, C. and R.B. Phelps (1995), 'Wood Flows and Storage after Timber Harvest', unpublished.

SAGyP (1994), 'Evaluación Financiera de Modelos Forestales para Producción de Madera y Fijación de Carbono en Regiones Seleccionadas', mimeo, Proyecto Forest AR, Secretaría de Agricultura, Ganadería y Pesca, República Argentina.

SAGyP (1995), 'Argentina: Desarrollo de Modelos' Forestales para Fijación de Carbono y Producción de Madera', mimeo, Proyecto Forest AR, Secretaría de Agricultura, Ganadería, y Pesca, República Argentina. (Paper to be presented at the XX World Meeting of the IUFRO in Finland, August 1995.)

Sedjo, R.A. and A.M. Solomon (1989), 'Climate and Forests', in *Greenhouse Warming: Abatement abd Adaptation,* ed. N.J. Rosenberg, W.E. Easterling, P.R. Crosson, and J. Darmstadter. Washington, DC: Resources for the Future.

Sedjo, R.A., J. Wisniewski, A. Sample, and J.D. Kinsman (1995), 'Managing Carbon via Forestry: Assessment of some Economic Studies', *Environmental and Resource Economics,* 5, (8), forthcoming.

Critical Reviews in Environmental Science and Technology, 27(Special): S193–S211 (1997)

THE ECONOMIC AND ENVIRONMENTAL IMPACT OF PAPER RECYCLING

STIG BYSTRÖM[1] and LARS LÖNNSTEDT[2]

[1]MoDo Pulp and Paper Inc., S-891 01 Örnsköldsvik, Sweden and the Department of Forest-Industry- Market Studies, the Swedish University of Agricultural Sciences, Box 7054, S-750 07 Uppsala, Sweden; [2]The Department of Forestry Economics, the Swedish University of Agricultural Sciences, S-901 83 Umeå, Sweden

ABSTRACT: The Optimal Fibre Flow-model, a combined optimization and simulation model, calculates the optimal combination of energy recovery and recycling of waste paper for paper and board production. In addition, the environmental impact is estimated by using an Environment Load Unit-index (ELU-index). The ELU-index assigns an environmental load value to emissions and to the use of non-renewable resources such as oil and coal. If the Scandinavian forest industry uses hydroelectric energy in pulp and paper production, and the utilization rate of waste paper increases, the ELU-index increase. However, a political decision that would force the forest industry to decrease its utilization rate and thus favor a 10% increase in energy recovery would result in a marginal loss for the Scandinavian forest industry of about 5.5 USD/ton. On the other hand, if the electricity is produced from fossil fuel, the ELU-index is initially quite stable, but later increases. A minimum is found at a utilization rate of about 25 per cent. Here, a forced decrease in the utilization rate, along with a 10% increase in energy recovery, would cause a marginal loss for the Scandinavian forest industry with about 6.5 USD/ton.

KEY WORDS: systems analysis, model, policy analysis, life cycle analysis, waste paper, energy.

1. INTRODUCTION

Concern about the environment, demands for economizing in the consumption of resources and an increased interest in reuse and recycling philosophy are strong currents in modern society. The targets for environmental criticism are often the throw-away mentality of consumer society and the large-scale material and energy requirements of the processing industry. The forest industry has thus become a convenient target for the environmental ambitions of consumers and politicians. In countries like Germany, Sweden and the USA, this has led to demands for changes in the industrial forest production system. The forest industry of the Nordic countries fears that political decisions made by the European Union or by individual countries will force a fixed utilization rate for waste paper in paper and board producing. There is a risk that such decisions will lead to a sub-optimal use of waste paper if the environmental impacts of alternative uses are not considered. Important alternative uses in this regard are recycling for production of paper, energy recovery and landfill. Figure 1 gives a principle outline of linkages of the fibre flow when Western Europe is divided into two regions.

The benefits of paper recycling have not been fully analyzed (Grieg-Gran, 1994), through increased recycling is generally assumed to be desirable and necessary. Waste management policy in a number of countries is characterised by a prevailing hierarchy of options in which waste minimisation, reuse and recycling are all considered preferable to energy recovery, which in turn is considered superior to landfills.

Any assessment of recycling should compare the impacts of recycling and their associated costs and benefits with those of alternative options for waste disposal. A key issue is the impact on energy use in manufacturing. Processing waste paper for paper and board manufacture requires energy that is usually derived from fossil fuels. In contrast to the production of virgin fibre-based chemical pulp, waste paper processing does not yield a thermal surplus and thus thermal energy must be supplied to dry the paper web. If, however, the waste paper was recovered for energy purposes the need for fossil fuel would be reduced and this reduction would have a favourable impact on the carbon dioxide balance and the greenhouse effect. Moreover, pulp production based on virgin fibres requires consumption of round-wood and causes emissions of air-polluting compounds.

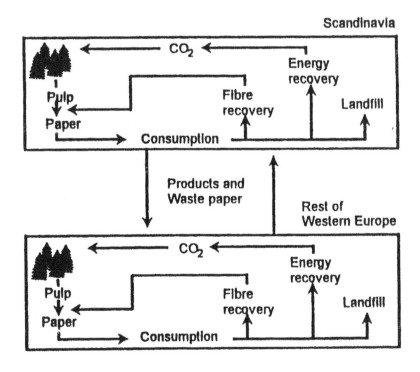

FIGURE 1. Principle flow of fibres in Western Europe.

2. APPROACH

2.1. Literature Reviewed

Examples of early economic studies of demand for, supply of or trade with waste paper are Grace et al. (1978) and Yohe (1979). They examined international trade and its importance to prices. Price expectations and the effect of price changes have been analyzed by Edwards (1979), Deadman &Turner (1981), and Kinkley & Lahiri (1984). Gill & Lahiri (1980) and Edgren & Moreland (1989) have found low price elasticities for waste paper, a finding which indicates that price subsidies are not to be recommended for stimulating increased use. In a Swedish study of the printing industry, Rehn (1993) shows that the uses of pulp, pulp wood and waste paper are each sensitive to their own price changes so that they will likely substitute for each other.

Lately, systems analysis and extensive modelling approaches have been used for studying the waste paper problem. Colletti & Boungiorno (1980) and the NAPAP model (Ince, 1994; Zhang et al., 1993) concentrate on production and economic aspects of waste paper recycling. Virtanen and Nilsson (1992) incorporate environmental aspects of recycling into their study.

A comprehensive review of existing information on the paper cycle from forestry through to recycling, energy recovery, and waste paper disposal has been prepared by The International Institute for Environment and Development (The Sustainable Paper Cycle, 1995). It is worth noting that consultancy companies have done some interesting analyses. Examples are Virta (1993) and FAO (1994) which give, respectively, valuable data and an analysis of the consequences of increased recycling in four different countries.

2.2. Aim and Methodology

In this paper, a modeling approach considers both an economic and an environmental measure of waste paper recycling. Characteristics of the model are:

- A simultaneous treatment of the following sectors:
  - Energy and fibre
  - Environment and economy
  - Quality and waste paper admixtures
- Simple system limits, i.e., an inclusion of most of the fibre production and fibre use in Western Europe, and
- A user-friendly model called the Optimal Fibre Flow Model.

The following questions are answered by the Model:

- What are the business economic costs of different recovery requirements?
- What are the environmental impacts of different recovery requirements?

The dynamics and development of the fibre cycle are analyzed using a combined optimization and simulation model. An engineering approach is taken to describe the production processes. However, both economic and environmental aspects are considered. The Model generates optimal flows of fibres under different assumptions. Consideration of the affect of all relevant processes and transports on the environment are included in the Model; for example, the carbon dioxide balance is calculated in the system. Thus, the Model not only includes calculations of the industry-related fibre cycles but also the role of forestry and forest products in the climatologically important circulation of carbon dioxide. The environmental effects of the different activities in the total system are also added using the same methodology as used in some life cycle analyses, where each individual emission and each use of non-renewable resources, such as oil and coal, is given an environmental load index value (ELU-index) taken from a system for Environmental Priority Strategies in Product Design, the so-called EPS-system (Steen & Ryding, 1994). The system is based on the willingness of people to pay for avoiding the consequences of different emissions. We register all production processes and emissions instead of as in the established methodology of life cycle analysis concentrating on just paper and board production. An additional reason for including all production processes and emissions is to demonstrate that the forest industry is not the large and dangerous poluter it is often perceived to be.

3. MODEL

3.1. Structure

Western Europe is divided into two regions, Scandinavia (Finland, Norway and Sweden) and Continental Western Europe and the U.K. (Compare Figure 1). Each region has production resources and a market for paper products and energy. The products produced are delivered either to the domestic market or to the export market. After end-use, paper is recycled for production of paper and board and/or recovered for energy use. If recycled, the waste paper is recovered, sorted, baled and transported to paper mills in one of the regions for production of recycled pulp. If recovered for energy use, the waste paper is assumed to follow the normal waste-handling system. It then replaces oil or coal. Eventually, the collected paper is exported to Scandinavia or the rest of Western Europe. The production value of waste paper depends on the price of fossil fuel and round timber. The higher the price of oil, the more waste paper is recovered for energy purposes. Waste paper which is not reused has no value in the Model. Thus, in the Model all waste paper is recovered. It is assumed that enough capacity exists for deinking and energy production.

Twelve different paper qualities are produced in the Model: Newsprint, SC paper, LWC, office paper (wood-free), coated paper (wood-free), tissue, white lined chipboard, "return fibre chipboard", wrapping paper, white liner, kraft-liner and fluting. Recipes specifying the need for fibres, filler and energy are given for each product. The Model chooses between virgin fibres and recycled fibres in keeping with the quality expected of the products. Five different flush pulps and market pulps are included. Dried pulp in sheets is delivered from Scandinavian producers to non-integrated paper mills in other parts of Western Europe. The need for pulp wood (short and long fibres) and energy is specified for each of the pulp qualities. Surplus energy from pulp processing is used in the paper production. Electricity can be produced from backpressure power or in condensation power stations that burn coal, oil, wood or waste paper.

Costs connected to the different processes are considered. The age distribution of the fibres in each product is calculated (Byström & Lönnstedt, 1995; also compare Göttsching, 1993). The model includes the yields in different processes, and these can be made age-dependent. Furthermore, the energy needs for production of chemicals are included. Different types of emissions to the atmosphere and water, except those from chemical plants, are calculated and later converted into comparative environmental indexes. Below, we describe the different subsystems that make up the Model.

The Forestry part of the Model describes how the forest absorbs carbon dioxide. Timber harvest and transport cause energy consumption and costs. Energy used in producing fertilizers is also considered.

The Pulp mill module describes the production of pulp using wood as the raw material. Apart from wood, use is made of electricity, thermal energy and chemicals. Excess energy in the pulp mill can be used in the paper mill. Electricity can be produced by back pressure steam turbines or by condensing turbines.

In the Deinking mill module, waste paper pulp is produced from recovered paper. The Model calculates the consequences of poor quality waste paper material. In other respects, the calculations are the same as for the pulp mill. The yield in the process that produces recycled pulp and the energy value in waste paper are calculated on the basis of the fibre composition of each individual product. In additon, the effect of filler is considered. The efficiency in the recycled pulp mill and the thermal energy recovered from burning paper are dependent on the composition of the paper.

The paper mill module of the Model describes how paper is produced from virgin pulp and waste paper pulp. In addition, use is made of different types of energy and fillers. Emissions to the atmosphere and to the water are registered. The paper products can be produced with different amounts of recovered paper from different products. Restrictions in the Model prohibit, however, incorrect combinations. Wood containing paper is not used, for example, when making wood-free qualities.

In the Model, collection of waste paper requires energy in the form of diesel fuel and other resources represented by variable costs. Standard emissions to the environment are considered. The need for resources varies depending on the product and the region. The resources needed (energy and money) to collect paper are progressive. For example, depending on quality, the resources needed to collect the last 30% of the consumption are 3 to 6 times higher then those needed to collect the first 30%. It is assumed that sufficient industrial capacity exists to recover waste paper as fibres or as energy.

All processes, including transport, require energy. All energy in the Model is generated in an energy plant where emissions to the environment are also calculated. Energy can be purchased on the market. Some forms of energy cannot be substituted, e.g., all transports are based on diesel fuel. Electricity and heat, on the other hand, can be generated both by fossil fuel (oil or coal) and by combustion of fibre products (including wood). In Scandinavia, electricity can be produced from water and fossil fuels, whereas the rest of Western Europe must rely on electricity generated by fossil fuels. Naturally, emissions are affected.

3.2. Mathematical Expressions

Incorporating considerations of demand (demand and prices for the different products are given), maximum existing capacity, production costs, transport costs and availability of raw material, the Model maximizes the profit, or using an economic terminology the producer surplus, by determining the product flows between different regions (export, import) and the waste paper's distribution between recycling for production and energy recovery. Finding a solution is somewhat tricky since many processing parameters depend on the composition of the waste paper. The composition, in turn, depends on the product flows and the distribution of the waste paper in different fields of use.

The linear programming problem is generated from a specially designed system, called GHOST, that includes functions for a simple generation of parts of the LP matrixes describing individual pulp and paper processes. In principle, the submatrixes describing the processes of each mill are generated first. The different processes are linked to each other with references in plain language, for example, a market product from one process can be raw material for another process. As hydroelectric energy and energy produced from fossil fuels are viewed as products, they are generated internally by the Model. The most important relationships are described by Equations (1) to (4).

$$R_{pi} = \sum C_{ijk}P_{pjk} \qquad (1)$$
$$P_p = \sum P_{pjk} \qquad (2)$$

where R_{pi} is raw material i used in process p, C_{ijk} is the recipe coefficient for raw material i, market product j and production product k, and P_{pjk} is product j from process p and product k. Note that for each market product, j, a number of different production products exist, designated by k, representing different ways to produce the same market product using different recipes, for example admixture of waste paper.

$$O_p = \Sigma\, R_{pi}Q_i \qquad (3)$$

where O_p is a part of the goal function and Q_i is the price of bought raw material.

$$C_p \geq P_p \qquad (4)$$

where C_p is maximum capacity for market product j from process p.

Relationship (5) shows that demand for a specified product can be supplied by all producers. Recycled raw materials (6) used for energy or production purposes can, at most, be collected up to the consumed volume. In the model,

$$D_{mj} = \Sigma\, P'_{mpj} \qquad (5)$$
$$D_{mj} \geq \Sigma\, W_{mpj} \qquad (6)$$

where D_{mj} is the demand for product j from market m, P'_{mpj} represents the shipments of product j from process p to market m, and W_{mpj} is the recycled flow of product j from market m to process p.

$$O_m = \Sigma\, T_{pj}P'_{pjk} + T'_{pj}W'_{mpj} \qquad (7)$$

where O_m is transportation costs for domestic and imported products for market m. The collection of raw materials is quite complex as the Model distinguishes different collection levels that affect the costs. However, in equation (7) collection is represented by one single coefficient, T. Transports affect the energy balance for each mill, i.e., indicate consumption of diesel oil and emissions. In the Model all emissions are viewed as flows of raw materials from the production processes.

The model follows this procedure for finding the optimal solution:

1. Assumptions are made about the processing factors, for example, the yield when producing paper from waste paper.
2. Based on these constants, an economic optimization is made that gives all relevant flows of material.
3. Knowing the flows of the different products, the age structure of the fibres and the distribution of the material are calculated.

4. Knowing the age and material distributions, an exact calculation of the processing factors can be made.
5. The newly calculated factors are compared with the previous ones. If differences exist, a new set of processing factors is calculated, for example, as an average of the two previous sets. Usually it takes just a few runs to find a stable solution.

Unlike other algorithms, the Model starts by calculating the processing factors based on assumed age and material distributions. In the second step an optimization is performed. Alternating between simulation and optimization, a stable solution is quickly found. The procedure for calculating Step 3 (i.e., the age structure and material composition of fibres) is described in an Appendix.

3.3. Environmental Load Unit Index

To calculate the environmental impact of the pulp and paper production, the use of non-renewable resources and the effects of emissions are added together. In the Model this is done by using an Environment Load Unit-index (ELU-index) (Steen&Ryding, 1994). For example, the "value" of the use of 1 kilogram of fresh water in areas with a water deficiency is 0.003 ELU. As can be seen in Table 1, the "punishment" for destroying 1 m^3 of oil is 360 ELU. The ELU-index for non-renewable resources reflects the market value of the resources, i.e., the demand and supply conditions. This explains why oil has a higher value than, for example, coal. Depending on use and emissions a value may be added. If oil is used as fuel, emissions from incineration are added. The Model only takes into account the ELU-index for important emissions and for the use of non-renewable raw materials. Effects on biodiversity caused by forest management and similar environmental impacts have not been assigned an ELU-index in this Model.

TABLE 1
ELU-Index used in the Model

Non-renewable resources

Emissions	Measure	ELU-index
Fossil oil	ELU/m³	360
Diesel oil	ELU/m³	336
Fossil coal	ELU/m³	100
CO_2	ELU/ton	88.9
CO	ELU/kilo	0.269
NO_x	ELU/kilo	0.217
S	ELU/kilo	0.1899
COD	ELU/kilo	0.0016

3.4. Data

The input data for the Model are prices, efficiencies and costs of production and transport. The fibre furnish and energy needs for each paper quality are specified. For each type of pulp, the need for wood and energy and the emissions to the environment from the production process are specified. Sources for the data are the Swedish Pulp and Paper Research Institute (Haglind et al., 1994) and databases at the MoDo Company. Data from Germany is assumed to reflect the situation in the whole of Western Europe. Vass and Haglind (1995) have made a Swedish litterature review of the environmental consequences of utilizing waste paper. The review contains valuable data about sludge, chemical use, transports, use of energy and emissions to air and water. The data for Sweden are reliable while the quality of data for the rest of Western Europe can be improved.

An extensive collection of data on the production of and trade in Western European forest products for 1990 has been carried out (Byström&Lönnstedt, 1995). This was the year for which the most up-to-date data for the countries studied could be found.

4. RESULTS

4.1. Economic use of Waste Paper

Using a production value of waste paper based on the prices of fossil fuels and round timber at the beginning of 1995, the most economical utilization rate for the Scandinavian forest industry is 57 per cent and 84 per cent for the forest industry in the rest of Western Europe. Thus, the economic optimization shows that from the perspective of business economics, a high utilization rate is profitable. If hydroelectric power is used in pulp and paper production and the Scandinavian forest industry is forced to decrease the utilization rate by 10 per cent, the marginal loss is about 5.5 USD/ton. On the other hand, if the electricity used is produced from fossil fuels, the loss is about 6.5 USD/ton. While mandatory utilization rates are set, the Model is free to optimally determine the admixture of waste paper for different products.

If the price of waste paper is defined by the prices of fossil fuels and round timber, the answer to one of our initial questions of view is obviously that a high utilization rate is to be preferred from a business point. However, from an environmental perspective, the answer is quite different.

4.2. Environmental Impact

The Model, accounting for all process restrictions, always maximizes the marginal revenue for the forest industry. It can, however, be forced to take additional restrictions into account. In the following examples the Model is forced to save forests in Scandinavia by recycling fibres. Given the restrictions, the Model optimizes the use of waste paper for energy recovery and recycling.

Figure 2 shows the use of fossil oil in the whole system when the utilization of recycled fibres in Scandinavian paper and board production changes. If electricity is produced from fossil fuels, the curve, with increasing utilization rates, first decreases and later increases. It has its minimum at a utilization rate of about 25%. If the paper production is based on hydroelectric energy, oil consumption increases continuously.

As oil consumption only tells part of the story, Figure 3 shows the total environmental impact measured as change in the ELU-index per ton of paper and board produced in Scandinavia. Furthermore Figure 3 illustrates the difference in load depending on whether the electricity in Scandinavia is hydroelectric or based on fossil fuels. During the first part of the 1990s, 47.5 per cent of Swedish electricity was produced at hydroelectric power stations.

As the de-inked paper utilization rate increases from 5% to 55%, the increase in load is about 100-150 ELU/ton paper. The disadvantage of recycling is greater if the electricity is hydroelectric. However, if electricity is produced from fossil fuels the ELU-index, with increasing utilization rates, decreases up to a utilization rate of about 25 per cent. After this, irrespective of the energy base, increased utilization of recycled paper is harmful for the environment.

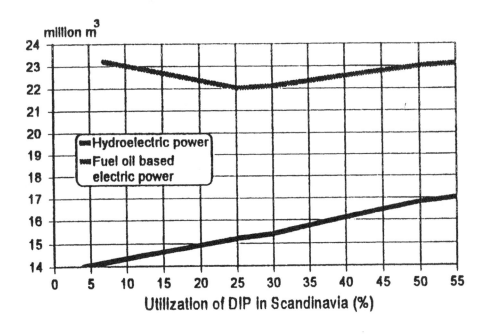

FIGURE 2. Consequences for oil consumption of a forced increase of Scandinavian utilization rates of De-Inked Paper (DIP).

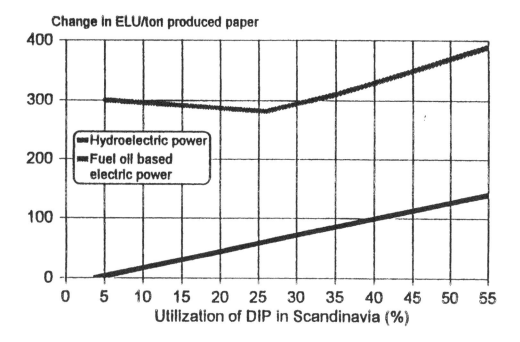

FIGURE 3. Consequences for the environment (measured by the ELU-index) of a forced increase of the Scandinavian utilization rates of De-Inked Paper (DIP).

In another example, use of waste paper is forced when producing newsprint and office paper. As above, a comparison is made between hydroelectric power and fossil electricity. The results are summarized in Table 2.

TABLE 2
Increase in Environmental Impact in ELU/Ton Product for Every 10% Increase in Recycled Fibres

	Hydroelectric power		Fossil electricity
	Newsprint Office paper	Office paper	Newsprint
Non-renewable resources	14 10	10	— 1 0
Emissions	18 10	13	+ 2
Total	32 20	23	— 8

S203

If electricity produced by hydroelectric power is used for newsprint production, the increased use of recycled fibres has an adverse effect on the environment. The index increases by 32 ELU/ton newsprint for every 10 per cent increase in recycled fibres in the product.The excess energy from the pulp mill (about 6 GJ/ton pulp) can be used for drying the paper (Figure 4). However, a lot of electric energy is used; 10.5 GJ/ton must be added. This corresponds to approximately 20 GJ of heat energy per ton if a condense turbine power plant is used. If newspaper after consumption is used for energy recovery 11.5 GJ of heat per ton will be produced.Thus, in total the energy saving for the whole system when TMP is used is about 7 GJ of heat per ton.When paper is recycled as fibres the thermo-mechanical pulping process no longer converts electricity into heat energy which is needed in the paper-making process. This energy loss is compensated for by fossil (oil) energy which is the cheapest available alternative. In addition to being a non-renewable resource, the burning of oil for energy produces emissions, of which carbon dioxide is the most important.

If, on the other hand, the electricity for newsprint production is generated from fossil fuels, an increased utilization rate has a favourable environmental impact. The explanation is that in this case the electricity, used for producing TMP, is produced from fossil fuel in condensing power plants with low efficiency (40%). If the electricity is not needed, as is the case when waste paper is used, it is more efficient to produce this heat directly from fossil fuel.Thermal energy must be added to the deinking process (2 GJ/ton) and to the paper mill (5.1 GJ/ton). Thus, instead of getting 6 GJ of thermal energy as in the first case, 7.1 GJ of heat per ton

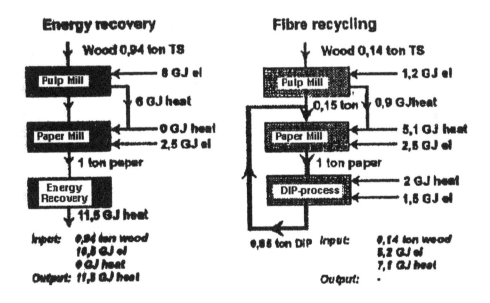

FIGURE 4.   Use and production of energy in newsprint production.

must be added (Figure 4). However, the use of electric energy is only 5.2 GJ/ton which is 5.3 GJ/ton less than when TMP is used. Differences in the type of transport of wood and waste paper contribute only slightly to the differences in the ELU-index. However, the index does indicate a small increase in environmental impact, mainly due to larger emissions.

In Table 2, figures describing the environmental impact caused by changing the utilization rate of recycled pulp in woodfree office paper are also included. The chemical pulp process converts about half of the wood used into pulp. The remainder of the wood raw material can be burned to create excess thermal energy that is normally used in the paper-making process. The total use of electric energy when office paper is produced from virgin pulp is about 5 GJ/ton (Figure 5). Excess energy from the pulp mill (about 7 GJ/ton pulp) can be used for drying the paper. If waste paper is used for energy recovery, about 11.5 GJ of heat per ton will be produced. If recycled fibres are used, the energy required for recycling plus the energy lost when the waste paper is not burned, must be derived from other sources, such as oil or coal. Thermal energy must be added to the deinking mill (2 GJ/ton) and to the paper mill (about 5.95 GJ/ton). Thus, instead of getting 11.5 GJ of thermal energy per ton as in the first case, 7.95 GJ of heat per ton must be added to the 4.38 GJ of electric energy. This increases the need for fossil fuel quite substantially, even if minor amounts of electric energy can be saved. Even if clean electricity is used, increased fibre recycling has a negative impact on the environment.

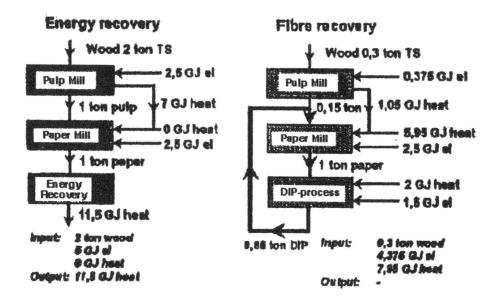

FIGURE 5. Use and production of energy in office paper production.

5. CONCLUSIONS

In the introduction to this paper we wrote that waste management policy in a number of countries is characterized by a hierarchy of options in which waste minimization, reuse and recycling are all considered preferable to energy recovery which in turn is considered superior to landfills. The issues are highly complex and the science for assessing them, life cycle analysis (LCA), is still in its infancy. Little analysis of the benefits of paper recycling has been made. Using the principles of LCA and a systems analysis approach simultaneously, we have studied the different alternatives.

The ELU-index gives no evidence that increased recycling is an environmentally-friendly policy (compare Kärnä et al. 1993). The result supports energy recovery from waste paper as a substitute for fossil fuels. This substitute will diminish the greenhouse effect. However, one consequence of replacing fossil fuels with energy recovery from waste paper may be a reduction in actual or potential profit levels for the forest industry.

It is of vital interest for humankind to decrease carbon dioxide in the atmosphere to avoid global warming. Maximum energy recovery of waste paper would only marginally influence the carbon dioxide balance of Western Europe (a few percents of the total fossil fuel use in Western Europe). Increased production of pulp based on wood, or use of waste paper as fuel, are both examples of a development that leads to replacement of fossil fuels and a consequent decrease in the release of anthropogenic carbon dioxide. Increasing the land area holding growing forests, which absorb carbon dioxide, is another way.

A major problem in many countries, and a driving force behind legislation, is the large volumes of paper and paper products in household waste, and the scarcity of room for landfills. These factors make for a strong argument for waste paper collection, especially in the densely populated countries of Western Europe. The question of whether the collected paper is recycled as raw material for paper or for energy production is secondary to the importance of using landfills efficiently.

The economic and political aspects of this question, however, are critical. Price relations between different forms of energy rule market demand, and energy prices are influenced by political decisions. The linkage between the different decision levels are complex. One important goal for the industry is to maximize profits. National governments determine national environmental policies, which affect the decisions taken by the industry. On the third international level, for example, the European Union, the policies of national governments need to be coordinated and formulated as an international environmental policy capable of dealing with the intensive trade in forest industrial products and the fact that emissions move over national borders. The conditions under which the model operates can be varied to account for, for example, for legislated requirements for recovery or admixture of waste paper, and the costs involved for virgin fibres, old fibres and fossil fuels.

The results of the project provide important data for decision-making among politicians, business management and environmental groups. Quantitative estimates are presented that can be used instead of qualitative judgements and general thinking. Hopefully the results will influence the debate in Europe regarding the use of waste paper.

For follow-up research on this issue we recommend that consideration be given to dynamic effects as changes in production capacities and product prices. As such, several interesting research topics exist:

- Include the markets, demand for and supply of forest industrial products, round timber and waste paper. This will allow a calculation of consumer and producer surplus. It is important to include social impacts such as employment rates.
- Analyze the consequences for the structure of the Western European forest sector of increased waste paper recovery and an increased utilization rate. Where will the new paper and board capacity be located? How will the quality of waste paper be affected? What effects will an increased use of waste paper have on the fibre flow and qualities?
- Given different assumptions about consumption, technological development and institutional changes (new laws), analyze when a balance between use of new and old fibres will be achieved.
- Change the boundaries of the system to include bioenergy as an alternative for fossil fuels.

For future studies of this kind, access to reliable data may prove a serious problem. Finland and Sweden have accessible data bases. The rest of Western Europe, however, has another tradition. Much would be gained if shared accessible data bases were developed in more countries or by a European organization. It is important that data acquisition, transformation and storage in data banks are made compatible. In this way, decisions of great consequence to our environment could be made on a more knowledgeable basis.

REFERENCES

Byström, S. & L. Lönnstedt. 1995. Fibre Age Distribution in Western Europe. Manuscript under preparation. 7 p.

Byström, S. & L. Lönnstedt. 1995. Waste Paper Usage and Fibre Flows in Western Europe. Forthcoming in *Resources, Conservation and Recycling*.

Colletti, J.P. & J. Buongiorno. 1980. Forecasts of Wastepaper Supply and Consumption in the United States to 1985. *Wood and Fiber*. 12: 233-243.

Deadman, D. & R.K.Turner. 1981. Modelling the Supply of Wastepaper. *Journal of Environmental Economics and Management*. 8: 100-103.

Edgren, J.A. & K.W. Moreland. 1989. An Econometric Analysis of Paper and Wastepaper Markets. *Resources and Energy*. 11: 299-319.

Edwards, R. 1979. Price Expectations and the Supply of Wastepaper. *Journal of Environmental Economics and Management*. 6: 332-340.

Food and Agricultural Organization of United Nations, Wood Industries Branch, Forest Products Division, Forestry Department. 1994. Paper Recycling Scenarios. FO:MISC/94/4. 49 p.

Gill, G. & K. Lahiri. 1980. An Econometric Model of Wastepaper in the USA. *Resource Policy*. pp 434-443.

Grace, R., R.K. Turner, and I. Walker.1978. Secondary Materials and International Trade. *Journal of Environmental Economics and Management*. 5: 172-186.

Grieg-Gran, M. 1994. Cost Benefit Implications of Paper Recycling Policies. Published in What is Determining International Competitiveness in the Global Pulp and Paper Industry. Proceedings, Third International Symposium, September 13-14, 1994, Seattle, Washington. University of Washington, College of Forest Resources, Center for International Trade in Forest Products, SP-17. p 19.

Göttsching, L. 1993. Waste Paper Recycling and Management of Paper Mill Residues. In Recycled Fibres - Issues and Trends. Food and Agricultural Organization of United Nations, Wood Industries Branch, Forest Products Division, Forestry Department, FO:MISC/93/10. pp 105-126.

Haglind, I., R. Lindström, A.M. Parming Vass & L. Strömberg. 1994. Skogsindustrins Tekniska Forskningsinstitut (STFI) baut eine Datenbank für Ökobilanzen in der Zellstoff- und Papierindustrie auf. (Also translated into French) *Der industriella Umweltschutz/La protection industrielle de l'environment*. 1:128-132.

Ince, P.J. 1994. Recycling and Long-range Timber Outlook. Background Research Report. Research Paper FPL-RP-534. Madison, WI: US Department of Agriculture, Forest Service, Forest Products Laboratory. 23 p.

Kinkley, C-C. & K. Lahiri. 1984. Testing the Rational Expectations Hypothesis in a Secondary Materials Market. *Journal of Environmental Economics and Management*. 11: 282-291.

Kärnä, A., J. Engström & T. Kutinlahti. 1993. Life Cycle Analysis of Newsprint. Publication 4-7, October. Finnish Pulp and Paper Research Institute, Espoo, Finland. 8 p.

Rehn, M. 1993. Produktionsteknologi i svensk tryckpappersindustri - en empirisk analys 1980-1990. Working Paper No. 174. Dept. of Forest Economics, The Swedish University of Agricultural Sciences, Umeå. 26 p.

Steen, B. & S-O. Ryding. 1994. Valuation of environmental impact within the EPS-system. Published in Integrating Impact Assessment into LCA. Proceedings of the LCA symposium held at the Fourth SETAC-European Congress, 11-14 April 1994, The Free University, Brussels, Belgium. 6 p.

The Sustainable Paper Cycle. Phase 1 Review Report. January 1995. Second Draft. World Business Council for Sustainable Development. Prepared by The International Institute for Environment and Development (IIED). 3 Endsiegh Street, London, WC1H 0DD, Unitied Kingdom.

Trømborg E. & B. Solberg. 1995. Economic Consequences of Increased Use of Recycled Fibre in the Newsprint and Magazine Paper Production in Norway. Paper presented at the workshop "Life-Cycle Analysis - A Challenge for Forestry and Forest Industry. 3-5 May. Hamburg, Germany. 11 p.

Vass A.M. & I. Haglind. 1995. Environmental consequences of waste paper usage and handling (in Swedish). Swedish Waste Research Council. AFR-Report 56. 65 p.

Virta, J. 1993. World-Wide Review of Recycled Fibre. In Recycled Fibres - Issues and Trends. Food and Agricultural Organization of United Nations, Wood Industries Branch, Forest Products Division, Forestry Department, FO:MISC/93/10. pp 15-44.

Virtanen, Y., and S. Nilsson. 1992. Some Environmental Policy Implications of Recycling Paper Products in Western Europe. IIASA, Laxenburg. 39 p.

Yohe, G.W. 1979. Secondary Materials and International Trade: A Comment on the Domestic Market. *Journal of Environmental Economics and Management.* 6: 199-203.

Zhang, D., J. Buongiorno, and P.J. Ince. 1993. PELPS III: A Microcomputer Price Endogenous Linear Programming System for Economic Modeling (Version 1.0). USDA Forest Service. Research Paper FPL-RP-256. Madison, WI: U.S. Department of Agriculture, Forest Service, Forest Products Laboratory. 43 p.

APPENDIX

Step 3, i.e., finding the age and material distributions, in the procedure for finding an optimal solution is described (compare section 3.2).The flow of products from producers in region 1 to consumers in region 1 is denoted as P_{11}, a vector with N different products. Correspondingly, the import of paper and board products from region 2 to consumers in region 1 is denoted as P_{21}. The flow of waste paper from consumers in region 1 to producers in region 1 is denoted as R_{11}, a vector with M different waste paper grades. Correspondingly, the import of waste paper grades from region 2 to producers in region 1 is denoted as R_{21}. Correspondingly, P_{22}, P_{12}, R_{22} and R_{12} are defined.

A_{R11} and A_{R21} define the conditions of the fibres in the two respective flows. The data are organized as a cube (Figure 6). (It is not necessary for the number of elements in each dimension to be the same). This method of organizing the data is used throughout the Model.

Given a certain waste paper grade, the resulting matrix, a level of the cube, describes different age classes and different material compositions (type of fibres

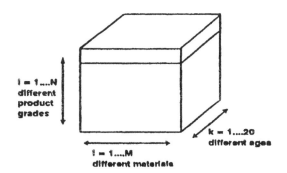

Each horizontal level in the cube describes a
product standardized so that $\Sigma A_{ki} = 1$
$i = 1...N$
$k = 1..20$
$l = 1...M$

FIGURE 6. Cube describing the flow of products.

and fillers). For this type of problem, the number of age classes are often restricted to between 10 and 20. The share of fibres belonging to the higher age classes is exceedingly small. In the calculations, these shares are added to the highest age class.

Prior to the deinking process, the flows of waste paper from the domestic and foreign markets are mixed. The volumes are used as weights. The conditions of the fibres in the mixed flow is described by a new cube.

$$A_{R1} = SUMAGE\ (R_{11},\ A_{R11},\ R_{21},\ A_{R21})$$

As a result of the deinking process, the proportions of chemical and mechanical fibres change. Filler and coating materials partly disappear. The resulting conditions of the fibres after deinking is described by a new cube:

$$A'_{R1} = YIELD\ (A_{R1}\ )$$

The virgin fibres from the pulping process, i.e., fibres with no circulation, and other raw materials such as fillers are described by the cube A_{Rm}. Based on the recipes used for different products, virgin and older fibres are mixed. The recipe used for each product is described by mix1. The conditions of the fibres in the mixed flow are described by a new cube:

$$A_m = RECIPE\ (mix1, A_{Rm}, A'_{R1})$$

When the fibres pass through the paper machine, mechanical damages occur. Once again the condition of the fibres is changed. The result is described by the following cube:

$$A_{P11} = A_{P12} = PAPER\ (A_m)$$

A_{P11} describes the condition of those fibres in paper and board products shipped to the domestic market and A_{P12} the condition of fibres exported to the other region. It should be noted that market 1 can import paper and board products from region 2. The conditions of those fibres are described by the cube A_{P21}.

After end-use, the paper and board products are recovered. Using the volumes of paper and board products shipped from region 1 and imported from region 2, a weighting is done for calculating the new cube.

$$A_{k1} = SUMAGE\ (P_{11}, A_{P11}, P_{21}, A_{P21})$$

The recovered paper is sorted into different grades. The sorting descriptions are given by coll1. The conditions for the fibres in the different waste-paper grades are described by the following cube:

$$A_{R11} = AGE\ (col11, A_{R11})$$

The flows for region 2 are defined in the same way. The steps in the algorithm are as follows:

1. Guess the initial values of the A_R-cubes, denoted as $A^{(1)}_R$.
2. Calculate $A^{(1)}_P = f(A^{(1)}_R)$
3. Calculate $A^{(2)}_R = f(A^{(1)}_P)$
4. Compare $A^{(2)}_R$ with $A^{(1)}_R$. If they differ, put $A^{(1)}_R = A^{(2)}_R$ and repeat the calculation, otherwise stop. Experience shows that 30 iterations are enough to find a stable solution. The calculation takes just a few seconds on a powerful PC.

Critical Reviews in Environmental Science and Technology, 27(Special): S213–S226 (1997)

FOREST/BIOMASS BASED MITIGATION STRATEGIES: DOES THE TIMING OF CARBON REDUCTIONS MATTER?

Gregg Marland,[1] Bernhard Schlamadinger,[2] and Paul Leiby[1]
[1]Oak Ridge National Laboratory, Oak Ridge, TN 37831-6335, USA; [2]Joanneum Research, Elisabethstrasse 11, A-8010 Grar, Austria

ABSTRACT: Many of the strategies proposed for using forest managememt to mitigate the increasing concentration of atmospheric CO_2 are characterized by C emissions reductions which are not uniform over time. A simple model of the carbon budgets of different forest management schemes illustrates the possibilities and poses the question whether the timing of emissions reductions matters.

KEY WORDS: forest management, carbon reductions, emissions reduction.

1. INTRODUCTION

A number of recent studies have pointed out that, if we choose to reduce net CO_2 emissions to the atmosphere, the costs and benefits of reductions will depend not only on how much we reduce emissions, but when. Richards (1994) notes that, so long as we have discounting, current reductions have greater value than future reductions. Manne and Richels (1992), on the other hand, suggest that there is value in flexibility in timing and that it may be less expensive, in the face of uncertainty, to initiate reductions in the future than to initiate them in the present. Richards (1994) suggests that with discounting there is value in temporary storage of carbon. Temporary storage may provide the added benefit of delaying change while we increase understanding of the consequences. Conversely, Manne and Richels (1992) suggest that when there is a possibility of technical improvement or substitutability between inputs of capital, labor, and energy; there may be advantage in borrowing emissions now in order to increase reductions later.

One of the alternatives frequently suggested for reducing net emissions of CO_2 is management of forest lands and forest resources (see, e.g. Hall et al., 1990). We have developed a simple model of C flows to examine some of the possibilities for mitigating net C emissions via forest management. Previous analyses with this model have shown that different alternatives for forest management can have very different paths for the net C flux as a function of time (Schlamadinger and Marland, 1995a and 1995b). The time paths of both implementation costs and the benefits of reduced atmospheric CO_2 will vary with the strategy chosen. If the value of emissions reductions is a function of their timing, it may be important to look not just at the net C offset of forestry projects at some time in the future, but at the pattern of reductions over time. In this paper we examine the pattern of change for C flows on a single parcel of land and then the cumulative flows over a number

1064-3389/97/$.50

of parcels when a forest management program is phased in over a larger area. We also comment on the relative merits, in terms of the net C balance, of different forestry alternatives if the value of emissions reductions is a function of their timing. Our intent has been to examine the timing and consequent path of benefits of carbon reductions and the extent to which the initial condition and phase-in phenomena might influence the nature or magnitude of the flow of benefits. The model tracks the net flux of C to the atmosphere but does not represent the implementation costs of the various changes in land use nor the revenues generated by selling wood products or biofuels.

2. THE MODEL

The model we use for this analysis is a spread-sheet model that provides a simplified description of carbon stocks and flows associated with management of forests, We try to recognize the full system that is impacted when the forest management regime is changed. The model calculates carbon accumulation in trees, in long-lived wood products, in short-lived wood products, in fossil fuels that are not burned because wood products are used instead of more energy-intensive materials, and in fossil fuels saved because wood is used as a fuel instead. The model represents the changing stock of carbon in soils and forest litter. It considers the energy inputs that are required for forest management and harvest and the energy required to produce and deliver alternate products or energy carriers. The structure of the model is illustrated in Figure 1. The model is described in detail in Schlamadinger and Marland (1995a and 1995b).

We summarize briefly some of the critical parameters used in the scenarios below. Full justification of these selections is provided in Schlamadinger and Marland (1995a). We assume that the maximum sustainable biomass on the site modeled is 160 Mg C/ha. Biomass growth will be at a constant rate (in Mg C/ha/yr) until half of this value is achieved and then at a gradually decreasing rate that is a function of the then-current status (Marland and Marland, 1992). Thus the maximum sustainable biomass is not related to the growth rate, although the time required to approach the maximum is. Thirty percent of harvested material is used to make long-lived products, 20% is used for short-lived products, and 5% is used for very short-lived products. Twenty-two percent of the harvest (logging and milling residues) is burned to produce energy and 23% of the harvested material is left on-site where it decays instantly to CO_2. Long-lived products are assumed to decay exponentially with a mean lifetime of 40 years while short-lived products have a mean lifetime of 10 years and very short-lived products decay instantly. We assume, optimistically perhaps, that 30% of products are used for energy at the end of their useful life. For short-rotation plantations, 80% of the harvest is burned as a fuel and 20% is lost during harvest and haul and decays instantly.

We have defined two important parameters that describe, 1. the amount of C emission from fossil fuels that is avoided when biofuels are used instead (direct energy substitution), and 2. the amount of fossil C not oxidized because wood products are used instead of products from other, more energy-intensive materials like concrete and steel (indirect energy substitution). Both of these displacement factors have units of Mg C/Mg C and represent the net amount of fossil fuel C not oxidized because 1 Mg of biomass C is used for energy or is stored in wood products. The displacement factor for fossil fuel (D_f) is defined as:

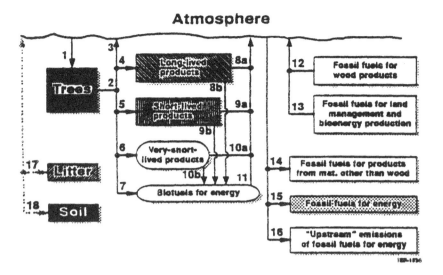

FIGURE 1. Carbon pools and fluxes as described by the model. Flux 1 is the net annual flux from the atmosphere to the forest and fluxes 2–7 are fluxes that occur whenever there is a harvest Fluxes 8–11 occur over time, depending on the lifetime of the respective products. Fluxes 14–16 are hypothetical fluxes, away from the atmosphere because they are fossil-fuel fluxes avoided as a result of the use of bioenergy or wood products. Fluxes 17 and 18 are net fluxes to or from litter and soils and, at this stage of model development, are assumptions based on the nature of the management change being modeled.

$$D_f = \frac{\text{efficiency of bioenergy system}}{\text{efficiency of displaced fossil system}} \times$$

$$\frac{C \text{ emission per J of fossil fuel}}{C \text{ emission per J of biofuel}}$$

Our scenarios assume that biomass is used for electric power generation but has a higher C emission rate and lower combustion efficiency than would the mix of fossil fuels for which it substitutes. With current technology 1 Mg of C in wood fuel can displace about 0.6 Mg of C in fossil fuel ($D_f = 0.6$) (Marland and Marland, 1992), but 1.0 might be approached if wood can substitute for coal and achieve the same net thermal conversion efficiency. The displacement factor for products (D_p) is defined as:

$$D_p = \left(-E_w \times F_w + E_{NW} \times F_{NW} \times M_{NW/W} \times L_{W/NW}\right) \times \left(1/C_W\right)$$

where

E_W = energy required to make I Mg of wood products (GJ/Mg),

F_W = C emission rate for fuel used to make wood products (Mg C/GJ),

E_{NW} = energy required to make non-wood products (GJ/Mg),

F_{NW} = C emission rate for fuel used to make non-wood products (Mg C/GJ),

$M_{NW/W}$ = ratio of the mass of non-wood material to wood material to make products that provide the same service (Mg/Mg),

$L_{W/NW}$ = ratio of the mean service lifetime of products from wood and non-wood materials (yr/yr), and

C_W = C content of dry wood (Mg C/Mg wood).

For now we assume that the two industries use a similar fuel mix ($F_W = F_{NW}$, which has an average value of about 0.016 Mg C/GJ for the U.S. economy [Marland and Pippin, 1990]) and that the alternate products have comparable service lives ($L_{W/NW} = 1$). C_W is about 0.5.

Worrell et al. (1994) report that the energy input to produce wood materials in Europe ranges from 2.71 GJ/Mg for European softwoods to 13.94 GJ/Mg for plywood. Comparable values for plastics are 60–80, for steel 20–25, and for aluminum some 190 GJ/Mg. Ignoring the final step of producing products from the raw materials, we assume for this comparison that wood products which require 10 GJ/Mg for their production will replace a mix of products which requires an average of 30 GJ/Mg. The value of D_p then depends strongly on $M_{NW/W}$, the mass of non-wood products which will be substituted by 1 Mg of wood products. This will clearly depend on the material being substituted and the application intended and much more work is required to use this factor in a detailed analysis. Our value of 0.5 for long-lived products implies that 1 Mg of wood products displaces about 0.85 Mg of alternate products while our value of 0.25 for short-lived products implies that 1 Mg of wood products displaces 0.59 Mg of alternate products.

Burschel et al. (1993) made a survey of data and recommended that for solid wood used in construction an average of 0.28 Mg C of embodied energy was displaced for each m^3 of wood used. This corresponds to a value of D_p slightly greater than 1, suggesting that our value of 0.5 may be conservative for some applications. The numbers from Burschel et al. (1993) are close to those from Buchanan (1991) who compared the production of light beams, heavy beams, and house framing from wood versus steel. Buchanan and Honey (1994) analyzed complete buildings in New Zealand and reported results for construction of industrial buildings, office buildings and a multi-story hostel which would compute to values of 2.3 to 0. 9 for our parameter D_p.

One can argue whether production of biomass fuels in fact saves fossil fuels (see e.g. ETSAP, 1994). The argument is especially interesting in developing countries where energy demand exceeds energy supply and the availability of additional biomass fuel might expand energy consumption rather than displace current fuel use. On the other hand, the growth in energy use in developing countries is generally constrained by capital requirements and the availability of a new fuel source might indeed provide for fuel substitution. We take the position here that the marginal fuel today, globally, is a fossil fuel, that provision of a biomass fuel fills a demand that would eventually have been met with fossil fuels, and our scenarios assume that biomass fuel substitutes for fossil fuel, recognizing that with realistic, current, conversion efficiencies 1 kg of C in biomass substitutes for 0.6 kg C in fossil fuel.

We have tried to select reasonable, representative parameters for the model runs shown here, but we caution that these scenarios are largely illustrative and cannot capture the complexity of real systems or the wide variability of site-specific parameters that will be so important on individual projects. In particular, the indirect fuel substitution factor is a poorly known member that has a great impact on some of our scenario outcomes. This number depends on how wood products are used, what alternative materials would be used in their absence, and the efficiency of substitution. The impact of this component in our current scenarios suggests the importance of more detailed examination and understanding.

3. SCENARIOS FOR REDUCING NET EMISSIONS OF CO_2

We are interested in the net change in C stocks that can be achieved with a change in land management and in the time path of those changes. For simplicity we consider 4 types of land management and the effect on net C stocks of changing from one to another. The 4 possibilities include: a. land that is not forested or for which the forest is badly degraded and the total on-site C storage is very small, b. land that is in mature forest and on-site C storage is large and not changing with time, c. land that is being managed in short-rotation forestry to produce an energy crop, and d. land that is in second-growth forest with or without periodic harvest to produce conventional wood products.

With these 4 possible types of land management, there are 12 possible changes in management type (a–b, a–c, etc.), 7 that are of particular interest for their potential to yield a decrease in the net flux of C to the atmosphere. We will not, for example, pursue here changes to type a or away from type c. We show diagrams for the 3 possibilities involving forestation on unforested land (a–b, a–c, and a–d) and the 2 cases involving clearing and reforestation of mature forest for production of traditional and/or energy crops (b–c and b–d). For each case we produce a diagram showing the impact on net C stocks for a single parcel of 1 ha. For some cases, a second diagram is used to show the cumulative net effect when the change in land management is phased in over 42 years to eventually involve 100 ha and produce a constant annual output of products and/or fuels.

Because our interest is in examining patterns and principles, we have made some of the parameter selections for reasons of computational and illustrative simplicity. For short-rotation management to produce an energy crop, we assume a harvest rotation cycle of 7 years. The rotation time for long-rotation management scenarios is taken as 42 years, a multiple of 7. To describe the phase-in of a new management type over a large area, we consider that 1 parcel is changed each year until all parcels are converted and a balanced-age rotation achieved. For simplicity, we use 42 years as the phase-in time when the management type is changed. For example, when mature forest is converted to a long-rotation harvest sequence (transition b–d above), this means that 1 parcel of mature forest is harvested and replanted at the beginning of each of the years 1 to 42 and the first parcel is harvested again in year 43, and so on. When mature forest is converted to a short-rotation plantation (transition b–c), 1 parcel is harvested and replanted with fast-growing trees at the beginning of years 1 to 7 and in each of years 8 to 14 there is 1 parcel of mature forest harvested and replanted while 1 parcel of the new plantation is also harvested and replanted, and so on. Results are normalized to 42 parcels that comprise a total of 100 ha.

4. RESULTS FOR ONE PARCEL OF LAND

For land that is not now forested, we have the option of planting it to trees that are intended to grow and store carbon for a long time, planting it to trees that will be put in a long-rotation cycle to be harvested for traditional wood products and energy, or planting it to a short-rotation energy crop. Without providing a quantitative description, Marland and Marland (1992) showed that the best choice (based only on C flows) will depend on site-specific characteristics such as the expected growth-rate and the accessibility for and efficiency of harvest. Consequently, the three scenarios shown in Figure 2 have site and species parameters appropriate for the option selected and should be compared with each other only with caution. For example, the short-rotation plantation is assumed to have intensive management and species selected for high juvenile growth rates and the net annual harvestable increment is taken to be 6 Mg C/ha/yr. The other two scenarios have smaller annual inputs to management and the juvenile growth rates are 2 Mg C/ha/yr.

Examination of Figures 2a, 2b, and 2c shows that all 3 scenarios on unforested land provide net carbon sequestering from the beginning, all 3 have carbon accumulation in sods and litter (the longer the rotation length, the greater the accumulation), and that each harvest cycle results in some loss to the atmosphere of C that was collected in the growing trees. The magnitude of the effective C release at each harvest depends on the efficiency with which wood products are produced and the efficiency with which the forest harvest provides direct or indirect displacement of fossil fuels.

In the reforestation scenario (Figure 2a) the dominant effect on the carbon cycle is accumulation of C in trees, while in the short-rotation scenario (Figure 2c) the dominant effect is the direct displacement of fossil fuels by biomass fuels. In the long-rotation scenario (Figure 2b), direct fossil fuel displacement is important as is the least-well-understood component in our scenario, the indirect displacement of fossil-fuel use because wood products are used in place of more energy-intensive products. The over-all result of this latter scenario (Figure 2b) depends heavily on the way we have parameterized this indirect fuel displacement. By the end of 100 years the C stored in biomass and wood products is at or approaching steady state in all 3 scenarios and continued growth in cumulative net carbon sequestered is a function of the extent to which fossil fuel is being displaced. At 100 years more C is stored in trees in the reforestation scenario than has accumulated in trees, forest products, and displaced fuels in the long-rotation forest; but the difference shrinks as the forest growth rate declines toward maturity while fossil-fuel displacement continues with each harvest cycle.

If the beginning state involves a standing forest (Figure 3), a change in management type includes an initial harvest, with allocation of the harvest to the various product and fuel categories. The net release of C to the atmosphere is a function, again, of the efficiency with which the harvest displaces fossil fuel or enters durable products. In the scenarios illustrated, the net initial release to the atmosphere is clearly a function of the way we have allocated the harvested material among the different product categories and of the factors for direct and indirect displacement of fossil fuels. There could, in theory, be circumstances with no net release (or even a net C benefit) if forest products displaced extremely energy-intensive alternative products and/or if biofuels were used very efficiently to displace inefficient fossil energy systems. Our choice of parameters results in a C release that amounts to about 15% of the C contained in the harvested biomass.

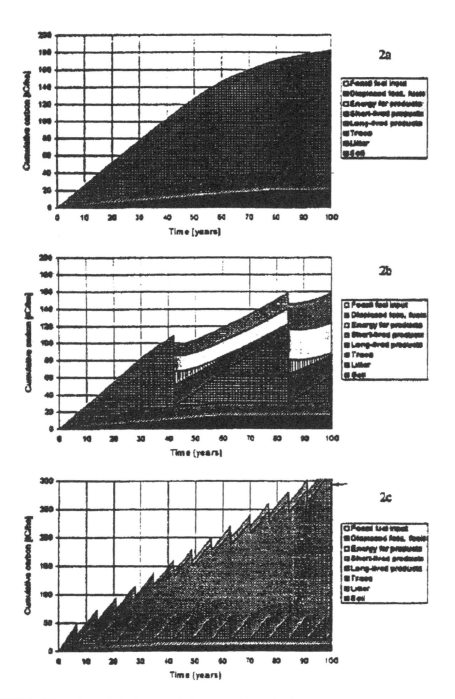

FIGURE 2. The net cumulative increase in the respective pools of C that would otherwise accumulate in the atmosphere. These are the accumulations of C when a 1 ha parcel of land that was not previously forested is planted to trees that will be protected to store C (Figure 2a), be put in 42-year rotation to produce a mix of wood and energy products (Figure 2b) or, be put in 7-year rotation to produce an energy crop (Figure 2c). All components do not appear on all diagrams. The representation for fossil fuel input is the C emissions from the fossil energy which is required to convert wood in the field to a fuel at the furnace less the comparable C emissions from converting coal. This value is often negative. In these cases the pattern is assumed to be transparent and the net balance of saved C emissions is represented not by the top of the pattern in the figure but by the dark line designated by the arrow to the right of the diagram.

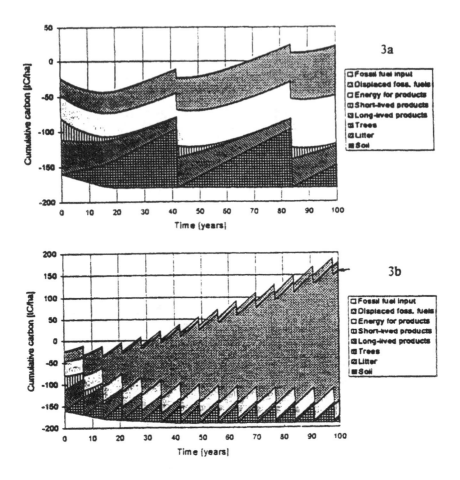

FIGURE 3. The net cumulative change in C pools when mature forest is harvested at time = 0 to initiate a forest plantation that produces a mix of products on a 42-year harvest rotation (3a) or a wood energy crop on a 7-year harvest rotation (3b). The initial drop to − 160 Mg C/ha represents harvest of the initial stand. Conversion to forest products, fuel displacement, and forest regrowth are represented by positive values above this baseline. Drops in the baseline below −160 Mg C/ha represent loss of C from soils and litter. As in Figure 2c, in 3b the net cumulative C emission avoided is represented by the line indicated with the arrow to the right of the diagram.

These plots of the net carbon impact of managing 1 ha previously in old-growth forest show that it takes about 68 years for the long-rotation plantation and about 30 years for the short-rotation plantation to return the C balance to the initial condition with respect to emissions to the atmosphere. That is, although C is released in the initial harvest, the net effect on the atmosphere is neutral after 68 or 30 years because of C in durable products, unburned fossil fuels, and C uptake in regrowing forests. Again, the difference in break-even times is largely because the short-rotation plantation is implemented on different lands or under different conditions and enjoys higher growth rates due to species selection, management intensity, and/or site selection.

Harmon et al. (1990) have noted that when an old-growth forest is harvested, the time required to return to the initial state of C storage may run to hundreds of years. Our conclusions are not substantially different except that we suggest the balance should acknowledge not only on-site and product storage of C — as done by Harmon et al. — but should include the effect of harvested C in the energy system.

5. RESULTS FOR 100 HECTARES AND FOR GROWING SYSTEMS

The two scenarios of Figure 3 involve initial clearing of existing forest and early debits with respect to the net flux of C to the atmosphere. If we look beyond a single parcel of land, to a larger system phased in over time to provide a larger eventual mitigation potential and a constant annual flow of forest products, phasing in of the initial loss can result in a net debit that lasts for a longer time. In Figure 4, 42 parcels of land, comprising a total of 100 ha are phased in as described in Section 3 above. The time to return to the initial C balance with respect to the atmosphere is now 52 years for the short-rotation plantation (Figure 4b) and nearly 90 years for the long-rotation system (Figure 4a). In these two scenarios, it is the direct and indirect displacement of fossil-fuel use that provide the dominant C benefits over the long term.

Recognizing that the initial C deficit of these scenarios can be regarded as a capital investment in future sequestration, we have looked one step further at the implications of system growth. The 100 hectare scenarios were defined with the idea that a useful plantation system has to maintain a constant flow of products and these scenarios achieve such a flow after 42 years. If the demand for forest products is growing so that there is continuing pressure to clear standing forest and place the land in a productive cycle, net C emissions accompanying harvest may be the dominant factor and the time to pay back initial C losses very large.

Chapman (1975) raised similar questions about nuclear power when it was perceived that the initial energy investment was so large that perhaps the entire output of existing plants could be consumed in bioding new power plants, if the growth rate was too great. Chapman concluded that for nuclear power this is not a problem until extremely high growth rates and that capital problems are sure to precede energy balance problems. Roth (1994) calculated the energy balance for growing energy systems and concluded that an energy system with an expotential increase in its size would only yield net energy outputs if the doubling of the system size did not occur within the energetic payback time of a single power plant. Figure 5 shows the C flows when mature forest is cleared to implement a long-rotation forest-management system with product output growing linearly. One additional of the 100 ha systems shown above is added every 5 years. The capital investment, in terms of C emissions, is large with respect to the forest growth rate, and the time required to recover the original C investment is thus very long when there is continuing growth in product demand.

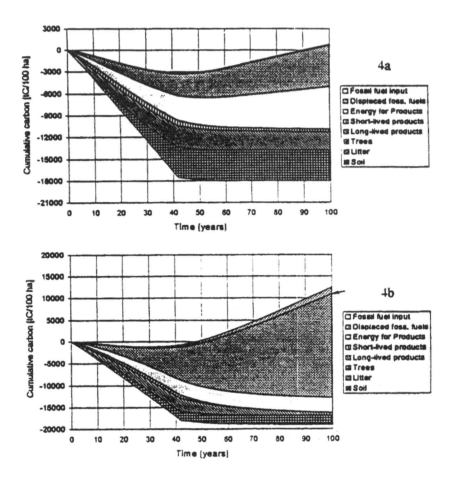

FIGURE 4. The net cumulative increase in C pools for the same scenarios as in Figure 3 except that 42 identical parcels totalling 100 ha have been phased in over a period of 42 years to produce a constant annual flow of products.

6. THE PRESENT VALUE OF FUTURE REDUCTIONS IN EMISSIONS

The series of scenarios described above all lead eventually to net C benefits, although this may involve very long times in an expanding system that starts on forested land. Nonetheless, it is clear that the scenarios differ not only in the amount of C eventually sequestered but also in the pattern of sequestration over time. Attention should be paid to the differences between forest management strategies in terms of their short-run and long-run effects on the carbon balance. Reforestation and protection (strategy 2a) offers large initial reductions in net C emissions, but reaches its steady-state C storage and offers little further C reductions after 75 to 100 years. In contrast, a strategy based on clearing existing forest in order to establish a fast-growing plantation (strategy 3b) releases C initially, but offers a steady flow of incremental C reductions thereafter. Which should be preferred?

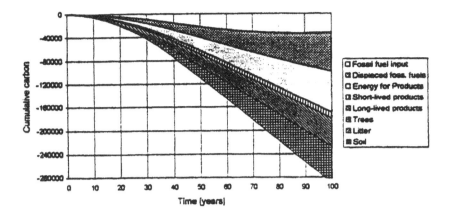

FIGURE 5. The net cumulative increase in C pools for the same scenario as in Figures 3a and 4a except that there is a continuing increase in the flow of forest products. The flow of products increases linearly with 1 additional unit of product flow every 5 years and so does the area in the system (thus the system size does not appear on the y-axis as in Figures 3a and 4a).

Admittedly, considering the C reduction path of forest management strategies without considering their management costs or fuel and product benefits is incomplete. Nonetheless, the discounted (present value) of C reduction paths is informative since it indicates cases where more costly management strategies may be worthwhile. It can also highlight, for example, cases where, on the basis of net carbon balance alone, strategies involving the clearing of standing forests to allow more intensive, short-rotation harvesting may or may not be effective. Depending on the rate of time-tradeoff between C emissions now and later, such a strategy may yield greater net carbon reduction benefit than either leaving mature forests in place or reforesting currently unforested, but less productive, land. The tradeoffs will depend on many site and situation-specific parameters but we can illustrate the point with the scenarios described above.

The rate of time-tradeoff between C emissions now and later can be called the "carbon discount rate." This rate differs from the usual discount rate applied to financial transactions because the financial discount rate reflects the opportunity cost of money. The opportunity cost of money is derived from the technical opportunities to turn current wealth into greater wealth later by investment now. The carbon discount factor should reflect the marginal rate of substitution between damages from emissions now versus damages from emissions later (see e.g. Lind et al., 1982; Arrow et al., 1995). The carbon discount factor will depend on the time path of stock accumulation and decay of C in the atmosphere, the relation between C stocks and subsequent climate changes, and the marginal damages (in lost utility) of climate change given changing income and adaptation technology. The carbon discount rate is the time derivative of the carbon discount factor.

Since the carbon discount rate depends on the time path of global C emissions, assumptions must be made to proceed. One assumption is that an optimal time path will be followed. In this case, the results of a global climate change model such as DICE (Nordhaus, 1994) can be used to estimate both the optimal time path and an appropriate discount rate consistent with the model's world view. A similar method was adopted by Reilly and Richards (1993) and Richards (1994). An interesting result of such an exercise is that the implied discount rate on C emissions is very low, less than the social rate of time preference (assumed by Nordhaus to be 3%) and that the discount rate is time dependent, increasing from 1.2% in the first decade to 2.8% in the tenth decade. This is because C stocks are increasing and the marginal damage of emissions is rising over time, although marginal damages are rising at a diminishing rate. A similar result is reported by Richards (1994, p. 41) in the case where atmospheric stocks are growing faster than linearly and damages are non-linear in the level of stocks. (Richards also performed numerical experiments and found that for linear growth in atmospheric stocks, and C increments that are small in comparison to initial stocks, the linear approximation of damages is reasonable and C can be discounted at the social rate of time preference with little loss of accuracy.) Applying this discount rate to the time path of C emissions reductions for different forest management strategies, we find that the difference in C paths leads to some very different net present values of C reductions (Table 1). Table 1 examines 3 of our scenarios that have approximately equal values for the net savings of C emissions (with no discounting) after 100 years (scenarios 2a, 2b, and 3b) and shows how the net present value of C emissions varies with the discount rate.

7. SUMMARY AND CONCLUSIONS

There are a number of forest management strategies which offer opportunities to mitigate net emissions of CO_2 to the atmosphere. This is particularly clear if one considers the full range of impacts which forest management and forest products have on the flows of C and is demonstrated with the series of scenarios modeled here. The magnitude of C benefits is very dependent on the rate of forest growth and the manner and efficiency with which wood products and biomass fuels are substituted for alternate materials or fossil

TABLE 1
Present Value of Future C Reductions. The C Discount Rate used Accounts for the Changing Marginal Damage of Emissions Over Time, as Well as the Valuation of Financial Damages at Different Times. (*Based on the DICE Model of Nordhaus, 1994)

Management scenario	Carbon discount rate			
	0%	1%	DICE rate*	5%
Reforest and protect (scenario 2a)	182	128	105	49
Reforest for long-rotation management (scenario 2b)	157	112	94	48
Clear forest for short-rotation plantation (scenario 3b)	159	76	40	−9

fuels. The flow of C-reductions benefits will also depend on the way in which a management strategy is phased in and how the land was being used before the strategy was implemented.

The series of scenarios described here illustrate that different strategies for using forest management to mitigate net emissions of CO_2 to the atmosphere can yield very different time paths for emissions reductions. Different strategies may offer permanent C sequestration, temporary C storage, or may borrow emissions now to increase future reductions. Temporary storage implies emitting one unit less now and one unit more later. Temporary storage has a positive value if the present value of marginal C reductions (discounted at the social rate of time preference) is lower, the further out in time these reductions are (Richards, 1993). In this most likely case, the C discount rate is positive, C "borrowing" (emitting one more unit of C now and one unit less later) has a negative value. However, what might be called C "investing" (emitting one unit more now so that later emissions are reduced by more than one) may still be worthwhile, depending on the biophysical and engineering parameters of the system and on the carbon discount rate. When a lower discount rate is applied, the value of temporary storage is diminished and the value of a C investment strategy is enhanced.

Table 1 shows that forest management plans which appear very similar at the end of 100 years in undiscounted terms (a commonly suggested comparison) may indicate very different contributions to the costs of climate change once their time path of C emissions is discounted.

Recognition that future reductions in emissions may be valued differently than current reductions has particular importance for strategies which involve clearing an existing forest in order to gain additional future reductions (C investing). With high rates of forest productivity and high efficiency in using the forest harvest, these investment strategies can yield positive net current value for C reductions at low discount rates. The net current value of reductions is negative at high discount rates (Table 1). For lower rates of forest productivity and less efficient utilization of the forest harvest (see, for example, Marland and Marland, 1992) these scenarios will fail to yield positive values for C sequestering even at low discount rates. On the other hand, these investment strategies could, over a sufficiently long time horizon, be preferable to a strategy of protecting existing forest or reforesting currently unforested but less productive land if the productivity is high, the harvest is used efficiently, and the discount rate is low. Numerical tests suggest that if C atmospheric stocks are rising and damages are increasing super-linearly with stocks, then a discount rate below the social rate of time preference is appropriate.

ACKNOWLEDGMENTS

Four anonymous reviewers offered suggestions which contributed significantly to the precision and clarity of this paper. Gregg Marland and Paul Leiby were supported by the U.S. Department of Energy, under contract DE-AC05-84OR21400 with Lockheed Martin Energy Systems, Inc. B. Schlamadinger's work was carried out within the Environment Research Program of the European Union with funds from the Austrian Ministry of Science and Research.

REFERENCES

Arrow, K. J., W. R. Cline, K. Maler, M. Munasinghe, and J. E. Stiglitz (1995), Intertemporal Equity and Discounting, Intergovernmental Panel on Climate Change, 1995 Report, Chapter 4 draft.

Buchanan. A. H. (1991), Building Materials and the Greenhouse Effect. *New Zealand Journal of Timber Construction* **7(1),** 6–10.

Buchanan. A. H. and B. G. Honey (1994), Energy and Carbon Dioxide Implications of Building Construction, *Energy and Buildings* **20,** 205–217.

Burschel, P., E. Kürsten, B. C. Larson, and M. Weber (1993), Present Role of German Forests and Forestry in the National Carbon Budget and Options to its Increase, *Water, Air and Soil Pollution* **70,** 325–340.

Chapman, P. F. (1975), Energy Analysis of Nuclear Power Stations, *Energy Policy,* Dec. 1975, 285–298.

ETSAP (1994), With CO_2 Restrictions is Energy Saved Energy Spared? *ETSAP News* no. 4, p. 1–2, Energy Technology Systems Analysis Programme, Netherlands Energy Research Foundation, Petten.

Hall, D. O., H. E. Mynick, and R. H. Williams (1990), Carbon Sequestration Versus Fossil Fuel Substitution — Alternative Roles for Biomass in Coping with Greenhouse Warming, Princeton University CEES Report 255, Princeton, NJ 08544.

Harmon, M. E., W. K. Ferrell, and J. F. Franklin (1990), Effects on Carbon Storage of Conversion of Old-Growth Forests to Young Forests. *Science* **247,** 699–701.

Lind, R. C., K. J. Arrow, G. R. Corey, P. Dasgupta, A. K. Sen, T. Stauffer, J. E. Stiglitz, J. A. Stockfisch, and R. Wilson (1982), *Discounting for Time and Risk in Energy Policy,*. Johns Hopkins University Press, Baltimore, MD.

Manne, A. S. and R. G. Richels (1992), Reducing U.S. CO_2 Emissions — The Value of Flexibility in Timing, presented at Symposium on Global Climate Change — A Petroleum Industry Perspective, 12–15 April 1995, Rome, Electric Power Research Institute, Palo Alto, Calif.

Marland, G. and A. Pippin, (1990), United States Emissions of Carbon Dioxide to the Earth's Atmosphere by Economic Activity, *Energy Systems and Policy* **14,** 319–336.

Marland, G., and S. Marland (1992), Should We Store Carbon in Trees? *Water, Air and Soil Pollution* **64,** 181–195.

Nordhaus, W. D. (1994), *Managing the Global Commons,* MIT Press, Cambridge, MA.

Reilly, J. M. and K. R. Richards (1993), Climate Change Damage and the Trace Gas Index Issue, *Environmental and Resource Economics* **3,** 41–61.

Richards, K. R. (1994), Valuation of Temporary and Future Greenhouse Gas Reductions, paper presented at International Atomic Energy Agency Symposium, 10 Nov. 1994, Battelle Pacific Northwest Laboratories, Washington, DC.

Roth E. (1994), Vollständige Energieszenarien bei Kraftwerkszubau (Complete Energy Scenarios for Phase-In-Program of Power Plants), *Brennstoff Wärme Kraft* **46(3),** 77–81.

Schlamadinger, B. and G. Marland (1995a), The Role of Forest Bioenergy in the Global Carbon Cycle, *Biomass and Bioenergy,* in press.

Schlamadinger, B. and G. Marland (1995b), Carbon Implications of Forest Management Strategies, in M. J. Apps and D. T. Price (eds.), *The Role of Global Forest Ecosystems and Forest Management in the Global Carbon Cycle,* NATO ASI Series. vol. XX, in press.

Worrell, E., R. J. J. van Heijningen, J. F. M. de Castro, J. H. 0. Hazewinkel, J. G. de Beer, A. P. C. Faaij, and K. Vringer (1994), New Gross Energy-Requirement Figures for Materials Production, *Energy* **19,** 627–640.

Critical Reviews in Environmental Science and Technology, 27(Special): S227–S244 (1997)

FORESTRY OPTIONS FOR SEQUESTERING CARBON IN MEXICO: COMPARATIVE ECONOMIC ANALYSIS OF THREE CASE STUDIES

OMAR MASERA, MAURICIO R. BELLON, and GERARDO SEGURA

Centro de Ecología, National University of Mexico (UNAM), A.P. 152 Pátzcuaro 61609 Michoacán, Mexico

ABSTRACT: Benefit cost analyses for three case studies which avoid carbon emissions and increase carbon sequestration in Mexican forests, a pulp wood plantation, a native temperate forest and a native tropical evergreen forest, are performed. Management of native temperate and tropical forests offer the most promising alternatives for carbon sequestration. The cost effectiveness of commercial plantations critically depends on very high site productivity.

KEY WORDS: Carbon sequestration, cost benefit analyses, native forests, plantations, forestry, global climatic change, Mexico, developing countries.

1. INTRODUCTION

The rapid advance of forest degradation and deforestation at a global scale results in significant carbon dioxide emissions to the atmosphere among many other local and regional problems. Currently, land use changes constitute the second largest source of carbon dioxide emissions to the atmosphere, the first in Latin America, with estimated gross emissions of 1.6 gigatons of carbon (GtC) per year (IPCC, 1992) and a net flux of 0.9 GtC/yr (Dixon et al., 1994).

However, if adequately managed, forests may serve as net carbon sinks, contributing to significantly offset emissions from the energy sector (Trexler and Haugen, 1993). Also, several options to sequester carbon in forests appear competitive with carbon saving measures in the energy and other economic sectors (Sedjo et al., 1995; Andrasko et al., 1991). Examining the cost effectiveness of forest management options that lead to a reduction in carbon emissions and to increasing carbon sequestration is particularly relevant for Mexico. Currently forests constitute about 40% of total CO2 emissions in Mexico (Masera et al, 1992). Also, making forests become net carbon sinks may help buy time to develop low carbon emission alternatives in the energy sector, which currently depends heavily on fossil fuels.

This paper performs a benefit cost analysis of three case studies in Mexican forests, including a commercial plantation, a native temperate forest and a native evergreen forest. The cases selected for management of native forests are examples of the new management technologies and of the associated social organization that are currently developing in the country. While the experience with pulp wood plantations is very limited in the country, this option is currently strongly promoted by the Mexican government. The first two sections present the context of the Mexican forest sector. Subsequent sections show the long-term potential for carbon sequestration in the country and the benefit cost analysis of the case studies.

2. THE MEXICAN FOREST SECTOR: GENERAL ASPECTS

Mexico has nearly 50 million hectares of natural closed forests that represent about 25% of the country's land surface. Approximately half of these forests are tropical (including evergreen and deciduous forests) and the other half are conifer and conifer-broadleaf forest associations (referred to as temperate forests). Approximately 2.6 million ha of closed forests are currently protected (1.9 million ha of tropical and 0.7 million ha of temperate forests). In addition there are at least 18 million hectares of forest lands currently heavily perturbed and degraded (SARH, 1992).

The main features of the forest sector in Mexico could be summarized as follows: (i) *Forests are highly diverse*. It is estimated that approximately 10 percent of the world's biodiversity is concentrated in Mexico (Flores and Gerez 1988); (ii) *A large proportion of the demand for forest products comes from subsistence uses*. The apparent consumption of wood products (production plus imports minus exports) is estimated at 47.6 million m^3 for 1990. Fuelwood accounts for 78% of total demand, the rest being for industrial use (10.1 million m^3). Sixty percent of industrial demand for wood products goes for timber products and 40% for pulp and paper products (CNIF, 1992); *(iii) Commercial forest harvesting is conducted almost exclusively on native forests*; *(iv) Most forest lands are common property*. Approximately 80% of the commercial forest lands in Mexico are collectively owned by communities and *ejidos*, and the remaining 20% is private property (Lara, 1992). It is estimated that about 10 million people live in forest lands, and at least 22 million rural inhabitants depend on fuelwood and other subsistence uses from the forests (Lara, 1992; Masera, 1993); *(v) The forest sector is undergoing a severe economic crisis*. During the last decade, the production of the timber and the pulp and paper industries decreased 24% and 32% respectively between 1985 to 1990, imports have increased in almost 90% since 1988, and there has been a reduction of 15% in the contribution of the forest sector to the GNP since 1987 (CNIF 1992). The causes of the crisis of the Mexican forest sector are rooted in three main factors that are closely interrelated: (i) an institutional and economic framework biased against the forest sector and forest owners; (ii) pres-

sures for conversion of forests to agriculture and pasture lands, and (iii) ineffi-
ciency of the forest industry and forest management practices.

3. DEFORESTATION AND CARBON EMISSIONS

Mexico has lost a large fraction of its original forest coverage. It is estimated that
tropical evergreen forests currently are restricted to 10 percent of their original area
(Rzedowski 1978). Since the sixties, the deforestation process has been particularly dramatic. In the late sixties and during the seventies, "development projects"
—many of them funded by multilateral lending agencies— and important subsidies for cattle ranching provided the basis for extensive clearing of forested areas.
Deforestation continued during the 1980s, fostered by the country's economic
crisis and a deepening of rural poverty.

Estimates of deforestation are not very reliable (they range from 370,000 ha/
yr to 1.5 million ha/yr) with a best estimate of 670,000 ha/yr for the mid 1980's
(Masera et al., 1992). Of these, 170,000 ha are from temperate and 500,000 ha from
tropical forests. The causes of deforestation vary according to forest type. The most
important cause for deforestation, particularly in tropical forests, has been the
conversion of forest to pasture resulting from an aggressive expansion of cattle
ranching activities since the early 1940's. Forest fires, shifting agriculture, and
timber mining, which generally lead to permanent changes in land use patterns, are
other common causes of deforestation, mainly in temperate forests.

Besides local and regional problems, it has been estimated that the current
pattern of deforestation may result in emissions of 53.4 million tons of carbon
(MtC), or about as much as 41% of total carbon emissions, in Mexico (Masera et
al., 1992). Conversion to pasture could be responsible for as much as 60% of the
total annual carbon balance. While accounting for only one-fourth of the area
affected by deforestation and forest fires, tropical evergreen forests are estimated
to be responsible for about 50% of total annual carbon balance. About 84% of total
emissions are estimated to come from tropical forests.

4. OPTIONS FOR CARBON SEQUESTRATION IN MEXICAN FORESTS

The current pattern of forest resource use in Mexico is largely non-sustainable. It
is leading to the depletion of natural forests, results in large net CO_2 emissions,
does not provide with adequate economic returns for local owners, and does not
meet the demand for subsistence and industrial forest products in the country.

Despite these problems, recent work has shown that the potential exists for
Mexican forests to become large carbon sinks. Detailed scenarios show that the
undertaking of several forest management options for both the conservation of

forests and for increasing the forested area may result in cumulative carbon sequestration amounting from 3.5 to 5.4 GtC, in the long term (Masera et al., 1995). Prominent among these options are the management of native forests and, to a lesser extent commercial pulpwood plantations, which may account for almost 50% of total carbon sequestration (Table 2). Several actions are needed at the policy and technical levels for the identified carbon sequestration potential to actually be tapped. Important among these actions is a better understanding of the costs, benefits and the comparative cost effectiveness of alternative forestry options. Given the specificity of situations in the Mexican forest sector, the suggested analysis should start with a detailed economic examination of representative case studies.

5. BENEFIT COST ANALYSES OF THREE CASE STUDIES

Three case studies where reliable information could be obtained were selected as case studies: (i) an industrial plantation; (ii) a managed native temperate forest; and (iii) a managed native evergreen tropical forest.

The analysis of carbon sequestration is conducted estimating the unit costs and benefits of carbon sequestration. The financial analysis uses roadside costs and benefits and was conducted for five real discount rates (0%, 1%, 3%, 5%, and 10%), normalizing the results per unit area. For brevity, only the results for the 10% discount rate are discussed here. The analysis also assumes that the options will remain in place in perpetuity (i.e., costs and benefits are calculated for infinite rotations and/or infinite harvesting). For plantations, the net carbon sequestration in vegetation is based on the average standing biomass per unit area (i.e. half of total biomass at rotation age). For management of native forests, carbon savings are estimated as the avoided emissions from deforestation per unit area. In both cases, the carbon pools of decomposing matter and forest products are also estimated. Because of lack of information, the soil carbon pool is not incorporated into the analysis.

The net present cost of carbon sequestration (or savings) (NPCCS) (in U.S. dollars per ton of carbon) is calculated according to the formula:

$$NPCCS = NPV/ NCS$$

where NPV is the net present value of monetary flows (used for both costs and net benefits) and NCS is the net carbon sequestration or avoided carbon emissions.

5.1. Management of Native Temperate Forests: The Case of San Pedro El Alto, Oaxaca

Almost 95% of commercial forestry is conducted in native temperate conifer forests in Mexico (SARH, 1992a). As stated before, most forests resources are

TABLE 1

Long-Term (2090) Carbon Sequestration Potential for Mexican Forests

Option	Incremental Area (million ha)	Unit Costs ($/tC)[a]	Unit Carbon sequestration (tC/h)[b]	Total sequestration (GtC)[c]
Conservation				
Natural Protected Areas	6.0	7–14	40–130	0.5–0.6
Management Native Forests	18.7	0.3–3	84–130	1.5–2.3
Improved woodburning cook stoves	n.a.	8–14	1.3tC/stove/yr	0.05–0.3
Afforestation				
Restoration Plantations	16.6	6.9	50–150	1.3–1.8
Commercial Pulp Wood Plantations	1.6–3.0	5–11	25–55	0.1–0.2
Total	42.7–44.3	3.4–5.2		

Notes: Source: Masera et al. 1995.

[a] Only initia or establishment costs are considered (range shown indicates differences by forest type), therefore the stated figures for unit costs of carbon underestimate actual costs.

[b] Only carbon sequestered in vegetation and soils is included.

[c] Total carbon sequestration calculated as unit carbon sequestration times area by forest types. Carbon sequestration estimates assume the option will remain in place in perpetuity. Refer to Bellon et al. (1994) for a detailed discussion of the assumptions used.

owned by *ejidos* and communities. Also, during the last 20 years an increasing number of these *ejidos* and communities have got in charge of the management of their forest resources through "communal forestry enterprises." A large percent of Mexico's commercial temperate forests are highly productive under natural conditions and can significantly increase this productivity when properly managed.

The *ejido* of San Pedro el Alto is located in the state of Oaxaca. The main economic activity is forest production, contributing 80% of the *ejido* income. The forest is composed of 77% of pines and 23% of oaks (Table 2a). The end products are timber, paper, and fuelwood, which mostly supply the forest industry of Oaxaca. Until 1984 the forest was given in concession to the Compañia Forestal de Oaxaca; since then, the forest exploitation has been controlled by the *ejido* members (SARH, 1990). They are the workers and managers of the forest, although they use the services of outside consultants for technical matters.

These consultants have developed a new improved management system that should increase the productivity from the current 3 m³/ha/yr to 6.8 m³/ha/yr, during a rotation period of 30 years, for an average productivity of 4.8 m³/ha/yr.[1] The new forest management system is based on changing the age structure and composition of the forest towards younger age classes and a higher percentage of pines. This system also aims at **increasing** the biomass stock at the end of the rotation period. The relevant information for the economic and carbon analysis is summarized in Table 2a.

For all discount rates the forest management provides positive net present values, which decrease with an increasing discount rate. Establishment costs are $341 and the NPV is $491/ha at 10% d.r. The project unit carbon sequestration was estimated as the avoided emissions from deforestation, noting that the area would likely become a degraded pasture had the project not been undertaken. A total of 64 tonC/ha are sequestered in this option. At 10% d.r. roadside NPCCS are $8/tonC (Table 3).

5.2. Management of Native Tropical Forests: The Case of Plant Piloto Forestal

Although currently only 5% (0.9 million ha) of commercial logging happens in tropical evergreen forests, these forests present 5.9 million ha with commercial value or 36% of Mexico's total commercial forest area. As in other tropical evergreen forests of the neotropics, no effective silviculture systems for timber production have been developed to ensure sustainability in Mexico. One of the main complexities for the implementation of a sustainable management system in these forests is the large diversity of tree species, of which only a small proportion has commercial value.

In the last 10 years the "Plan Piloto Forestal", a project to develop and implement sustainable strategies to manage and conserve the tropical forests of

---

[1]   Commercial timber productivity from management of native temperate forests ranges from 1 to 3 m³/ha/yr in natural conditions to 7 m³/ha/yr with more efficient management in Northern Mexico. Temperate forests of Central Mexico with warmer and more humid climates, can produce more than 15 m³/ha/year in natural conditions (Jardel 1986) and probably up to 20 m³/ha/year if they were more efficiently managed.

TABLE 2a
Main Parameters Management Temperate Forests

A. Economic Parameters

Total Area	27,400 ha
Rotation	30 yr
Average yield	4.8 m³/ha/year
Commercial Stock	144 m³/ha
Road side cost	$12.33/m³
Mill site cost (incl. transportation and loading)	$29.95/m³
Average price of wood (mill site)	$44.10/m³

B. Carbon Related Parameters

Emittable pre-management carbon (tC/ha)	64	Using forest stock at the end of rotation, adjusted by combustion efficiency and assuming the alternative use is a degraded pasture with 5 tC/ha
Tree species	Pinus spp. & Quercus spp.	SARH (1990)
Wood Density (ton/m³)	0.50	Echenique (1982)
Carbon content (%)	50%	Assumed
Harvestable biomass at rotation age (m³/ha)	144	SARH (1990)
Ratio commercial /total biomass	2.0	Cannell (1982)
Soil Carbon (t/ha)	n.d.	
Average NPP (tC/ha/yr)	2.01	Calculated from above parameters
End use product	Timber (50%), paper (45%) and fuelwood (5%)	
Duration end use product (yr)	10, 1, 1	Assumed
Decomposition period (yr)	8	Assumed

southeast Mexico, has been conducted in the state of Quintana Roo in Southern Mexico with the assistance of the German government. This project covers a large area and numerous *ejidos* throughout the state. As part of this project, an experimental silvicultural system has been developed in which detailed inventories of commercial and non-commercial species have been created. Approximately, sixteen species are harvested. They are divided in precious, soft and hard woods. The first category includes mahogany and Honduras mahogany, which are fundamental for the economic viability of the system, due to their high value and good market. The other species have lower values and limited markets. Although still using a selective cutting system with a rotation age of 75 years, extraction operations are conducted in a way that relatively large gaps (1000-2000 m^2) are created to ensure the regeneration of mahogany. After the operation, these gaps are reforested with mahogany.[3]

The BCA presented is based on data from the Zona Maya, in the central region of Quintana Roo, that includes 19 *ejidos*, for a total forested area of 150,000 ha. The main parameters used for the economic and carbon analysis are presented in Table 2b.[4]

In all cases the option provides positive net present values, which decrease with an increasing discount rate. Unit annualized costs are very low ($15/ha/yr to $3/ha/yr at 10% d.r.) as well as annualized net benefits ($5 to $1 at 10% d.r.). These low values are the result of the very low commercial productivity of the forest (0.82 m^3/ha/yr) (Table 2b). The project unit carbon sequestration was estimated as the avoided emissions from deforestation. The most probable alternative land use in this area is a degraded pasture. Net carbon sequestration reaches 81 tonC/ha. At 10% d.r. roadside NPCCS are $0.8/tonC (Table 3).

5.3. Pulp Wood Plantations: The Case of La Sabana

Official sources estimate that Mexico has as much as 22 million ha of land appropriate for commercial pulp wood plantations (SARH 1992b). However, the stated figure largely overestimates the actual potential for this forest option, as such a large area will likely conflict with other priority land uses (specifically with food production), and marginal areas may not have the potential productivity to sustain a commercial enterprise. The Mexican experience with pulp wood plantations is very limited. Only one plantation project has actually been operating commer-

---

[2]   Two additional important aspects of this experience that should be noted are: (a) a previous forest concession was canceled and harvesting rights were given to the owners of the land (mainly *ejido* members); (b) it has had political support from the state government (A. Ehnis, pers. comm.).

[3]   In this system, as well as in the case of the management of native temperate forests, transportation and hauling are the two main components of mill site costs, representing from 55% to 66% of total costs per cubic meter.

cially, with negative results (see below). Currently, however, pulp wood plantations have been strongly advocated by the Government. As of 1994, projects covering 1.0 million ha have been launched in different parts of the country, with 15 thousand ha. already established.

La Sabana Plantation, located in the State of Oaxaca, constitutes one of the oldest and best studied plantation projects in Mexico. The only end-use product for the plantation is paper. Three species of tropical pines are used, *Pinus caribaea* (var. *caribaea* and *hondurensis*), *P. oocarpa* (var. *ochoteranai*), and *P. tropicalis,* and managed in a 15-16 year rotation period. Most of the area where the plantation was established corresponds to shifting agriculture and induced grassland for cattle, in a previously tropical evergreen forest. The plantation was established mainly in communal and *ejido* lands.

Due to poor management, the actual average plantation yield was 3.7 m³/ha/yr which results in net economic losses for all discount rates.[4] A sensitivity analysis was performed in order to test the response of the BCA to changes in productivity. In the ranges of productivity recorded for the plantation (3.7 to 7 m³/ha) and for the four discount rates there were always net losses. We thus performed the BCA assuming an optimistic average yield of 17.2 m³/ha/yr for the plantation, which constitutes the government's goal for pine plantations established in temperate forests.

Table 2c presents the main economic and carbon related parameters used in the analysis. Achieving a commercial stock of 275 m³/ha at the rotation period, the plantation results in a net sequestration of 63 tonC/ha. Establishment costs are $1234/ha or $20/tonC (Table 3).[5] Net benefits are positive for the lower discount rates, but become negative for the 10% d.r. ($-419/ha, Table 3). The NPCCS is $-7/tonC. These results indicates that even under very optimistic assumptions regarding the plantation yields, it is not profitable and net transfers will be needed to achieve the potential carbon sequestration.

Because achieving very high yields is such a key issue for the profitability (and the carbon sequestration implications) of the plantation, it has to be located in very fertile lands in order to be a good business (a yield of at least 20 m³/yr/ha is needed for NPV to be positive). This goes against the common notion that pulp wood plantations are a good option for marginal lands, and also suggests that plantation establishment will have to compete with other profitable options (e.g., agriculture, cattle ranching, coffee production). Another important result is the fact that establishment costs account for a small proportion of total costs, 10% including roads cost and 5% without them. Since in many studies of CO_2 sequestration these costs are the most frequently used as a baseline (e.g., Dixon et al., 1991), our results suggests that these studies may seriously underestimate costs. The cost figures

[4] Apparently the reported low productivity for the plantation were related to a poor selection and management of the pine germplasm (M. Kanninen, pers. comm.).

[5] Average estimated establishment costs for other commercial plantations in Mexico range from US$580/ha to US$700 for Pine plantations and US$387/ha for Eucalyptus plantations (Bellón et al., 1994).

TABLE 2B
Main Parameters Management Tropical Forests

A. Economic Parameters

Total Area	150,000 ha
Rotation	75 yr
Average yield	0.82 m³/ha/year
Commercial Stock	61.6 m³/ha
Road side cost	
Precious	$41.89/ m³
Soft	$30.51/m³
Hard	$31.02/ m³
Mill site price of wood	
Precious	$179.64/m³
Soft	$69.55/m³
Hard	$62.32/m³

B. Carbon Related Parameters

Emittable pre-with	80	Using Cannell's (1984) formula
management carbon (1988),		forest inventory from Martin
(tC/ha) and		adjusted by combustion efficiency
		assuming the alternative use is a degraded pasture with 5 tC/ha
Tree species	Mahogany, Honduras Mahogany, Manilkara	Plan Piloto Forestal (1993)
	Zapota, and 13	other species
Wood Density (ton/m³)	0.71	Using Echenique (1982), for a weighed average of most
common		
		species harvested
Carbon content (%)	50%	Assumed
	Commercial stock at	61.6 Estimated from PPF
(1993) rotation age (m³/ha)		
	Ratio commercial /total	1.3 Cannell (1982)
aboveground biomass		
Soil Carbon (t/ha)	n.a.	
Average NPP (tC/ha/yr)	0.51	Calculated from above parameters
End use products	Furniture (87%), Paper (13%)	
	Duration end use	
products (yr)	20,1	Assumed
Decomposition period (yr)	5	Assumed

TABLE 2C
Main Parameters Pulp Wood Plantation

A. Economic Parameters

Total Area	6,238 ha
Rotation	16 yr
Average yield (actual)	7.12 m³/ha/year
Average yield (assumed)	17.20 m³/ha/year
Commercial Stock (using assumed yield)	275 m³/ha
Road side cost	$10.76/ m³
Mill site cost	$13.09/m³
Mill site price of wood	$54.71/m³

B. Carbon Related Parameters

Emittable pre-plantation carbon (tC/ha)	5	Assumed, no data available
Tree species	Pinus Caribea, Oocarpa, and Tropicalis	Alho et al. (1990)
Wood Density (ton/m³)	0.47	Echenique (1982)
Carbon content (%)	50%	Assumed
Commercial stock at rotation age (m³/ha)	275	Alho et al. (1990)
Ratio commercial/total biomass	1.5	Cannell (1982)
Soil Carbon (t/ha)	n.a.	
Average NPP (tC/ha/yr)	6.06	Calculated from above parameters
End use product	Paper	Alho et al (1990)
Duration end use product (yr)	1	Assumed
Decomposition period (yr)	8	Assumed

presented for the scenario (which are based mostly on initial or establishment costs) must then be regarded as well below the actual cost of carbon sequestration.

5.4. Comparative Analysis

As we noted above, the three case studies analyzed are representative of some of the most important response options for carbon sequestration of Mexico's forests. However, the economic attractiveness and the viability of each of these options is intimately related to the local biophysical and socioeconomic conditions of the area where it is implemented. For this reason, a careful analysis should be conducted before generalizations are made about the applicability of any of the proposed options at the country level. Below, we provide some elements that might guide the undertaking of such detailed analyses.

(i) *Commercial Pulpwood Plantations: limited applicability linked to very high yields and short rotation periods* Our results show that pulp wood plantations, even if resulting in moderate net carbon sequestration (when established in agricultural or pasture lands), tend not to be economically feasible in areas with commercial productivity lower than 20 m^3/ha/yr, or where rotation ages are long (15-20 yr). It has been suggested that large areas of the southeastern states of Tabasco, Campeche and Veracruz, in Southern Mexico, are appropriate for highly productive, short rotation, pulpwood plantations that can reach productivity as high as 50 m^3/ha/yr (Zobel pers. com.). For these plantations to be profitable, however, they need to be established in compact and well communicated areas of at least 20,000 ha, with a relatively mild topography. Plantations also exhibit the largest annualized costs and need the largest initial investment.

(ii) *Management of native tropical forests: high carbon sequestration but marginal net economic benefits.* This option provides the highest carbon sequestration potential of all three. However, given its very low unit net benefits, its economic sustainability, tends to be fragile. The low net benefits results from the fact that timber markets are limited to very few commercial species that occupy a small proportion of the forest biomass, and because the average commercial forest productivity tends to be relatively low (<1 m^3/ha/yr). Transportation costs also account for a large proportion of total wood costs at mill site. In the particular case of southern Quintana Roo other important factors such as a low opportunity cost of the land (cattle ranching and agriculture are not very attractive in the region), a low population density, the existence of large forest properties with a high stock of commercial species, and the subsidized technical assistance provided by the government and other organizations have also played an important role in the success of this forest management system. Unless some incentives or subsidies are provided to increase unit revenues, this option will not be attractive in regions of tropical evergreen forests with higher opportunity costs for land, and/or reduced stock of commercial species (like the evergreen forests of the State of Chiapas).

(iii) *Management of native temperate forests: larger applicability with intermediate carbon sequestration and larger local benefits.* Management of temperate native forests provides intermediate carbon sequestration and is the most profitable of the three options. Thus, it could be the most promising option to sequester carbon in many areas of the country, mainly because it is the one that offers the largest benefits to land owners. Several factors are associated with the profitability of this option. In contrast with tropical forests, for example, temperate forests have a much larger percent of their total timber volume in commercial species, which represents a larger capital stock to justify investment in forest management and infrastructure. Also, unlike the tropical forest of Quintana Roo, the temperate forest of Oaxaca, which are representative of many mountainous areas of central Mexico, are highly productive and the silvicultural systems applied to them are effectively improving their natural productivity. At a substantially lower annualized unit cost than plantations, and given the social ownership of most native forests in the country, the management of native temperate forests

TABLE 3
Economic and Carbon Indicators Case Studies

Option		Native Pulp Wood Plantation	Native Temperate Forest	Tropical Forest
Establishment Cost ($/ha)		1,234(1,234)	341(341)	43 (43)
NPV($/ha)		1,298(–419)	967(491)	121(63)
Net Carbon Sequestration (NCS) (tC/ha)	Vegetation	49	33	36
	Decomposing Matter	16	22	36
	Forest Products	3	9	9
Establisment Cost ($/tC)		20(20)	5 (5)	0.5(0.5)
NPCCS ($t/C)		21(–7)	15(8)	1.5(0.8)

Source: Adapted from Masera et al., 1995. The economic analysis assumes a real discount rate of 5% and 10% (10% discount values shown in parentheses). NPV is calculated for perpetual rotations, according to the formula: $NPV_R = \Sigma\ [\ B(t) - C(t)\ ]\ /\ (1+r)^n$, and $NPV = NPV_R\ /\ (1 - (1+r)^{-R})$, where "r" is the discount rate, and "R" the rotation age in years (EAP, 1995).

provides a good opportunity for strengthening local villages through the creation of "social enterprises" for forest management (i.e. enterprises owned and operated by *ejidos* and communities). It should be noted, however, that, at the stated commercial yields, the annualized net benefits per unit area of these systems (about $50/ha/yr are still low compared to local average wages (approximately $2,000/yr), meaning that the option will look attractive only to villages with large forest resource endowments. Technical assistance from the government and NGO's, and the improvement in road infrastructure (currently transportation costs account for more than half of total costs mill site) are essential to actually achieve the potential benefits.

6. EVALUATION OF RESPONSE OPTIONS: IMPROVING BENEFIT-COST ANALYSES

A fundamental assumption of benefit cost analysis is that the level of economic welfare experienced by individuals can be measured in terms of market prices (Hufschmidt et al., 1983). Nevertheless, market prices do not always reflect the true opportunity costs of a commodity or a resource for society, since market imperfections are common.

In the Mexican forest sector, the problem of imperfect markets is prevalent. The timber market has been characterized by a few large buyers that purchase and market most of the timber. Clearly, these small number of buyers greatly influence the price of wood. Historically, forests have been allocated for exploitation through concessions, that in most cases have responded more to political pressures or favors than to efficiency concerns. The concession allocation mechanism has been criticized because it fails to combine forest tenure with capture of stumpage value, which in turn prevents the government or concessionaires from detecting and responding to rising stumpage values, and translates into artificially low revenues for the government, and reduced potential funds for forest management (Vincent, 1992). An important component of both the San Pedro el Alto and the Plan Piloto projects has been that concessions were cancelled, and the harvest of the forests were given to the landowners.

Finally, most forested lands in Mexico are under the *ejido* or communal land tenure, these lands cannot legally be rented, sold or mortgaged,[6] being outside the land market and with no prices to guide the process of land allocation among alternative uses. Since these lands cannot be mortgaged, *ejido* or community members are restricted in their ability to secure credit and financial backing, having

---

[6]   The Agrarian law has recently been changed, allowing, under certain circumstances, *ejido* land to be converted into private property (but not *bienes comunales*), and hence be sold, rented or mortgaged, and therefore enter into the market.

a limited access to credit markets. The government has provided some credit, but in many cases with discretionary and non-economic criteria. Furthermore, in peasant and communal lands, subsistence production is important, and markets in general are not well developed. These conditions suggest the possibility of under-estimating the costs and benefits associated with land use changes based only on available prices. This may not be such a serious problem in well developed areas, where markets are better developed and private property is common. Hence, in peasant and communal areas, unless special care is put to correct for market imperfections, the results of a benefit-cost analysis may seriously underestimate the costs of carbon sequestration, and their results should be interpreted only as a minimum.

Evidence that this may be the case, is the fact that *ejidos* and communities that have been faced with the option of entering into an agreement to use their land for forest plantations have generally opposed them on the grounds that the rent offered for their land does not compensate for their losses, and the risk incurred is high (Chapela, 1992; Rojas, 1991). This clearly suggests high opportunity costs for these people that the BCA is not capturing.

In order to improve BCA for forestry options under the circumstances de-scribed above, the followings aspects might be considered:

(i)   to include the loss of subsistence products due to a forestry option (e.g. forest plantation), including the cost of creating the new job opportunities that will provide the cash to purchase these goods (with similar or lower levels of risk than current subsistence production, since these people are usually risk-averse), as well as the institutions and infrastructure that will be required to provide these goods at levels and prices that do not make these people worse-off than before.

(ii)   the costs of creating "intermediate institutions" should be considered. An analysis of successful forest projects in Mexico, such as the Plan Piloto Forestal of Quintana Roo or San Juan Nuevo in Michoacán, suggests that an important part of their success is that institutions have been created to provide marketing support, lobbying with the government, technical sup-port, particularly to foster research and technological options to accomplish sustainable forest management. We have called these institutions *interme-diate*, because they are between local ones, such as the ejidos, and regional and national ones, such as the state and national governments. There are several reasons for these intermediate institutions. First, we do not know enough about techniques in tropical forests, which makes research indis-pensable. Second, to decrease the costs of technical consultants, several *ejidos* may need to pool together to create an economy of scale that makes attractive for highly trained and expensive technicians to work with them. Third, marketing is expensive and requires the pooling of resources to gain information and to promote sales. Fourth, to deal with the government, it is necessary to be strong, which may only be accomplished by pooling resources and presenting a united front. Unfortunately, we still do not know how much does it cost to supply these intermediate institutions, but their costs, both economic and non-economic, should be estimated to provide a more accurate idea of the costs of sustainable forests options.

(iii)   overall concerns regarding forest use need to be better integrated into the analysis of carbon sequestration. The maximization of only one objective (carbon sequestration) might not result automatically in the best case scenario due to other forest uses and concerns. For example, the use of intensive management methods which rely on application of herbicides and chemical fertilizers, while certainly providing larger carbon sequestration, can result in local pollution problems.

7. CONCLUSIONS

While currently a net source of carbon emissions to the atmosphere, the Mexican forest sector has the potential to become a large carbon sink. The case studies analyzed suggest that sustainable management of native temperate and, to a less extent, of native tropical forest appear to be the most important alternatives in terms of both total carbon sequestration and its economic feasibility in Mexico. Because most forest resources are common property in Mexico, the sustainable— both environmentally and socially—harvesting of native forests may become a powerful instrument for the empowerment of local communities. Encouraging experiences of social forestry enterprises exist in Mexico, showing the viability of this approach. These experiences include the management of both temperate (San Juan Nuevo in Michoacan State, UZACHI in Oaxaca State) and tropical (Union de Ejidos Mayas in Quintana Roo State) native forests. Several lessons could be learned from these experiences, which help improve the prospects for local people-based sustainable forest management in the countryside.

The analysis also shows that evaluating the costs of carbon sequestration only on the basis of establishment costs tend to result in very optimistic figures. In countries like Mexico, where the forest sector lacks efficient markets, important limitations arise to benefit-cost analyses. The development of shadow prices, although feasible is very difficult due to the interconnectedness of economic and social aspects common in this sector. In all these cases, the results of the benefit-cost analysis presented provide just a minimum estimate of a project's worth.

ACKNOWLEDGMENTS

Funding for this work was provided by the Climate Change Division of the U.S. Environmental Protection Agency. We also thank Yamil Bonduki and two anonymous reviewers for their valuable comments on previous manuscripts.

REFERENCES

Alho, P., R. Alvarez Reyes, E. Ekola, H. Granholm, L. Jylhä, M. Katila, E. Korhonen, O. Mikkonen, P. Muuttomaa y V. Voipo (1990). *Plan de Manejo Integral de La Sabana: Informe Principal.* Programa de Cooperación Científica y Técnica en el Sector Forestal entre México y Finlandia, Mexico City.

Andrasko, K., K. Heaton, and S. Winnett (1991). "Estimating the Costs of Forest Sector Management Options: Overview of Site, National, and Global Analyses", Presented at the "Technical Workshop to Explore Options for Global Forest Management", Bangkok, April 24-30.

Bellón, M.R., O.R. Masera, and G. Segura (1994). "Response Options for Sequestering Carbon in Mexican Forests", LBL Report, Energy and Environment Division, Lawrence-Berkeley Laboratory, Berkeley (unpublished).

Cannell, M.G.R. (1984). Biomass of Forest Stands. *Forest Ecology & Management* 8:299-312.

Cannell, M.G.R. (1982). *World Forest Biomass and Primary Production Data.* Academic Press, London.

Chapela, F. (1992). El proyecto de La Sabana Mixe. *El Cotidiano* 48: 58-61.

CNIF (1991 & 1992). Memoria Económica. Cámara Nacional de la Industria Forestal. Mexico City.

Dixon, R.K., S. Brown, R.A. Houghton, A.M. Solomon, M.C. Trexler, and J. Wisniewski (1994). "Carbon Pools and Flux of Global Forest Ecosystems". *Science* 263:185-190.

Dixon, R. K., P. E. Schroeder y J. K. Winjum (eds) (1991). *Assessment of Promising Forest Management Practices and Technologies for Enhancing the Conservation and Sequestration of Atmospheric Carbon and their Costs at the Site Level.* US Environmental Protection Agency, Corvallis, Oregon.

Echenique, R. (1982). "Características de la Madera y su Uso en la Construcción", Instituto Nacional de Investigaciones Forestales, Mexico City.

Ehnis, A., (1993). Personal communication.

Energy Analysis Program (EAP) (1995). "Guidance for Mitigation Assessments, Version 2.0", Report LBL # 36387. Prepared for the "U.S. Support for Country Studies to Address Climate Change", Lawrence-Berkeley Laboratory, Berkeley, California (March).

Flores-Villela, O. and P. Gerez (1988). *Conservación en México: Síntesis sobre Vertebrados Terrestres, Vegetación y Uso del Suelo.* Instituto Nacional de Investigaciones sobre Recursos Bióticos-Conservación Internacional, Mexico City.

Hufschmidt, M. M., D. E. James, A. D. Meister, B. T. Bower y J. A. Dixon (1983). *Environment, Natural Systems, and Development: An Economic Valuation Guide.* Johns Hopkins University Press, Baltimore.

Intergovernmental Panel on Climate Change (IPCC) (1992). *Climate Change 1992: The Supplementary Report to the IPCC Scientific Assessment.* Cambridge University Press, Melbourne.

Jardel, P.E. (1986). "Efecto de la explotación forestal en la estructura y regeneración del bosque de coníferas de la vertiente oriental del Cofre de Perote". Masters Thesis. INIREB. Jalapa, Veracruz.

Kanninen, M. (1993). Personal communication.

Lara, P.Y. (1992). "Posibles impactos de las reformas al artículo 27 sobre los recursos forestales". *El Cotidiano* 48:13-20.

Marland, G. (1988). "The Prospect for solving the CO2 problem through global reforestation. Department of Energy (DOE), Office of Energy Research Report DOE/NBB-0082, TR039. U.S. Government Printing Office, Washington D.C.

Martin, P. (1988). "Biomassen Inventur im Tropischen Feuchtwald", M.S. Thesis, Albert Ludwigs University, Germany.

Masera, O.R., M.R. Bellón, and G. Segura (1995). "Forest Options for Sequestering Carbon in Mexico's Forests". Forthcoming in *Biomass and Bioenergy.*

Masera, O.R. (1993). "Sustainable Fuelwood Use in Rural Mexico, Volume I: Current Patterns of Resource Use", Report # LBL-34634, Energy and Environment Division, Lawrence Berkeley Laboratory, Berkeley, California (April).

Masera, O.R., M.J. Ordóñez, and R. Dirzo (1992). "Carbon Emissions from Deforestation in Mexico: Current Situation and Long-term Scenarios"; in W. Makundi and J. Sathaye (eds.) "Carbon Emission and Sequestration in Forests: Case Studies from Seven Developing Countries: Summary." Lawrence Berkeley Laboratory Report # LBL-32665, University of California, Berkeley, California, (August).

Mc Neely, J.A. and K. Miller (eds.) (1983). *National Parks, Conservation and Development*, IUCN, Smithsonian Institution Press, Washington D.C.

Plan Piloto Forestal (1993). "Costos de Producción y Extracción". (unpublished).

Rojas, R. (1991). "La empresa Simpson Investment negocia ya con ejidatarios". *La Jornada* 12/6/1991.

Rzedowski, J. (1978). *Vegetación de México*. Limusa, Mexico City.

SARH (1992). "Inventario Forestal Nacional de Gran Visión 1991-1992". Reporte Principal, Subsecretaría Forestal, Mexico City.

SARH (1992a). "Informe Anual. Dirección General de Política Forestal". SARH-Subsecretaría Forestal, Mexico City.

SARH (1992b). "Programa Nacional de Reforestación: Plan Rector de la Reforestación". SARH-Subsecretaría Forestal, Mexico City.

SARH (1990). "Estudio de Manejo Integral Forestal para la Comunidad de San Pedro el Alto, Distrito de Zimatlan de Alvarez, Oaxaca", Subsecretaría Forestal, SARH, Oaxaca, Mexico.

SARH (1989). "Informe técnico sobre auditorías operacionales de unidades para el desarrollo forestal". SARH-Subsecretaría Forestal, Mexico City.

Sedjo, R.A., J. Wisniewski, A.V. Sample, and J.D. Kinsman (1995). "The Economics of Managing Carbon via Forestry: Assesment of Existing Studies". *Environmental and Resource Economics* 5: 1-27.

Trexler, M.C., and C.A. Haugen (1993). *Keeping it Green: Using Tropical Forestry to Mitigate Global Warming.* World Resources Institute, Washington D.C.

Vincent, J. R. (1992). "The tropical timber trade and sustainable development". *Science* 256: 1651-1655.

Zobel (1993). Personal communication.

Critical Reviews in Environmental Science and Technology, 27(Special): S245–S257 (1997)

THE PHYSICAL RISKS OF REFORESTATION AS A STRATEGY TO OFFSET GLOBAL CLIMATE CHANGE

ROBERT J. MOULTON and JOHN F. KELLY
USDA Forest Service

ABSTRACT: Assessment of physical risks is important with respect to costs and carbon yields from tree planting. Plantation losses were estimated for the Southern U.S. Commercial timber harvests are the principal cause of losses. Excluding harvesting losses, the annual survival rate is 98.4 percent. Wildfire, insects, and inclement weather are not major factors; diseases are the leading cause of mortality, but affect only a small number of plantations.

KEY WORDS: Global climate change, carbon sinks, tree planting, tree plantation survival, risks.

1. INTRODUCTION

"How will the emission reductions be recovered if the trees die, if, for example, there is a hurricane and all the trees are blown over?"

—Question from a member of Congress on tree planting as a response strategy in the President's Climate Change Action Plan (November, 1993).

Quantification and documentation of anticipated survival and losses due to physical disturbances are important with respect to tree planting as a mitigation strategy for climate change. This is true for at least two reasons: first, from the perspective of science and analysis, losses of tree plantations affect the quantity and unit cost of carbon dioxide extracted from the air and stored as solid carbon in forest vegetation and soils; second, the perception of tree planting as a risky endeavor may materially influence its acceptance by public and private sector policy-making officials as a means to address global climate change.

The Climate Change Action Plan (CCAP) (1993) outlines a total of 46 actions that, collectively, are intended to reduce U.S. green house gas emissions to 1990 levels by the year 2000. The Plan was developed to fulfill U.S. obligations as one of 161 nations that signed the Framework Convention on Climate Change at the United Nations Conference on the Environment and Development (UNCED), or Earth Summit, held in Rio de Janeiro in June, 1992. The CCAP also served to fulfill a promise that then-candidate Bill Clinton made to address global warming at an Earth Day celebration during his Presidential campaign.

1064-3389/97/$.50

Central to this paper is Action #44 in the CCAP: Accelerate Tree Planting in Nonindustrial Private Forests. This action was prepared by the senior author, as a member of the Sinks Group, which developed and evaluated a number of forestry and agricultural actions that would sequester carbon and/or reduce the emission of green house gases. The specific tree planting program was principally based on earlier publications by Moulton and Richards (1993), Richards, Moulton and Birdsey (1993) and Moulton and Richards (1990). Allowances for plantation losses were applied in estimating program costs and carbon yields for Action #44 based on information assembled and developed by the authors. This paper, however, is the first published documentation of the basis for survival and loss rates incorporated into tree planting in the CCAP.

Specifically, the objective of the research effort described in this paper was to estimate plantation survival and losses through the use of USDA Forest Service Forest Inventory and Analysis (FIA) data for 12 states in the Southern United States (Table 1). Quantification of plantation survival and losses is especially important in the South, as 90 percent of the acreage to be planted to trees under the CCAP is located in this region.

In total, there are some 90,000 FIA plots laid out in a grid across the United States. Initially, it was proposed that plantations be studied for the entire country. This proved to be impractical, however, as plantations are either too few in number, stand origin (plantation vs. natural) has not been recorded at all, or not for successive surveys (required for trend analysis) for states outside of the South.

This research and the results reported in this paper, are a component of a larger study being conducted by the authors and Donald Hodges, W. F. Watson, and Keith Bellie of Mississippi State University. That study will consolidate information from a variety of sources. For example, it will look at the effects of fire control on plantation survival and the findings of other studies of smaller scale.

One observation that can be made with respect to the larger study is that while there is a very large literature on individual damaging agents—such as wildfire and individual insects and diseases—the focus of most studies is on the damaging agents themselves, and perhaps, their effects on individual trees. While such studies are of value in their own right, and have numerous applications, they are of limited usefulness for evaluating aggregate effects of physical disturbances on forests over large areas for long periods of time, and hence, are difficult, if not impractical, to apply in designing long-term national and regional forestry programs to mitigate global climate change.

We also reviewed forest economics literature to determine how physical risks are handled with respect to financial investments in forest plantations. Here it was found that most often physical risk: (1) was not mentioned, which by default implies an assumption of zero loss, or (2) was covered by assumption, i.e., an assigned rate without accompanying documentation. Foresters and forest economists might argue that this simply means that physical risks in plantation forestry are so minor that the subject does not warrant much consideration. Perhaps so. But such a response would probably not satisfy the Congressman who asked the question that begins this paper, nor other policy officials who are faced with making funding decisions for large scale climate change programs.

2. METHODS

2.1. General Approaches

Evaluation of plantation survival and loss rates, and the subsequent application in analyses, lends itself to two general approaches. The two approaches differ with respect to the use of:

1. A single overall rate
2. Multiple rates related to life stages of plantations
   a. Initial establishment
   b. Post establishment

Studies that have evaluated the retention and condition of tree plantations established on private lands through USDA cost-share assistance programs (Kurtz, Alig and Mills, 1980; Alig, Mills and Shackelford, 1980; Kurtz et. al., 1994) are based on field observations of plantations made 10 or more years after the trees were planted. These studies provided direct evidence of overall rates of survival; however, to the extent that plantations of different ages are included in a single study, or that the different studies pertain to similar populations of plantations, inferences can also be drawn of survival rates related to plantation age. Adjustments for survival/losses of plantations using a single overall rate are commonly made on the output (yield) side of the equation.

In contrast, there are numerous studies of tree seedling survival. These are short term studies, usually made one to two years after an area is planted, and look directly at the initial survival of plantations (item 2a). Since these losses occur early in the life of plantations, it is convenient to apply allowances for losses in program planning as an adjustment to costs through the use of a cost multiplier of the form $1/1\text{-}r$, where r is the average annual rate of loss. As an example, if $r = 0.10$, then the cost multiplier is 1.11 (1/1 - 0.1). This assumes that both the original planting and all subsequent plantings are subject to the same average annual rate of loss, which is a fairly reasonable assumption for large scale programs in which tree planting would be done each year for a decade or more. As an example of the use of the cost multiplier shown above, if a program called for the planting of 1,000 acres (405 ha) of trees, costs would be based on 1,111 acres (450 ha) of trees (1,000 x 1.11). This is the sum of the original planting (1,000) plus the first replant of 100 acres, a second replant of 10 acres, and final replant of 1 acre.

The final class of studies (2b) looks at post establishment survival of plantations. This paper evaluates plantations subsequent to their identification as plantations, hence, it is concerned solely with post establishment survival to be applied on an area (acreage) basis as an adjustment to expected yields (output).

The FIA program is administered by a Washington (DC) Office staff and six units within individual Forest and Range Experiment Stations. Plot locations established to assess forest resources are remeasured on an average national cycle of approximately 10 years; in the Southern States, the average cycle has been shorter over the past decade, approximately 7-9 years. Inventories are conducted and analyzed on a state-by-state basis (for example, McWilliams, 1992, and Thompson and Johnson, 1994). The data collected by the FIA program afford the best opportunity to assess plantation survival for large regions of the country.

Plantations are routinely identified in the course of FIA assessments in the South. However, plantation data covering at least two complete inventories have been collected by only two units--both in the newly-designated Southern Research Station. These units cover the 12 Southern States, from Virginia to Texas. Plantations are an important element of the resource in these states.

Data from the latest inventories were utilized in this study. The most recent inventory is often referred to here as the final measurement; while the previous inventory is often referred to as the initial measurement. Most of these Southern States have, however, had at least six inventories conducted by the FIA program. Changes in sample designs, variables collected, and data limitations preclude the examination of all inventories for this topic. For some states, three inventories are available for analysis; future plans include the examination of additional inventory cycles where possible.

The FIA data analyzed here consists of both plot-level and tree-level information. Plot-level data includes such items as ground use, site class, stand origin, stocking level (an estimate of site occupancy), plot expansion factors, owner category, and forest type. Tree-level data includes such items as species, tree history, volume, cause of mortality, and damage to live trees. Various inventory attributes have been calculated, including growth, removals, and mortality volumes.

2.2. Application of FIA Data to Plantation Survival Analysis

The revisit of sample plots is an important element allowing assessment of survival for previously established plantations. However, because the inventory procedures were designed to assess a wide range of items, they were not specifically tailored to plantation survival. Thus, a cruiser records the stand origin for individual plots at the time of each plot visit. Sometimes a plantation may lose its obvious identity as artificially regenerated. This is possible for a variety of reasons, including secondary and understory growth, and intermediate management activities. The loss of the artificial origin identity may be especially prominent for stands that were not originally planted or seeded in straight rows. While the FIA inventories include seeded stands in the category with planted stands as artificially regenerated, it is likely that few stands seeded are actually recognized as being artificially regenerated. Seeding is not used widely in the South, as planting is the preferred method of forest managers for artificially regenerating stands.

Because of the potential loss of the planted origin of stands in subsequent inventories, the definition of surviving plantations used here was very specific, and includes harvesting history, volume trend, and the age of the stand, in addition to the recorded stand origin. To qualify as surviving plantations, two conditions had to be met. First, the plots had to be to be coded as plantations in the remeasurement (final) visit by the cruiser, or (in the case where the plot was coded as natural) timber removals were limited to no more than one-half the initial volume, while showing an overall volume increase. Second, age at remeasurement had to indicate an older stand than was previously recorded (this condition was necessary to ensure that previous plantations were not harvested or otherwise destroyed and subsequently replaced with young plantations).

Several data anomalies had to be dealt with in preparation of the data set for analysis. The first anomaly deals with age. A small number of plots were classed as mixed-aged in Midsouth data, due to inventory procedures and the inclusion of non-artificial components on these plots. For these plots, a specific age was assigned using a volume-age relationship.

A second anomaly deals with the sample representation. Although the FIA program uses permanent plots, the number of acres represented by each sample location is not constant from plot to plot, or over subsequent inventories for the same plot; this number of acres is referred to as the expansion factor. The changing values for expansion factors may occur for a variety of reasons. One prominent reason is that sample locations may be added from time to time. In order to keep the analysis compatible for successive inventories, a constant expansion factor was used for each plot--the factor for the initial inventory. The maintenance of a constant expansion factor is valid in this analysis because of the focus on subsequent survival of the plot characteristics from one inventory to the next. Since the focus of the analysis is on resulting status of plots in the initial inventory, the original sample representation is appropriate. The concern here is not with assessing the current extent of plantations, but rather survival in the most recent periods available.

Other data characteristics must be recognized in order to adequately understand the analysis and its limitations. Since cruisers visit plots only at specific time points, interim occurrences are not well documented by FIA procedures. While a cruiser may observe and record a tree being cut and used, there is usually no knowledge regarding the condition of the tree just prior to its being cut. For example, a tree that dies because of some insect or disease and is subsequently cut may likely not be correctly assessed as salvaged mortality--especially if the stand is clearcut or sustains a heavy partial cut. Furthermore, damage agents recorded (or not recorded) as affecting live trees in the initial inventory may not (or may) be important to the subsequent history of the tree or stand. This is true because of the rather lengthy time between inventory visits.

Survival rates were calculated using times between inventories for individual plots, not the nominal inventory dates. Since the expansion factors vary considerably, the calculations accounted for these differences by weighting plots according to the value of their expansion factors.

2.3. Commercial Harvests

Commercial harvests are commonplace in Southern forests. Often, the harvest of mature timber is the only cause of significant disturbance in many forest stands. Given the importance of commercial harvests in forest management, this must be recognized as a component of management schemes. For application in this study, an objective definition (as opposed to a subjective field classification) of commercial harvest was developed that requires either (1) that harvests remove at least 1,500 cubic feet per acre of growing stock volume or (2) the age at harvest be at least 25 years for sites capable of producing at least 50 cubic feet per acre at the culmination of mean annual growth (for sites capable of producing less that 50 cubic feet, age was set at 30 years). In addition, no mortality was to have been recorded, other than a possible small amount caused by the logging operation or subsequent management activities, and more than one-half of the initial basal area was to have been harvested.

2.4. Regression Analysis of Plantation Survival

A logistic model was postulated to estimate the probability of plantation survival. The logistic model predicts the probability of plantation survival over one period. In this study one period is defined as the length of time between the initial and final plot visits by FIA crews; it averages 7.5 years. The general specification for the predictive model is:

$$Ps = f(OWN, FTYPE, SKGS, AGE, DR) \qquad (1)$$

where

Ps   = survival of plantation plot for one period, 1 = survived, 0 = otherwise
OWN   = owner class, 1 = public or forest industry, 0 = other owner
FTYPE = forest type, 1 = pine or oak-pine, 0 = other
SKGS = stocking of growing stock trees, 1 = greater than 60%, 0 = equal to or less than 60%
AGE   = stand age
DR   = damage recorded, as a percent of initial live-tree basal area

The probability of plantation survival was analyzed using logistic transformation of plot data. The original model work included variables for both site quality and volume. Stand volume and age were, not surprisingly for plantations, found to be highly correlated. Stand volume was therefore dropped from the model as it did not add significantly to it. Site quality also was eliminated from the final model as it failed to contribute to the model. Model parameters were estimated using logistic regression, specifically the logistic procedure in SAS (SAS Institute Inc., 1990).

3. RESULTS

A total of 22,928,700 acres (9,279,118 ha) of plantations were identified and recorded by the initial inventories conducted from 1977 through 1986. Of these initial acres, 18,511,700 acres (7,491,582 ha) (81 percent) survived to the following inventories completed from 1987 through 1993, an average of 7.5 years later. The South-wide survival rate for plantations is 97.4 percent annually. When commercial harvests are excluded, the survival rate improves to 98.4 percent annually (Table 1).

The state-by-state results are very similar to the South-wide results. Annual survival rates for all plantations ranged from a low of 96.7 percent (Florida) to a high of 99.5 percent (Oklahoma); excluding harvests increased annual survival rates from a low of 97.6 percent (Texas) to 99.5 percent (Oklahoma).

Timber volumes, a proxy for carbon storage, increased at an average annual rate of 581 million cubic feet per year (6%/year) between measurements for the 22,928,700 million acres (9,279,118 ha) in the original surveys (includes volumes on both surviving stands and residual volumes on nonsurviving stands). Alternatively, the increase in wood volume was 497 million cubic feet per year (5%/year) when only volumes on surviving stands are considered.

3.1. Stand Characteristics and Survival Rates

The youngest age classes have the highest survival rates. For plantations 10 years or less when initially measured, the annual survival rate is 98.8 percent. Forest industry-owned and National Forest plantations have the highest rates of survival, after accounting for commercial harvests. Excluding commercial harvests, forest industry and National Forest plantations have annual survival rates of 98.7; non-industrial private plantations have a 97.9 percent survival rate. Most plantations (60 percent) are owned by forest industry.

While the vast majority of plantations in the South are planted with pine species, the forest type may change over time to oak-pine or a hardwood type due to natural invasion by trees of other species. The survival rates are the highest for plantations where the forest type is either pine or oak-pine. This is logical, since

TABLE 1

Area of Plantations on Timberland and Plantation Survival from Sample Plots Initially Identified as Planted during the Initial Inventory and Surviving to the Most Recent Inventory, by State, for 12 Southern States

| State and inventory dates | Area of plantations | | Average annual Rates of survival | |
	Initial Inventory	Surviving to most recent inventory	All plantations	Excluding commercial harvests
	——————thousand acres————		———percent———	
AL (82, 90)	2,721.0	2,185.2	97.7	98.3
AR (78, 88)	895.4	659.7	97.3	97.9
FL (80, 87)	3,449.4	2,571.1	96.7	98.3
GA (82, 88)	3,892.4	3,090.1	97.0	98.6
LA (84, 91)	1,964.9	1,660.8	97.9	99.1
MS (77, 87)	1,842.6	1,340.2	97.1	97.8
NC (84, 90)	1,697.0	1,557.9	98.7	99.2
OK (86, 93)	505.8	488.5	99.5	99.5
SC (86, 93)	2,147.6	1,775.1	97.3	98.3
TN (80, 89)	536.3	397.5	97.1	97.7
TX (86, 92)	1,792.0	1,414.1	96.8	97.6
VA (85, 91)	1,484.3	1,371.5	98.7	99.1
All states	22,928.7	18,511.7	97.4	98.4
	—————thousand hectares————			
	9,279.1	7,491.6		

pine plantations that are less successful in terms of survival are the most subject to invasion by the more tolerant hardwoods.

When commercial harvests are excluded, there appears to be little difference in survival rates between plantations with varying amounts of growing stock volume. Of course, when commercial harvests are considered, the survival rate diminishes as the volume per acre rises to reflect the commercial value of these stands to their owners.

While there are distinct limitations for the recording of damage by FIA field crews, the relationship between the amount of damage recorded and the survival rate is definite, if not particularly strong. For planted stands with recorded damage on more than one-third of basal area, the annual survival rate is 96.5 percent, while otherwise it is 97.7. The difference is much closer when comparing stands where commercial harvests are excluded (98.0 versus 98.5), likely reflecting factors involved in management decisions regarding which stands should be harvested.

3.2. Nonsurviving Plantations

While the majority (54 percent) of plantations not surviving to successive inventory periods are either commercially harvested or landcleared, a number dropout for various other reasons (Table 2). Primarily, these remaining plantations are either harvested prematurely or experience mortality (Table 3). Other plantations dropout because of low stocking and other, unexplained reasons.

Of all the nonsurviving plantations, a plurality of area is in the age class 26-30 at the time of dropout (Table 2). This age class is common for dropouts because of the preponderance of commercially harvested stands at this age. The plantations that are landcleared are fairly evenly distributed between all age classes, while other nonsurvivors are somewhat concentrated in the younger age classes.

Of the plantations where mortality was an apparent cause of nonsurvival, a plurality of the area - 33 percent - was primarily affected by diseases (Table 3). Management activity, including logging damage, also affected a notable area. Only a few of the stands, amounting to 161,500 acres (65,358 ha) were affected by insects, fire, or weather. This latter area represents 22 percent of the plantations not surviving due to mortality, and only 8 percent of all the plantations not surviving that were not commercially harvested or landcleared. It is important to remember that the utility of these estimates are constrained by the inherent limitations of FIA inventories, including infrequent visits to the plots and the collection of only limited information regarding specific agents affecting tree and stand survival.

Low stocking is potentially a factor for plantation dropouts, since the average stocking is lowest for other nonsurvivors, which doesn't include landcleared or commercially harvested plots. However, this isn't a pronounced characteristic, probably since natural components do contribute to overall stocking levels in plantations where survival of the planted component is low. It is interesting, but certainly not surprising, to note that commercially harvested plots have the highest stocking level of all categories, including surviving plantations.

3.3. RESULTS OF THE REGRESSION ANALYSIS

Logistic analysis of discrete data is based on the binomial distribution and not on the normal distribution, as is common for other analyses. Assessing the model usually follows from the evaluation of a likelihood ratio test (Hosmer and Lemeshow, 1989). A stepwise logistic regression procedure was used to evaluate seven independent variables; these were four discrete (dummy variables) and three continuous variables. Evaluation of the models suggested elimination of two of the original variables as discussed earlier. The likelihood of the final model below (equation 2) was significant at the $p = 0.0001$ level. Results of testing individual coefficients in the model were also significant at the $p = 0.0001$ level. Results of the multiple logistical regression yielded the following estimates of coefficients (asymptotic odds ratios are indicated in parentheses):

TABLE 2
Area of Plantations not Surviving and Disposition, by Age Class at Termination, 12 Southern States

Age class at termination	Non-surviving plantations	Land cleared	Commercial harvests	Other nonsurvivors
		Disposition of nonsurvivors		
		thousand acres		
1-5	61.4	31.7	0.0	29.7
6-10	475.8	73.9	13.1	388.8
11-15	411.5	93.0	5.7	312.7
16-20	438.7	63.7	71.3	303.7
21-25	944.1	54.4	367.1	522.6
26-30	1,007.9	72.8	766.1	169.0
31-35	490.8	5.6	388.7	96.5
36-40	295.0	22.3	180.4	92.3
41-45	147.9	23.3	78.9	45.6
46+	144.0	10.4	63.9	69.7
All ages	4,417.0	451.3	1,935.1	2,030.6
	thousand hectares			
	1,787.5	182.6	783.1	821.8

$$\log(Ps/1\text{-}Ps) = 1.799 + 0.138*OWN + 0.472*FTYPE + 0.257*SKGS + -0.058*AGE + -0.007*DR$$
$$(6.04) \quad (1.15) \quad\quad (1.60) \quad\quad\quad (1.29) \quad\quad\quad (0.94) \quad\quad (0.99)$$
$$(2)$$

Analysis of predicted probabilities and the associated observed responses indicated 74 percent were concordant. The signs of the coefficients were as expected from casual examination of the data. As values for the quantitative variables AGE and DR rise, the probability of survival decreases. With advancing age, the likelihood of merchantable harvests increase, and the more damage recorded, the likelihood of adverse health conditions increase.

While the probability of survival predicted by the model is for an inter-inventory period, the annual survival rate may be computed using the average interval for the plots used in the model estimation.

An alternative model was also estimated, excluding the commercially harvested and landcleared plantations from the analysis. Examples of results for pine plantations on nonindustrial private land with well-stock, undamaged stands are provided in Table 4. Even for plantations that are 30-40 years of age, the average annual loss rate is estimated at less than 2 percent, excluding land clearings and commercial harvests.

TABLE 3
Area of Other Nonsurvivors and Disposition, by Primary Agent Causing Mortality, Damage, or Removal, 12 Southern States

| | Disposition of nonsurvivors | | |
Primary agent affecting stands	Stands with removals, no mortality	Stands with mortality[b]	Other stands (No mortality or removals)[c]
	———thousand acres———		
None	0.0	0.0	205.9
Insects	0.0	34.5	12.1
Disease	0.0	245.7	62.3
Fire	0.0	43.5	2.3
Weather	0.0	83.5	6.7
Management[a]	0.0	149.5	8.3
Logging	698.2	0.0	0.0
Other agents	0.0	182.1	296.1
All agents	698.2	738.8	593.7
	———thousand hectares———		
	282.6	299.0	240.3

[a] Includes logging damage to live trees and trees intentionally killed by timber stand improvement measures.
[b] Cause of mortality for plurality of dead basal area.
[c] Agent of damage for plurality of live-tree basal area with recorded damage.

TABLE 4
Examples of Average Annual Loss Rates for Well-Stocked, Undamaged Pine Plantations on Nonindustrial Private Land, by Stand Age, as Estimated by the Logistic Model

| | Average annual loss rates | |
Stand age	All plantations	Excluding commercial harvests and landclearings
	——— percent ———	
5	1.3	0.9
10	1.7	1.0
15	2.1	1.1
20	2.7	1.3
25	3.4	1.4
30	4.2	1.5
40	6.0	1.8

4. DISCUSSION

The results here show a low risk of plantation losses. Excluding commercial harvests, which is a deliberative decision of the landowner (unless timber theft is involved), the annual survival rate of plantations in the South is 98.4 percent. Further excluding land use changes, which are also a deliberative decision of the landowner, the survival rate increases to 98.7 percent annually. Additionally excluding premature harvests, which are also a conscious decision by land managers, the annual rate of plantation survival goes to over 99 percent.

The occurrence of wildfire, insect attack, or inclement weather are not major factors in the survival of plantations. While disease may affect a few plantations, the biggest factor by far appears to be the actions of the land managers. The lack of influence of wildfire and other destructive agents on plantation survival probably speak to the success of programs that have been implemented to protect Southern forests.

These conclusions, however, must be tempered somewhat by recognition of FIA data limitations for assessment of the exact cause for nonsurvival of plantations. These data are useful for assessment of overall post-establishment survival and for description of general plantation characteristics, but are limited in their ability to discern exact causes of nonsurvival in all cases.

The logistic model developed does indeed include important and significant variables which are tracked by the FIA inventories. It's possible that other variables may be important for predicting plantation losses relating to commercial harvests and land clearings. These could include landowner attitudes, income utility, and the value of alternative uses of forest property.

While this initial part of this study indicates that few factors external to deliberate management decisions affect plantation survival, a more complete analysis is scheduled to be completed on the physical risks of reforestation as a strategy to offset global climate change.

ACKNOWLEDGEMENTS

Joseph F. Glover with the Forest Inventory and Analysis Unit, USDA FS Research Station in Asheville, NC provided inventory data for the Southeastern states. Charles E. Thomas with the Institute for Quantitative Studies, USDA FS Southern Research Station, New Orleans, LA provided assistance with the statistical model. Felicia L. Lockhart and Jeralyn D. Snellgrove, USDA FS Cooperative Forestry staff in Washington, DC, prepared the computer graphics presented at the Bergendal Conference. We are grateful for the assistance provided by these individuals.

REFERENCES

Alig, R.J., T.J. Mills, and R.L. Shackelford. 1980. Most Soil Bank Plantations in the South have been retained; some need follow-up treatments. Southern Journal of Applied Forestry. 4:60-64.

Climate Change Action Plan. 1993. The White House. Washington, D.C. 50 p. Hosmer, D. W. and S. Lemeshow. 1989. Applied logistic regression. New York: John Wiley & Sons. 307 p.

Kurtz, W. B., R.J. Alig, and T.J. Mills. 1980. Retention and condition of Agricultural Conservation Program conifer plantings. Journal of Forestry. 78(5) 273-276.

Kurtz, W. B., T.A. Noweg, R.J. Moulton, and R.J. Alig. 1994. An analysis of the retention, condition and land use implications of tree plantings under the Soil Bank Program, the Forestry Incentives Program and the Agricultural Conservation Program. Missouri Agricultural Experiment Station Report SR464. 86 pp. + app.

McWilliams, W.H. 1992. Forest resources of Alabama. Resour. Bull. SO-170. New Orleans, LA: U.S. Department of Agriculture, Forest Service, Southern Forest Experiment Station. 71 p.

Moulton, R J. and K.R. Richards. 1990. Costs of sequestering carbon through tree planting and forest management in the United States. U. S. Department of Agriculture Forest Service General Technical Report GTR WO-58. Washington, D.C. 46 p.

Moulton, R.J. and K.R. Richards. 1993. Accelerated tree planting on nonindustrail private lands. The White House Conference on Global Climate Change: Second Sinks Workshop. Washington, D.C. July 15.

Richards, K.R., R.J. Moulton, and R.A. Birdsey. 1993. Costs of creating carbon sinks in the U.S. Energy Conversion and Management. 34: 905-912.

SAS Institute Inc. 1990. SAS/STAT user's guide, version 6, fourth edition, volume 2. Cary, NC: SAS Institute Inc. 849 p.

Thompson, M.T. and T.G. Johnson. 1994. Virginia's forests. 1992. Resour. Bull. SE-151. Asheville, NC: U.S. Department of Agriculture, Forest Service, Southeastern Forest Experiment Station. 103 p.

U.S. Department of Agriculture, Forest Service. 1992. Forest Service resource inventories: an overview. Washington, DC: U.S. Department of Agriculture, Forest Service. 39 p.

Critical Reviews in Environmental Science and Technology, 27(Special): S259–S268 (1997)

CAN RECYCLING OF WASTE HELP US TO SEQUESTER CARBON IN FORESTRY? EXPERIMENTAL RESULTS AND ECONOMIC VISIONS

LARS OWE NILSSON

Swedish University of Agricultural Sciences, Department of Ecology and Environmental Research, P.O. Box 7072, S-750 07 Uppsala, Sweden.

ABSTRACT: This study deals with describing the carbon sequestration potential and the economic prerequisites when using waste water as fertilizer. There is a large potential to sequester carbon and at the same time purify waste water in Norway spruce. It is concluded that the costs for sequestering carbon through application with waste water can be lower than when using commercial fertilizers.

KEY WORDS: Carbon sequestration, waste water, fertilization, waste purification, cost-effectiveness.

1. INTRODUCTION

The scientific community is looking for economically efficient methods to sequester carbon via forestry with the aim to increase the flows of carbon dioxide from the atmosphere into forest ecosystems (Sedjo et al., 1995). At the same time, industrialized societies are facing problems with various forms of waste which gradually have grown to almost acute dimensions. The costs for landfills and purification of wastes are today high enough to motivate a search for alternative solutions of those problems. Gradually an awareness of the ecological and economical benefits of recycling of wastes is developing in the forest industry (Greger et al. 1995), agriculture and the society as a whole.

One benefit of some wastes are their relatively high content of nutrients. This is to be found among all in waste water and sludge (e.g. Perttu 1993). It therefore can be argued that if we could use those wastes as fertilizer we could spare the costs of purification. If we received increased production this would also reduce the amount of carbon dioxide in the atmosphere. And if the produced biomass was utilized for combustion to replace fossil fuels, this would act to reduce the net emissions of carbon dioxide to the atmosphere.

There are some environmental problems with the use of waste as fertilizer which up to date still lack a proper solution. One of those problems includes the content of heavy metals found in among all industrial waste. Even if there is waste with very low content of heavy metals the fact that the content sometimes is high leads to a hesitation to use it.

Obviously new techniques to easily purify waste from its heavy metal content here would be very useful. Such new techniques also are under development. Another environmental concern is the risk of leaching of nutrients due to waste application from the soil into ground water.

Several experiments to use vegetation as purification filters for wastes have been initiated during the last few decades. In the majority of those experiments wastes have been applied to crops or fast grown short rotation tree species. Today we lack information about the capacity of forests to accumulate nutrients from waste and improve the production (sequester carbon) as well as the economic prerequisites of such activities. From a cultivation point of view we also need more information on how to use wastes without negative ecological consequences (i.e. leaching of nutrients from the soil) and how wastes technically should be applied in terms of a rational economy.

The aim of this paper is to indicate the N uptake and C sequestration capacity in a Norway spruce ecosystem as a result of waste water application in south Sweden. The aim is also to compare the costs of sequestering carbon in forests using either commercial fertilizers or waste fertilizer.

2. MATERIALS AND METHODS

2.1. Site and Experimental Set-Up

During 1988 we initiated a field experiment where among all water and nutrient availability was modified in a Norway spruce stand. The experiment was part of a large experimental set-up including 10 different field manipulations. The research area is located at Skogaby in the south-western part of Sweden, about 16 km from the sea of Kattegat and 95-115 m above the sea level. The climate is maritime with a mean annual precipitation of about 1150 mm and an annual mean air temperature of 7.5°C. The vegetation period is approximately 200 days, starting in the beginning of April and terminating in early November. The deposition is 15-20 kg N ha^{-1}y^{-1} and 20-25 kg S ha^{-1}y^{-1}. The soil type is a Haplic podzol (FAO-UNESCO, 1988). The stand, planted in 1966, is relatively homogenous. In 1987 it had a mean of 2347 stems ha^{-1} and a mean basal area of 24.7 m^2 ha^{-1}. In terms of bedrock, soil, tree species and provenance, stand productivity, site history, stand management etc the research stand is considered to be representative for the region.

The experiment has a randomised block design with four replicates and each gross plot is about 2000 m^2 in size. Irrigation (I) was performed during the growing period as soon as a 20 mm water storage deficit in the soil had developed. Combined improvement of both water and nutrient availability (IF-treatment) was achieved by irrigation and application with liquid fertilizer containing a complete set of nutrients (100 kg N ha^{-1} y^{-1}), according to the Ingestad principle (Ingestad, 1988). The following addition rates of macro nutrients where used; N=100, P=17, K=48, Ca=6, Mg=6, S=9, kg ha^{-1}y^{-1}. The fertilizer also included a balanced set of micro nutrients. Irrigation was made similarly for the I- and IF- treatments. Treatment started in 1988 and was performed in similar manner through the experimental period each year up until 1993. The treatment is described in more detail by Nilsson (1995).

2.2. Dry Mass Production and N Uptake

Data on accumulation and total production/uptake of dry mass and N after 6 years of treatment are given in Table 1. Total above ground production for untreated trees was higher during 1987-90 with 28.4 Mg dry mass ha^{-1} compared with 21.8 Mg during 1990-93, respectively. Changes in total above ground accumulation were more pronounced with 22.5 and 13.1 Mg ha^{-1} for the early and late period, respectively. Litterfall increased over time from 5.8 to 8.7 Mg ha^{-1} from the first to the second 3-year period. IF-treated trees showed increases in total production from +54% vs C to +116% vs C during 1987-90 and 1990-93, respectively. Total above ground accumulation of dry mass increased with +64% vs C and +164% vs C during the 2 periods. Also needle litterfall for the IF-treated trees was increased with +16% and +49% vs C during the first and second 3-year-period of the experiment, respectively. Total above ground production for irrigated trees improved with +23% vs C and +85% vs C and, total above ground dry mass accumulation with +31% vs C and +145% vs C during the two experimental periods. Needle litterfall for irrigated trees, however, showed negligible differences compared to control trees. The reasons for the variation in production over time are discussed by Nilsson (1995).

Total above ground N accumulation over 6 years was 34 kg ha^{-1} for untreated trees, 109 kg ha^{-1} for irrigated trees and 347 kg ha^{-1} for fertilized and irrigated trees. The N content in litterfall over the 6-year experimental period was 173-174 kg ha^{-1} for untreated and irrigated trees, but markedly higher values, 312 kg ha^{-1}, were found for the IF-treated trees.

TABLE 1

Total Above Ground Dry Mass Accumulation (Sum of Stems with Bark, Needles, Living and Dead Branches), Litter Fall and Total Production (Mg ha-1) as well as N Accumulation, Losses of N Through Litter Fall, and Gross N Uptake (kg N ha-1) for 3 Different Periods During the Experiment. Production Figures are Given in Mg Dry Weight but Can Easily be Converted into Mg of C by Multiplication with 0.50. Values from the Same Period in the Same Column Marked with Different Letters Differ Significantly between Treatments ($p < 0.05$). Data from Nilsson (1995)

Treatm/ period	Total above ground accumul	% of C	Total litter-fall	Total net pro-duction	% of C	Total N accumul	Needle litter N	Gross N uptake
C 87-90	22.52a	100	5.84a	28.36a	100	38.0a	67.9a	108.5a
90-93	13.05a	100	8.71a	21.76a	100	-4.2a	105.2a	103.4a
87-93	35.54a	100	14.55a	50.09a	100	33.8a	173.1a	212.0a
I 87-90	29.39b	131	5.55a	34.94a	123	77.4a	61.7a	138.9a
90-93	31.97b	245	8.19a	40.16b	185	31.4a	112.6a	143.5a
87-93	61.25b	172	13.74a	74.99b	150	108.9b	174.3a	282.4a
IF 87-90	36.91c	164	6.72b	43.63c	154	224.3b	92.2b	318.3b
90-93	34.57b	265	12.36b	46.93b	216	123.2b	220.0b	344.9b
87-93	72.02c	203	19.08b	91.10c	182	347.4c	312.2b	663.2b

The gross uptake of N over 6 years was considerable higher for IF-treated trees (663 kg ha^{-1}) compared to untreated (212 kg ha^{-1}) and irrigated trees (282 kg ha^{-1}).

3. RESULTS

According to Perttu (1993) the mean nutrient composition in Swedish waste water is N=100, P=18 and K=64 (relative values in relation to N). This is fairly similar to the composition used in this study with: N=100, P=17 and K=48. In earlier studies on the same site by Nilsson and Wiklund (1992; 1994; 1995), where growth response and nutrient uptake was examined after 3 years of treatment with different nutrient and water availability regimes, was concluded, that N was the only growth limiting nutrient at the site. In treatment with P, K, Ca and Mg fertilization no growth response was achieved although the trees responded with marked increases in their uptake of those elements. On the contrary application with ammonium sulphate (no P, K, Ca or Mg) resulted in significant growth and N uptake increases. In earlier studies by Nilsson and Wiklund (1992) and Nilsson (1995) was also concluded that water availability is a highly growth limiting factor at the site. The results of those findings show similar results regarding N uptake and carbon sequestration as found in the cited study by Nilsson (1995) would be obtained if normal Swedish waste water was utilized instead of the Ingestad solution used in the IF-fertilizer and if irrigation was done in the same way.

4. DISCUSSION

4.1. Potential For N Purification and Carbon Sequestration

As the litterfall increases due to fertilization this means a gradual build up of carbon in the humus layer of the soil. Data from the Skogaby site also show a build up of the humus dry mass from 11.3 Mg ha^{-1} for untreated trees to 19.5 Mg ha^{-1} for IF-treated trees (Johan Bergholm, 1995 unpublished data). Irrigated trees, however, showed a small and insignificant reduction to 9.7 Mg ha^{-1}. Investigations of carbon pools in different layers of the mineral soil so far show no significant changes due to the treatments. In earlier long-term fertilization experiments of Norway spruce it has been shown that carbon is accumulating both in the humus layer and in the mineral soil (Berdén, 1994). This effect probably is caused by both increased litterfall and decreased microbial activity in a situation with high N availability (Nohrstedt, et al., 1989). Fertilization with N, thus, results in sequestration of carbon both in the trees and in the soil.

If we try to sum up the changes in dry mass pools due to IF-treatment we find an increase in the accumulation above ground of 36.5 Mg ha^{-1} and of 8.2 Mg ha^{-1} in the humus. In total this means 44.7 Mg dry mass ha^{-1}. It is reasonable to believe additional accumulation in the root system and in the tree stump below ground. In an earlier study by Persson et al. (1995) a dry mass pool of stump above and below ground was shown as, coarse and fine roots of 9.3 Mg ha^{-1} in untreated trees of the same site. If we assume similar increases in dry mass accumulation as for above ground compartments (about 100%) the total accumulation increase due to IF treatment would be in the order of 54 Mg ha^{-1}. Let us put those figures in the range of 50-55 Mg ha^{-1} due to uncertainties connected with root pool changes.

The N uptake above ground for IF-treated trees was 451 kg ha^{-1} greater than for untreated trees. 139 kg ha^{-1} of this increased uptake was lost back to the soil system through the litter fall. It is possible that already a fraction of the N in this litter may be recycled as new N uptake gradually starts to be mineralized. Let us for simplicity assume this fraction to be negligible. Majdi and Persson (1993) found an accumulation of N in living roots of 48 kg ha^{-1} for untreated trees of the same site. It is likely that increased N accumulation rates are to be found for IF-treated trees. If we assume this rate to be in the order of 50 kg N ha^{-1} it seems that the total increase in N uptake related to the IF-treatment was in the order of 500 kg N ha^{-1}.

If instead of waste water using combinations of sludge and bioash also a good nutrient availability regime is possible. According to Perttu (1993) however, we would then end-up with a surplus of all macro nutrients except N. Well balanced in terms of N addition this application form should also be useful for carbon sequestration and purification purposes in Norway spruce. In this application form however, there is no beneficial effect from irrigation.

4.2. Does Fertilization Increase the Growth of Forests?

A prerequisite for having a positive growth response in forests due to application with nutrient rich waste water or sludge is that the forests are nutrient limited in their growth. If we discuss waste water the water in itself besides the nutrients may improve the production of the forest. Looking globally we find that water and nutrient availability are two major factors controlling forest growth in most parts of the world. On the most forested areas net primary production (NPP) can be enhanced through fertilization and/or irrigation. There are several examples of enhanced forest productivity due to fertilization. Thus, in Portugal and Brazil eucalyptus have reached annual production of up to 40 Mg ha^{-1} y^{-1}. In Australia, Fife and Nambiar (1995) have shown clear responses of N fertilization on the growth (up to 99% growth increase) of *Pinus radiata* grown in a very dry environment. The authors also concluded that improved N status of the trees leads to a more efficient water utilization. In Sweden a series of fertilization experiments have shown clear growth response due to N fertilization on both Scots pine and Norway spruce (Tamm, 1991; Sune Linder, personal communication 1995). Similar positive results on production due to N-addition have been observed in a large number of fertilizer experiments (about 70%) in south Sweden by different units of the Swedish University of Agricultural Sciences and by the Forest Research Institute of SkogForsk (Hans-Örjan Nohrstedt, personal communication 1995). As pointed out in an overview by Davey (1990) there is large evidence for positive growth responses due to fertilization in both the US and Canada. Similar unexpected growth responses this time with liquid fertilization have been noted in North Carolina of the US (Lee Allen, personal communication 1995). Cole (1995) and Henry et al. (1994) showed a sustainable basal area increase of 4-500% after application of nutrient rich sewage sludge in a 45 year old Douglas fir forest in NE US. Perttu (1993) has indicated improved production for willow as a result of both application with waste water and sludge. In temperate and boreal regions it seems like N most often is the primary limiting nutrient for NPP in forest ecosystems. However, there are several areas where other nutrients are limiting forest production, mostly P, K and Mg, but also micro nutrients like, B, Cu, Mn or Zn. In some fertilization studies there results, large increases of the storage of carbon in the soil, primarily caused by a build up of the humus layer (Berdén, 1994). The questions of to which extent changes in allocation patterns between above and below ground biomass do contribute to a visible build up of above ground biomass pools (which mostly has been the measured one) is balanced with lower biomass allocation to the roots, still remains to be better answered.

Danish (Beier et al., 1995) and Swedish (Nilsson, 1995; Sune Linder, personal communication 1995) experiments have shown large growth increase due to irrigation alone despite a relatively humid climate. Positive NPP response due to irrigation has also been found among a series of experiments (EXMAN) in practically whole Europe. It is likely that water availability limits forest NPP in several of the worlds climatic zones.

4.3. Does Increased Production Mean Increased Carbon Sequestration?

One interesting analysis when considering carbon sequestration alternatives using commercial fertilizers is to investigate if the energy balance or the carbon balance resulting from those activities. How much energy is needed to produce the fertilizer? How much carbon is emitted due to this production? How much carbon and energy is required for transportation and application of the fertilizers? Studies have shown that a fraction of less than 10% of the carbon normally being fixed as increased production due to fertilization is lost from manufacturing, transport and, application of the fertilizer (David Hall, personal communication 1995). Other critical questions include: Does fertilization lead to increased emissions of other greenhouse gases like methane or N_2O so that we even may worsen the greenhouse effect? Some of the answers to all of those questions are answered by Parks et al. (1995 this volume).

4.4. Practical and Environmental Considerations

It seems most promising in a first development to use waste purification forests located relatively close to purification plants. However, if wastes are refined into similar solid form or chemical compositions as found in commercial fertilizers they could be transported in similar way as commercial fertilizers are done today without loosing any economic competitiveness. To obtain a high carbon sequestration and nutrient uptake potential we should utilize tree species with both high production and nutrient uptake potential. Short rotation forests may here be very good alternatives to normal forestry. Environmental considerations restricts us to use contaminated wastes with heavy metal contents above prohibited threshold values. To utilize such wastes, they need to be purified from its heavy metal content. Fertilization may also induce leaching of nutrients from the soil if being done in too high proportions. This is a rather negative environmental effect and cannot be tolerated. One other environmental concern deals with how biodiversity in a forest is affected by waste fertilization. Several different studies have shown a decreased biodiversity of both flora and fauna due to fertilization (Fogelfors and Steen, 1982; Wiklund et al., 1995). Large areas of the world are today being applied with commercial fertilizers. Those forests do face problems with their biodiversity in same way as normal agriculture does. In the future, however, a realistic development in highly populated areas include natural to semi natural forestry combined with highly intensive cultivations. In those intensive plantations the biodiversity requirements then naturally will be lower than for natural or semi-natural forests.

Application of waste should preferably be done during the vegetation period among all to prevent from risks for leaching of nutrients from the soil. How to do with waste being produced during winter time? One solution is that the purification plants during winter is continuing as earlier to handle this material. An alternative is refinement of the waste to become applicable fertilizer.This will solve the problem with storing of waste during the winter period. If the waste fertilizer is refined into similar forms as commercial fertilizers, the waste fertilizer could compete with commercial fertilizer which means that we do not only depend on near waste plant applications. Specific analyses are needed to clearly answer the economic prerequisites of this particular sub problem.

4.5. Economic Considerations of Sequestering Carbon Using Wastes

In an attempt to estimate the costs of sequestering carbon in forestry via fertilization Dixon (1995 this volume) have found an approximate cost in the range of $10-15 per tonne of carbon being sequestered. In those calculations the costs of buying the fertilizer is included. There are also costs for transportation and application of the fertilizers into the forests. Let us compare the costs if we shift from using commercial fertilizers into various forms of waste as fertilizer.

In a first qualitative comparison, the focus is on waste with so low concentrations of heavy metals that they are below the permitted boundary levels for application in forests. In this case we can exclude the costs of purification of waste which up to date is a significant one. We may then also exclude to use a large part of industrial wastes and concentrate on various forms of municipality wastes with so low content of heavy metals that it can be utilized as fertilizer without heavy metal purification. There are two types of waste in this category; waste water and sludge.

If instead of purifying the waste, the purification plant deliver the waste as fertilizer - it has the chance of saving significant expenditures. For example the costs in Sweden to purify waste water from its N content is estimated to be $5-10 per kg N (Kurth Perttu, personal communication 1995). If converting the 500 kg ha^{-1} in increased N uptake due to IF treatment into a purification value we find $2500-5000 ha^{-1} in purification value over the six year experimental period. Hahn (1992) attempted to estimate the purification value of sludge and found $445 per tonne. Comparison between those two estimates depends on amount of wastes being applied and its N content.

The Skogaby study also suggests that an additional 50-55 Mg ha^{-1} of dry mass was built into the ecosystem due to a hypothetical treatment with waste water, of which 36.5 Mg ha^{-1} was harvestable, thus representing a significant possibility to sequester atmospheric carbon dioxide into the forest ecosystem. The additional carbon sequestered could be used as bioenergy as substitution to fossil fuels. This additional biomass production also would represent an economic value namely the value paid by a market. However, costs are also involved in harvesting and transportation.

Before being applicable as fertilizer, however, refinements of the waste may be needed. For example sludge may be manufactured into pellets and waste water may be concentrated in order to lower transportation costs or increasing the flexibility when being applied. Few cost estimates from this type of handling are currently available. However, it seems reasonable to believe that the cost of transport and application is higher for waste water than for sludge as application of sludge may be done in higher doses and with lower intervals compared to waste water. In an attempt to estimate the cost of transporting pellets a distance of 20 km and then apply it on the ground Nilsson (1994) found a cost of $25-35 per tonne. Hahn (1992) estimated the fertilization value for a farmer to be about $500 per tonne of sludge. If applied in forests this value however should be somewhat lower. If we here add the non-payment value of sludge for the purification plant this was estimated by Hahn (1992) to be $445 per tonne.

If applying commercial fertilizer into a forest we have to face costs related to transportation and application. In case wastes are refined into the same form as found among commercial fertilizers, application costs would be the same for waste fertilizer as for commercial fertilizer. In this latter case probably transportation costs would be lower than for commercial fertilizers as the distance between factory and application site in most cases are lower compared to normal commercial fertilizer alternatives.

If summarising the non-payment benefits of not having to purify the wastes and reduce with the additional costs for refinement of waste into easily applicable forms, assuming waste fertilizer is manufactured in the same form as normal commercial fertilizer, and further note a similar application cost and lower transportation costs for waste fertilizer compared to commercial fertilizer it seems rather clear that the costs for fertilizing a forest with heavy metal free waste generally can be considerably lower compared to when using commercial fertilizers. Thus, the carbon sequestration costs from fertilization of forests estimated by Dixon (1995 this volume), $10-15 per tonne of carbon being sequestered, should be lower if we instead of commercial fertilizer used waste as fertilizer.

What to do with the additional biomass being produced? Several different species may be used. Fast growing short rotation forest may be used primarily for bioenergy production, and then serve as substitution to fossil fuel combustion. This would then represent a long-term sequestration of carbon into fossil fuels. Normal forestry species may also be cultivated, primarily such local species that respond well to fertilization and with promising economic prerequisites. This includes several normal utilizations like as pulp or timber or new ones as for example bioenergy.

It seems as a promising way of thinking to look for opportunities to combine the solution of several problems into one single policy. If we try to solve the problem with human wastes we might at the same time contribute to the solving of the problems related to the green-house effect. In trying to get rid of one waste category we will at the same time get rid of another. This is to move the development of our society into a 'recycling' society. Such a society would be more sustainable than the one we are currently developing. In economic terms this development seems more promising than when using commercial fertilisers when aiming to sequester carbon in forests. Further research and development will be required to better quantify its potential. Maybe the use of wastes as fertilizer in combination with biomass production programs with the aim of substituting fossil fuel use with renewable biomass will be profitable. In such a case we can put the carbon sequestration costs of this alternative to zero.

5. CONCLUSIONS

There is a large potential to sequester carbon through application with liquid waste water in Norway spruce. Over a six year period the trees took up about 500 kg N ha^{-1} as a result of an application of 600 kg N ha^{-1} at the same time as the ecosystem as a whole accumulated 50-55 Mg dry mass ha^{-1} more than untreated trees.

The costs for sequestering carbon through application with waste water can be lower than when using commercial fertilizers. The results open up visions for a future development where waste purification is combined with carbon sequestration. This development can result in ecologically sound cultivation; a move into the 'recycling' society and; at the same time it has good prerequisites for profitability for companies and from the national economy point of view.

ACKNOWLEDGEMENT

This study was financed by the National Swedish Environmental Protection Agency and the Swedish National Board for Industrial and Technical Development.

REFERENCES

Beier, C., Gundersen, P., Hansen, K, and Rasmussen, L. 1995. 'Experimental manipulation of water and nutrient input to a Norway spruce plantation at Klosterhede, Denmark. II Effects on tree growth and nutrition', *Plant and Soil* 168-169, 613-622.

Berdén, M. 1994. 'Ion leaching and soil acidification in a forest Haplic Podzol: Effecs of nitrogen application and clear-cutting', Doctoral Dissertation, Swedish University of Agricultural Sciences, Department of Ecology and Environmental Research, Rep 73.

Cole, D.W: 1995. 'Soil nutrient supply in natural and managed forests', *Plant and Soil* 168-169, 45-53.

Davey, C.B. 1990. 'Forest ferilization in the americas', In: Nutrition of trees, The Marcus Wallenberg Foundation, Symposia proceedings: 6, Lectures given at the 1989 Marcus Wallenberg Symposium in Falun, Sweden, on September 14, 1989, Strålins Tryckeri AB, Grycksbo, Sweden.

Dixon, R. K. (This volume) 'Silvicultural options to conserve and sequester carbon in forest systems: Preliminary economic assessment'

FAO-UNESCO. 1988. 'Soil map of the world', Revised legend, FAO, Rome.

Fife, D.N. and Nambiar, E.K.S. 1995. 'Effect of nitrogen on growth and water relations of radiata pine families', *Plant and Soil* 168-169, 279-285.

Fogelfors, H. and Steen, E. 1982. 'Vegetation changes during 25 years of in landscape conservation experiments near Uppsala', Sweden, SNV PM1623, (In Swedish).

Greger, M., Hamza, K. and Perttu, K. 1995. 'Recirculation of waste products from forest industry - a prestudy', Swedish Waste Research Council. Report 68.

Hahn, T. 1992. 'Avloppsvatten på åkermark eller soptipp? Ekonomiska konsekvenser för kommuner och lantbrukare' (Sewage sludge on farm land or dump? Economic consequences for communes and farmers), Swed. Univ. Agric. Sci., Fakta Ekonomi No. 1. (In Swedish).

Henry, C.L., Cole, D.W., Hinckley, T.M. and Harrison, R.B. 1994. 'The use of municipal and pulp and paper sludges to increase production in forestry', IUFRO Centennial Conference, Berlin, 1992.

Ingestad, T. 1988. 'A fertilization model based on the concepts of nutrient flux density and nutrient productivity', *Scandinavian Journal of Forest Research*, 3, 157-173.

Majdi, H. and Persson, H. 1993. 'Spatial distribution of fine roots, rhizoshere and bulk-soil chemistry in an acidified *Picea abies* stand', *Scandinavian Journal of Forest Re*search 8, 147-155.

Nilsson, A. 1994. 'Teknik och ekonomi för att behandla aska. Tema askåterföring' (Technique and economy for ash recycling), Symposia, Hudiksvall January 24-25, 1994. Bio-X, Gävle. (In Swedish)

Nilsson, L.O. 1995. 'Manipulation of conventional forest management practices to increase forest growth - results from the Swedish Skogaby project', *Forest Ecology and Management* (in press).

Nilsson, L.O. and Wiklund, K. 1992. 'Influence of nutrient and water stress on Norway spruce production in south Sweden - the role of air pollutants', *Plant and Soil* 147, 251-265.

Nilsson, L.O. and Wiklund, K. 1994. 'Nitrogen accumulation in a Norway spruce stand following ammonium sulphate application, fertigation, irrigation, drought and nitogen-free-fertilisation', *Plant and Soil* 164, 221-229.

Nilsson, L.O. and Wiklund, K. 1995. 'Nutrient balance and P, K, Ca, Mg, S and B accumulation in a Norway spruce stand following ammonium sulphate application, fertigation, irrigation, drought and N-free-fertilisation', *Plant and Soil* 168-169, 437-446.

Nohrstedt, H.-Ö., Arnebrant, K. Bååth, E. and Söderström, B. 1989. 'Changes in carbon content, respiration rate, ATP content, and microbial biomass in nitrogen-fertilized pine forest soils in Sweden', *Canadian Journal of Forest Research* 19, 323-328.

Parks, P., Hall, D., Kriström, B., Masera, O., Moulton, R.J., Plantinga, A.J., Swisher, J.N. and Winjum, J.(This volume). 'An economic approach to planting trees for carbon storage'

Perttu, K. 1993. 'Biomass production and nutrient removal from municipal wastes using willow vegetation filters', *Journal of Sustainable Forestry* 1(3), 57-70.

Persson, H., von Fircks, Y., Majdi, H. and Nilsson, L.O. 1995. 'Root distribution in a Norway spruce (*Picea abies* (L.) Karst.) stand subjected to drought and ammonium-sulphate application', *Plant and Soil* 168-169, 161-165.

Sedjo, R.A., Wisniewski, J., Sample, A. and Kinsman, J.D. 1995. 'The economics of managing carbon via forestry: Assessment of existing studies', *Journal of Environmental and Resource Economics*, 6, 139–165.

Tamm, C.O. 1991. 'Nitrogen in terrestrial ecosystems. Questions of productivity, vegetational changes, and ecosystem stability', Ecological studies 81, Springer-Verlag, Berlin, New York, London.

Wiklund, K., Nilsson, L.O. and Jacobsson, S. 1995. 'Effects of irrigation, fertilization and artificial drought on basidioma production in a Norway spruce stand', *Canadian Journal of Botany* 73, 200-208.

Critical Reviews in Environmental Science and Technology, 27(Special): S269–S277 (1997)

THE COST OF CARBON SEQUESTRATION IN FORESTS: A POSITIVE ANALYSIS

ANDREW J. PLANTINGA

Department of Resource Economics and Policy, University of Maine, Orono, ME 04469-5782, U.S.A.

ABSTRACT: Estimates of the cost of large-scale carbon sequestration programs are based on normative criteria. In this paper, carbon cost schedules are derived using land use elasticities from an econometric analysis of forest and agricultural land use. The results indicate that normative analyses underestimate the true costs of carbon sequestration programs.

KEY WORDS: Global change, carbon sequestration, forestry, agriculture, land use.

1. INTRODUCTION

Large-scale afforestation projects have the potential to significantly offset emissions of carbon dioxide from fossil fuel use and other sources (Harland, 1988). However, the decision to implement an afforestation program ultimately must be based on the cost of the program relative to other emissions reduction strategies. A number of estimates have been made of the costs of large-scale (e.g., national) carbon sequestration programs (e.g., Sedjo and Solomon, 1989; Moulton and Richards, 1990; Adams et al., 1993; Parks and Hardie, 1995). The most detailed studies identify marginal agricultural lands capable of growing trees, and then estimate the compensation required by landowners to convert the land to forest Compensation is provided for forest establishment costs and foregone agricultural rents (net of forestry returns in some cases). A total cost schedule for carbon sequestration is based on the cumulative cost of compensation, where lands are converted to forest in order of increasing cost.

Carbon cost estimates are based on the normative assumption that landowners will accept the compensation, as specified in the analyses, for converting their land to forest. Compensation rates are assembled from a variety of data sources and often represent averages over broad geographical areas. Ultimately, compensation rates are fairly subjective measures. Moreover, they do not account for many factors which may influence the decisions of landowners, including option values, private non-market benefits, and asymmetric information. Thus, normative analyses may underestimate or overestimate the true costs of a carbon sequestration program.

An alternative approach to estimating carbon sequestration costs is to determine how landowners actually respond to changes in the net returns to forestry and agriculture. In this paper, land use elasticities from an econometric study (Plantinga, 1995) are used to estimate the cost of sequestering carbon in forests in a fourteen-county region of Wisconsin. The elasticities are estimated using a flexible model which accounts, at least implicitly, for many of the factors influencing land use decisions. Marginal carbon cost schedules are derived using parameter values equal to those in normative studies. The normative estimates are found to be lo wer than the positive estimates, indicating that, at least in this region, the cost of carbon sequestration may be underestimated. The results suggest that factors which tend to increase required compensation may significantly affect the true cost of a program. Moreover, program costs may be higher due to asymmetric infoon between landowners and the government.

This paper has five sections. Section 2 discusses some of the methods which have been used to measure compensation rates and difficulties which arise with these approaches. In Section 3, the land use elasticities used in this study are discussed and the method of deriving cost estimates is presented. Section 4 presents the results of the analysis and a comparison to other studies. Conclusions are discussed in Section 5.

2. COMPENSATION FOR CONVERTING AGRICULTURAL LAND TO FOREST

This section evaluates the methods used to measure compensation for converting agricultural land to forest. Three studies of carbon sequestration programs in the United States, comparable to the analysis presented below, are considered.[1]

Moulton and Richards (1990) base their compensation measures on "maximum allowable rental rates" for the Conservation Reserve Program (CRP). The CRP is designed to take marginal agricultural lands out of production by compensating landowners for foregone agricultural returns and part of the cost of establishing or other permanent vegetative cover. The maximum allowable rental rates are the maximum payments for foregone agricultural returns which may be provided to farmers in the program. They are determined locally by committees of government officials and farmers in accordance with U.S. Department of Agriculture farm budgets and cropland rental rates. Adams et al. (1993) derive their estimates using a spatial equilbrium model of the U.S. agricultural sector (Chang et al., 1992). The model is constrained to allocate specific amounts of agricultural land to tree growing. Compensation equals the shadow price of the constraint plus forest establishment costs. The value of agricultural production is based on regional (State and substate) production budgets. Parks and Hardie (1995) use crop budgets representing average retums across major crops to determine compensation for the conversion of cropland. Compensation for pasture conversion is based on private market rental rates. Forest establishment costs and annualized forestry rents are also included.

The first issue that arises in determining compensation rates is what should payments cover? In all the studies, compensation is provided for tree planting costs and foregone agricultural rents. Forestry rents are excluded in Moulton and Richards and Adams et al.; however, in theory, forestry rents will be capitalized into the value of the land and may

---

[1]   Only analyses of national programs, with results comparable to those in this study, are considered. Stavins (1995) estimates the cost of sequestering carbon in forests for the Mississippi River floodplain region using an approach similar to the one in this study.

therefore decrease the level of required compensation. Due to a lack of data, private non-market benefits from forests have not been reflected in compensation rates, though numerous surveys of private forest landowners indicate these amenity values are significant (Kurtz and Crouse, 1981). Indeed, some non-industrial private forest landowners in the United States appear to own forest primarily for recreation and other reasons unrelated to commodity production.

In Adams et al. and Parks and Hardie, compensation is based on the conventional "net present value greater than zero" rule. That is, the discounted benefits from conversion must just exceed the discounted costs. However, to achieve long-term carbon storage, the conversion of forest land back to agriculture must be restricted to some degree, or participation in the program must be irreversible. If there is uncertainty over the foregone benefits and costs of agriculture, then there may be a value to delaying the irreversible enrollment decision. Landowners may need to be compensated for this option value, in addition to foregone agricultural returns and planting costs. There is considerable anecdotal evidence that option values play an important role in firms' investment decisions (Dixit and Pindyck. 1994).

Since landowners have different skills and knowledge, it is likely that rents and costs will vary across parcels of land, even if they have similar characteristics (e.g., productivity). All the studies measure compensation with rents and costs representing averages over geographical areas such as counties and States. Unless landowners are clustered at the mean, the average compensation may be too little or too much to induce conversion of the desired number of hectares. Moreover, even if the distribution of landowners by required compensation can be determined, the government may be unable to sort landowners according to minimum compensation rates (Smith. 1995). If there is asymmetric information between landowners and the government, landowners may have an incentive to claim more compensation than the minimum amount required. The government can insure that a specific carbon sequestration goal is reached by offering compensation at the highest marginal rate.

The true costs of a carbon sequestration program may be overestimated due to failure to account for forestry rents and private non-market benefits from forests, and underestimated due to option values and asymmetric information. The use of averages may understate or overstate the true costs. Moulton and Richards may avoid some of these problems by basing their compensation mtes on those used in the CRP. The enrollment goal of the CRP, 16 to 18 million hectares, is close to being met, indicating that the maximum allowable rental rates have been sufficient to induce many landowners to convert their land.

3. A POSITIVE APPROACH TO THE DERIVATION OF CARBON COSTS

This section presents the methodology used to derive positive estimates of the cost of sequestering carbon in forests. The basis for the estimates are land use cities in Plantinga (1995). The proportion of forested land in a survey unit (hereafter. county) can be represented generally as:

$$f_{it} = \sum_{j=1}^{J} \pi_{ijt} \theta_{ij} \tag{1}$$

where f_{it} is the forested proportion of the land in county i at time t, θ_{ij} is the proportion of land in county i in land productivity class j, J is the number of land productivity classes, and π_{ijt} is the probability that land productivity class j in county i is forested in time t. π_{ijt} is unobservable but may be specified as a function of county-level observable variables, X_{it}, and unobservable parameters to be estimated, β_j. The specification used in an application to Wisconsin is:

$$\pi_{ijt} = \frac{1}{1 + e^{-\left(\beta_0 + \beta_j SMP_{it}\right)}} \tag{2}$$

where β_0 is an intercept term and SMP_{it} is the stumpage to milk price ratio. The stumpage price is a weighted average across species and the milk price is an average across grades. The milk price reflects the influence of federal price support programs.

The model given by Equations (1) and (2) is flexible in that average rents, measured by average prices, influence the probability that land of a particular quality is put into forestry or agriculture. The probabilities are derived using a random utility framework which implicitly accounts for unobservable factors specific to individual landowners (e.g., Maddala, 1983). These may include departures from the average level of rents due to particular skills and knowledge and the effects of option values and private non-market benefits. The model accounts, at least implicitly, for many of the factors discussed in Section 2. As indicated, it is difficult to account for these factors using a normative approach since they must be specified explicitly.

The application of the model in Plantinga (1995) is to a 14-county region in southwestern Wisconsin for the period 1968 to 1983.[2] During this period, a large amount of marginal agricultural land, particularly pasture and wooded pasture, reverted to forest in the region. Elasticities are estimated at the mean of this period for four land productivity classes (that is, $J = 4$) (Table 1).[3] Across all classes, a. small change in the price ratio has an approximately unit elastic effect on the. probability that the land is forested. However, as more land is converted to forest the opportunity cost of agriculture rises and so the elasticity declines. Compared to other studies (e.g., Parks, 1986). the land use elasticities estimated for this region are high, indicating that landowners require relatively little stimulus to convert agricultural land to forest. Thus, the cost of sequestering carbon should be low relative to other regions.

The cost of carbon sequestration may be determined by estimating how much the stumpage price must increase to ·induce landowners to convert a given amount of land to

---

[2]   The area considered is the U.S. Forest Service's Southwest Survey Unit (Spencer et al., 1989). About one of the land in cbe region is in forests and tw is in cropind and pasture. Oak-hickory is the dominant forest and dairy farming is the main agricultural enterprise.

[3]   Data are available for each county and the years 1968 and 1983. The time series observations are pooled. yielding 28 observations. Equation (2) is substituted into Equation (1) and the βs are estimated using a maximum likelhhood approach.

forest. Landowners receive the nominal increase in the stumpage value in the present period.[4] Formally, for a change in forest area of ΔA, landowners are compensated by an amount $\Delta P_h^s V^s \Delta A$ where ΔP_h^s is the change in stumpage price and V'^s is the stumpage volume at harvest for a typical rotation.[5] The subscript on price denotes the hth increment in forest area of ΔA. Since land use elasticities are declining as more land is converted to forest, $\Delta P_h^s > \Delta P_{h-1}^s$. On an annualized basis, the compensation or subsidy is:

$$S_h = \Delta P_h^s V^s \Delta A \left[\frac{r(1+r)^n}{(1+r)^n - 1} \right] \tag{3}$$

TABLE 1
Maximum Likelihood Estimates for Southwestern Wisconsin

Productivity class[1]	$\hat{\beta}_j$	$\hat{\pi}_j$ [2]	$\hat{\xi}_{\pi_j, SMP}$ [3]
Intercept	-2.96		
1	-0.17	0.019	-0.96
2	0.19	0.134	0.95
3	0.74	0.794	0.89
4	0.43	0.385	1.53
All classes		0.273	1.032

Source: Plantinga (1995).

[1] The productivity classes correspond to U.S. Department of Agriculture, Soil Conservation Service soil capability classes. In particular, productivity class I corresponds to soil classes I and II, 2 corresponds to classes III and IV, etc.

[2] $\hat{\pi}_j$ is an estimate, at the mean, of the probability that land quality class j is forested.

[3] $\hat{\xi}_{\pi_j, SMP}$ is an estimate, at the mean, of the elasticity of π_j with respect to SMP.

---

[4] This payment is not discussed since the model parameters (the β_j) measure discounting terms. This approach is tantamount to subsidizing forestry, and creates an incentive for landowners to clear existing forest in order to receive subsidies. This problem may be ameliorated by requiring landowners to document previous uses of their land and restricting enrollment to parcels in agriculture from a specified date.

[5] A 65-year rotation and a harvest volume of 8.58 thousand board feet, corresponding to medium site oak-hickory stands (Birdsey, 1992), is assumed.

where r is the interst rate and n is the rotation length. Each hectare converted to forest sequesters carbon at an average annual rate of C.

Total costs are determined in two ways. First it is assumed that the government can pay compensation at different rates and, in particular, offer landowners their minimum required compensation. In this case, the total annual costs of sequestering $C\Delta AH$ units of carbon per year is $TC^L(H) = \Sigma_{h=1}^{H} S_h$ where H denotes the total number of ΔA hectare increments. $TC^L(H)$ is the lowest compensation the government can pay to achieve a carbon target of $C\Delta AH$. It is more likely that the government cannot identify landowners according to minimum required compensation, even if it knows the distribution of costs (that is, each S_h). In this case, the government can insure that the target of $C\Delta AH$ is met if it offers all landowners S_H, implying total annual costs of $TC^U(H) = S_H H$. $TC^U(H)$ corresponds to the highest cost the government may have to pay to reach a goal of $C\Delta AH$. Thus, the total annual costs for an actual program will be bounded by $TC^L(H)$ and $TC^U(H)$.

The marginal costs for the Hth increment of $C\Delta A$ units of carbon are given by:

$$MC^L(H) = \frac{TC^L - TC^L(H-1)}{C\Delta A} \qquad (4)$$

$$MC^U(H) = \frac{TC^U - TC^U(H-1)}{C\Delta A} \qquad (5)$$

Marginal cost schedules are constructed by plotting $MC^L(H)$ and $MC^U(H)$ against the corresponding total annual carbon flow $C\Delta AH$.

4. ESTIMATION RESULTS AND DISCUSSION

A carbon sequestration program involving the conversion of 263 thousand hectares of land to forest in Southwestern Wisconsin, a 36 percent increase, is evaluated. The forest area target corresponds to the region's share of the eligible pasture and forest in Wisconsin, as specified in Moulton and Richards. The share for the region is determined by current pasture and forest land use shares (Spencer et al., 1988). The emphasis is on pasture conversion and forest management activities since the estimated elasticities primarily measure the effect of stumpage price increases on pasture and wooded pasture reversion (see Section 3). The carbon cost estimates derived using these elasticities reflect only the opportunity cost of agriculture, since natural regeneration is costless. However, in many cases, the cost estimates in the normative analyses also include costs for artificial regeneration. If these planting costs are omitted, the normative cost estimates are comparable to the estimates in this study.

Carbon cost schedules are estimated for two combinations of the sequestration and interest rate, namely $(C,r) = (4.35, 0.10)$ and $(5.56, 0.04)$ (Table 2). C is measured in short tons of carbon per hectare per year.[6] The first set of values is similar to those used in

[6]   A short ton is equal to approximately 0.91 metric tons. The avoirdupois weights are used throughout the remainder of this paper so that the results are readily comparable to those in the normative studies.

Moulton and Richards and Adams et al. The sequestration rate (4.35) is a weighted average of the rates for wet and dry pasture, forest planting, and passive and active forest management in the Lake States (see Moulton and Richards, Table 1). The weights are given by the land area shares of each activity. The second set of values corresponds to the parameters used in Parks and Hardie. The subsidy ranges from $1.30 per thousand board feet (mbf) for the first 10 thousand hectares to $35.90 per mbf for the last. The latter corresponds to an approximate 45 percent increase on the early-1980's stumpage price of $80 per mbf. When all 263 thousand hectares are converted to forest, the annual carbon storage is approximately 1150 thousand tons for $C = 4.35$ and 1450 thousand tons for $C = 5.56$.

For $(C,r) = (4.35,0.10)$, the marginal costs of the program range from $18 to $38 per ton when all 263 thousand hectares are converted. The weighted average of the Moulton and Richards cost estimates for Lakes States pasture and forest is approximately $13. When planting costs are omitted, the estimate falls to about $9. The cost estimates in Adams et al. are reported only for four carbon targets. However, for the lower targets, corresponding to the conversion of agricultural lands with relatively low opportunity costs, the Adams et al. estimates agree with those in Moulton and Richards. Thus, the estimates in this study, which are also for low cost lands, may be compared to those in Adams et al. by way of the Moulton and Richards estimates. For $(C,r)=(5.56,0.04)$, marginal costs are between $6 and $13 when all hectares are converted. The annualized average total cost in Parks and Hardie is approximately $10 per ton for a program involving 9 million hectares of crop and pasture land. The estimate is about $6 if annualized planting costs equal to those in Moulton and Richards are omitted.

There is considerable disparity between the results derived using the positive and normative approaches. The Moulton and Richards and Adams et al. estimates are well below the range estimated in this study, even if planting costs are included. The Parks and Hardie estimates also appear to be too low. As indicated, the marginal cost estimates in this

TABLE 2
Estimates of Marginal Carbon Sequestration Costs for Southwestern Wisconsin

	Hectares converted to forest				
	61	111	162	212	263
$(C,r) = (4.35,0.10)$					
$C\Delta AH$[1]	264	484	704	924	1144
MC^L[2]	3.62	6.83	10.26	13.95	17.90
MC^U	6.70	13.41	20.88	29.10	38.21
$(C,r) = (5.56,0.04)$					
$C\Delta AH$	338	619	900	1181	1463
MC^L	1.20	2.27	3.41	4.63	5.94
MC^U	2.22	4.45	6.93	9.66	12.68

[1] The annual carbon flows are measured in 1000 short tons per year.
[2] All marginal costs are measured in U.S. dollars per short ton.

study should be low compared to other regions. Therefore, since the marginal costs of the program evaluated by Parks and Hardie are increasing, the average total cost for this larger program should lie above the marginal cost estimates in this study. However, the average cost estimates in Parks and Hardie, with and without planting costs, lie within the estimated range of marginal costs.

5. CONCLUSIONS

This study uses the results of a positive analysis of land use to test the accuracy of carbon cost estimates based on normative criteria. The marginal cost estimates derived using the positive approach are found to be considerably higher than the normative estimates. This suggests that the true costs of a carbon sequestration program will be significantly affected by factors which tend to increase program costs, such as option values. Factors which may decrease program costs, such as private non-market benefits, appear to be less important. In addition, the results indicate that asymmetric information between landowners and the government may have a significant effect on the costs of a program. Depending on the governments ability to sort landowners by minimum compensation rates, program costs may be as much as twice that under perfect information. In this extreme case, the positive estimates differ from the normative estimates by a factor of three.

Although the normative estimates for this region appear to be too low, estimates for other regions may be more accurate. As indicated, the compensation rates in Moulton and Richards are based on the maximum allowable rental rates for the Conservation Reserve Program. Participation in the CRP has been relatively low in Wisconsin: no more than 5 percent of the land in any county has been enrolled, in contrast to nearly 30 percent enrollment in some counties in other States. The results of this study suggest that CRP rates in Wisconsin have been too low to induce the conversion of land to forest. However, CRP rates clearly have: been high enough in areas with high enrollment, suggesting that carbon sequestration costs have been more accurately estimated for these regions. Positive estimates of carbon costs for other regions (e.g., Stavins, 1995) will help determine if normative estimates are systematically low. Another direction for future research is to use the information on CRP enrollment to improve the accuracy of normative cost estimates. For instance, compensation rates could be adjusted upward in regions with low enrollment using data on the percent of eligible lands enrolled.

ACKNOWLEDGEMENTS

The author wishes to thank Ralph Alig, Maria Enzer, Lars Lonnstedt, Robert Moulton, and Robert Stavins for helpful comments and suggestions.

REFERENCES

Adams, R. M., D. M. Adams, J. M. Callaway, C. Chang, and B. A. McCarl (1993), 'Sequestering Carbon on Agricultural Land: Social Cost and Impacts on Timber Markets', *Contemporary Policy Issues* **XI,** 76–87.

Birdsay, R. A. (1992), 'Changes in Forest Carbon Storage from Increasing Forest Area and Timber Growth', in N. R. Sampson and D. Hair, eds., *Forests and Global Change, Volume One. Opportunities for Increasing Forest Cover,* American Forests, Washington, DC.

Chang, C., B. A. McCarl, J. W. Mjelde, and J. W. Richardson (1992), 'Sectoral Implications of Farm Program Modifications', *American Journal of Agricultural Economics* **74**, 38–49.

Dixit, A. K. and R. S. Pindyck (1994), *Investment Under Uncertainty,* Princeton University Press, Princeton, New Jersey.

Kurtz, W. B. and C. K. Crouse (1981), 'Nonindustrial Private Forest Ownership Studies: A Bibliography', Council of Planning Librarians Bibliography No. 60, Chicago, Illinois.

Maddala, G. S. (1983), *Limited Dependent and Qualitative Variables in Econometrics,* Cambridge University Press, London.

Harland, G. (1988), 'The Prospect of Solving the CO_2 Problem through Global Reforestation', U.S. Department of Energy, Office of Energy Research, DOE/NBB-0082, Washington, DC.

Moulton, R. J. and K. B. Richards (1990), 'Costs of Sequestering Carbon through Tree Planting and Forest Management in the U.S.', U.S. Department of Agriculture, Forest Service, General Technical Report WO-58, Washington, DC.

Parks, P. J. (1986), *The Influence of Economic and Demographic Factors on Forest Land Use Decisions,* Unpublished PhD. Dissertation, University of California, Berkeley, California.

Parks, P. J. and I. W. Hardie (1995), 'Least-Cost Forest Carbon Reserves: Cost-Effective Subsidies to Convert Marginal Agricultural Land to Forests', *Land Economics* **71**, forthcoming.

Plantinga, A. J. (1995), *The Allocation of Land to Forestry and Agriculture,* Unpublished PhD. Dissertation, University of California, Berkeley, California.

Sedjo, R. A. and A. M. Solomon (1989), 'Climate and Forests', in N. J. Rosenberg, W. E. Easterling, P. R. Crosson, and J. Darmstadter, eds., *Greenhouse Warming: Abatement and Adaptation,* Resources for the Future, Washington, DC.

Smith, R. (1995), 'The Conservation Reserve Program as a Least-Cost Land Retirement Mechanism', *American Journal of Agricultural Economics* **77**, 93–105.

Spencer, J. S. Jr., W. B. Smith, J. T. Hahn, and G. K. Raile (1988), 'Wisconsin's Fourth Forest Inventory, 1983', U.S. Department of Agriculture, Forest Service, North Central Forest Experiment Station, Resource Bulletin NC-107, St. Paul, Minnesota.

Stavins, R. N. (1995), 'The Costs of Carbon Sequestration: A Revealed-Preference Approach', Paper Presented at the Murray S. Johnson Conference on Environmental Economics, University of Texas, Austin, April 29, 1995.

Critical Reviews in Environmental Science and Technology, 27(Special): S279–S292 (1997)

THE TIME VALUE OF CARBON IN BOTTOM-UP STUDIES

KENNETH R. RICHARDS

Pacific Northwest Laboratory,* 901 D St. SW, Washington, D.C. 20024

ABSTRACT: In many climate change mitigation cost-effectiveness studies that do not develop endogenous shadow prices for the value of carbon it is necessary to determine the time value of reductions in carbon dioxide emissions. This paper examines the implications for the time value of carbon of exogenously specifying (1) the path of marginal damages over time, (2) the path of emissions over time, and (3) that the emissions path follow an optimal trajectory over time.

KEY WORDS: carbon dioxide, cost-effectiveness, discounting, dynamic optimization, sequestration.

1. INTRODUCTION

Ideally, carbon sequestration cost studies would be conducted as cost-benefit analyses, providing a ratio between the costs and benefits of forestry activities, each denominated in dollars. However, because these carbon sequestration analyses are generally conducted in isolation from either studies of damages or economy-wide modelling exercises that provide shadow prices, cost-benefit studies have not been employed.

In the absence of shadow prices, bottom-up analysis of greenhouse gas abatement strategies, and of carbon dioxide abatement in particular, have generally been conducted in a cost-effectiveness framework. Richards and Stokes (1995) identified three methods that have been used to derive a "dollars per tonne" measure of the cost-effectiveness of carbon sequestration projects: the *flow summation* method, the *average storage* method and the *levelization/discounting* method. The flow summation method treats all carbon capture as equivalent, regardless of the time at which it occurs. In contrast, the levelization/discounting method gives more weight to carbon captured earlier than carbon captured later. The average storage method measures change in average carbon storage rather than carbon flows. The choice of method for calculating the summary statistic can make as much as an order of magnitude difference in the result (Richards and Stokes 1995).

---

* Pacific Northwest Laboratory is operated for the U.S. Department of Energy by Battelle Memorial Institute under contract DE-AC06-76RLO 1830.

1064-3389/97/$.50

The underlying issue is the definition of a ton of carbon used in the denominator of the cost-effectiveness measure. Because carbon emissions reductions and carbon sequestration occur over many years, it is important to decide how to account for the time value of carbon. Analysts could report results of their technology studies in terms of the costs of achieving specific streams of carbon reductions over time. However, policy makers have been disinclined to deal with this type of multidimensional result. Instead they seek simpler summary statistics, such as dollars per tonne of carbon sequestration.

Oftentimes readers, and sometimes the studies themselves, have not been careful about differentiating among the methods, or the circumstances under which one approach is more appropriate than another. This has led to confusion as policymakers have compared the results of studies that appear similar, but are fundamentally different. This study addresses the question of how the time value of carbon should be treated in cost-effectiveness studies.

For purposes of developing a common unit of measure to use in cost-effectiveness studies, analysts need to employ a value-based definition of carbon emissions reductions. To develop a common unit, analysts must understand the pattern or general path of shadow prices over time. It is not necessary, however, to know the actual magnitude of shadow prices associated with emission reductions, only to identify a relation in which the marginal benefit, MB(t), of a one unit reduction in emissions (or alternatively, of atmospheric stocks) of carbon at time t can be expressed in a function,

$$MB(t) = f(MB(0), t)$$

This study develops a framework within which the analyst's choice of a common unit for these bottom-up studies can be considered. It illustrates that the appropriate treatment of the time value of carbon is dependent upon the policy context of the analysis. Therefore, it is important to identify how much of the policy context is determined by exogenously specified elements of the analysis. Cost-effectiveness studies should assure that the policy context within which the analysis takes place is clearly identified, and that the "tons" in the denominator of the cost-effectiveness metric are appropriately defined for the specified policy context.

A practical question that has developed in carbon sequestration cost-effectiveness studies relates to whether there is a positive value to projects that capture carbon and store it temporarily, only to release the carbon to the atmosphere at a later date. The analysis in this study demonstrates that whenever there is a positive time value to carbon, that is carbon reductions achieved earlier are more valuable than reductions occurring later, there is a positive value to the temporary capture and storage of carbon.

2. THE BASIC MODEL

The treatment of carbon sequestration occurring at different points in time for bottom-up models can be illustrated using a simple model of production (Reilly and Richards 1993). Consider the production function

$$F(t) = F(G(t), E(t)) \qquad (1)$$

where F is economic production at time t, G(t) is the atmospheric stock of carbon dioxide at time t, and E(t) is the emissions of carbon dioxide at time t. Unless otherwise stated, in this model $F_G < 0$; i.e.,the marginal productivity with respect to atmospheric stocks of carbon is negative. Conceptually, this marginal productivity is the negative of the marginal damages associated with increasing atmospheric carbon stocks. Also $F_E > 0$, so the costs of emissions control are positive. To assure well-behaved functions, $F_{GG} \leq 0$ and $F_{EE} \leq 0$. $F_{EG} = 0$ provides for separability of the two variables in the function.

The rate of change of atmospheric stocks is determined by the emissions rate E and the dissipation of existing stocks.

$$\frac{\partial G}{\partial t} = E(t) - \delta G(t) \qquad (2)$$

where δ is the atmospheric dissipation rate for carbon dioxide. This is an extremely simplified model of the relation between emissions and atmospheric stocks of carbon dioxide. It does, however, capture the concept of natural removal of atmospheric carbon dioxide by terrestrial and oceanic sinks. The simplicity of the atmospheric model facilitates the focus of this analysis on the shadow price of carbon. The differential Equation (2) can be solved for the level of G at any time t.

$$G(t) = e^{-\delta t} \int_{s=0}^{t} e^{\delta s} E(s) ds + R e^{-\delta t} \qquad (3)$$

At t = 0 the first term on the righthand side is zero, so R = G(0). Equation (3) becomes

$$G(t) = e^{-\delta t} \left(G(0) + \int_{s=0}^{t} e^{\delta s} E(s) ds \right) \qquad (4)$$

The basic model in Equations (1) and (4) provides a basis for examining how the time value of carbon varies depending upon the policy context that is assumed.

3. THREE POLICY CONTEXTS

The appropriate treatment of the discount rate applied to reductions of carbon dioxide emissions depends upon the policy context for which the policy analyst is conducting a bottom-up study of costs of mitigation. The simple model expressed by Equations (1) and (4) provides the framework for examining policy contexts within which the bottom-up analysis might be conducted This section considers three policy contexts that vary according to the exogenously specified variables and constraints: (1) the path of marginal damages over time is exogenously specified; (2) the emissions path as a function of time and the damages function (economic damages as a function of atmospheric stocks) are exogenously specified; and (3) the damages function is exogenously specified and emissions follow an optimal path over time. The analysis indicates that the choice of whether and how to treat the time value of carbon emissions reductions depends very much upon the policy context for which the analysis is designed.

3.1. Path of Marginal Damages Specified

In the first policy context, the policy analyst takes the path of marginal damages, $-F_G(t)$, as exogenously specified. In this case, the present value of a unit reduction in emissions at time $t = 0$ becomes

$$MB(0) = -\int_{s=0}^{\infty} e^{-rs} e^{-\partial s} F_G(s) ds \qquad (5)$$

This expression is equivalent to that developed by Sathaye et al. (1993) for the value of a present emissions reduction. $F_G(s)$ is the negative of the instantaneous marginal damages at time $t = s$; e^{-rs} reflects the social discounting of future damages, and e^{-s} accounts for the dissipation of carbon over time. To compare the relative value of reductions at different points in time, this simple model must be extended. Assume that an action is not taken until time t. The marginal damages must be discounted to start at t and run to ∞. The dissipation, however, must be 0 at time t and run to ∞. The expression becomes

$$MB(t) = -\int_{s=t}^{\infty} e^{-rs} e^{-\delta(s-t)} F_G(s) ds = -e^{\delta t} \int e^{-(r+\delta)s} F_G(s) ds \qquad (6)$$

The questions about the path of the marginal value of carbon emissions reductions over time can now be addressed. Note that the term preceding the integral adjusts the benefits of delaying emissions because there is also a delay in the dissipation of the increment to the atmospheric carbon dioxide stock. For any given path of marginal damages, $-F_G(t)$, the marginal benefit of emissions reductions can be tested for $t = 0$ and $t > 0$ to develop the time value of reductions of emissions.

3.1.1. The Path of Marginal Damages

Consider the specific case in which the marginal damages are exogenously specified to increase at an exponential rate, α, such that

$$F_G(t) = e^{\alpha t} F_0 \tag{7}$$

where $F_0 = F_G(0)$ is the marginal productivity with respect to atmospheric carbon stocks (i.e., the negative of marginal damages) at time 0, so that $F_0 < 0$. Inserting (7) into (6) and solving yields

$$MB(t) = \frac{-e^{\delta t} F_0}{\alpha - r - \delta} \left(e^{(\alpha - r - \delta)s} \right)_{s=t}^{\infty} \tag{8}$$

The critical relation in this expression is the relative value of a, r, and . There are four ranges of value for a that should be considered.

No growth in marginal damages, $\alpha = 0$. Where marginal damages are constant, the present value of reductions in atmospheric stocks declines at the social discount rate.

Slow to moderate growth in marginal damages, $0 < a < r$. As in the case of $\alpha = 0$, later reductions in atmospheric stocks have less value than earlier reductions. The value of later reductions declines less rapidly, however, than in the case of $\alpha = 0$. In this case, the value of carbon dioxide emission or atmospheric stock reductions is discounted at a positive rate less than the social discount rate.

Marginal damages rise at the social rate, $\alpha = r$. When the marginal damages are rising at the social discount rate there is no greater benefit to achieving emission reductions sooner rather than later. Thus, the value of carbon should not be discounted for occurring later rather than sooner.

Rapid growth in marginal damages, $r < \alpha < r + \delta$ The expressions for marginal benefits and rate of change of marginal benefits is the same as for the case in which $\alpha < r$, as shown in Equation (10). The consequences, however, are different. In this range $MB(t) > MB(0)$ and $MB(t) \to \infty$ as $t \to \infty$. It follows that later reductions have higher value than earlier reductions.

3.1.2. The Value of Temporary Storage

The general expression for the value of temporary storage of carbon is

$$VS_{0,t} = MB(0) - MB(t) \tag{9}$$

where $VS_{0,t}$ is the value of storage of carbon from time 0 to time t in the future. Continuing with the example of marginal damages path specified in expression (7) above, the expression for the value of temporary storage can be derived as

$$VS_{0,t} = MB(0) - MB(t) = \left[\left(e^{\delta t} - 1 \right) e^{(\alpha - r - \delta)\infty} + \left(1 - e^{(\alpha - r)t} \right) \right] \left(\frac{F_0}{\alpha - r - \delta} \right) \quad (10)$$

As would be expected (10) provides the following general result:

$$\alpha \prec r \Rightarrow VS_{0,t} \succ 0$$

$$\alpha = r \Rightarrow VS_{0,t} = 0$$

$$\alpha \succ r \Rightarrow VS_{0,t} \prec 0$$

That is, temporary storage has positive value only if the level of marginal damages is increasing at a rate less than the social discount rate. When marginal damages are rising at a rate less than the social discount rate, so $\alpha < r$, temporary storage has a positive value because it delays damages. When $\alpha < r$, however, the benefits of delaying the onset of the damages are more than offset by the rise in the level of marginal instantaneous damages, so temporary storage has a negative value. The two effects are perfectly balanced when $\alpha = r$.

3.2. Path of Emissions and Damage Function are Specified

Suppose that emissions paths, rather than marginal damages, are exogenously specified. For example, the Intergovernmental Panel on Climate Change conducted several exercises in which it specified various emissions scenarios and asked participating analysts to explore a number of questions related to environmental impacts and costs of controls. In this case, developing an expression for the shadow price of carbon requires specifying an instantaneous damage function (Reilly and Richards).[2] It is then conceptually straightforward to derive the path of marginal damages. Taking the derivative of the damages function with respect to atmospheric stocks provides an expression for the marginal damages. Equation (4) provides the expression for translating the emissions path into a pattern of atmospheric stock over time. Combining the atmospheric stock pattern with the marginal damages expression yields a measure of marginal damages over time. While there are many plausible damage functions, this approach is illustrated below for

---

[2] The instaneous damage function is a measure of damages as a function of the atmospheric stocks of greenhouse gases.

two hypothetical damages functions - linear and quadratic (Reilly and Richards 1993).

3.2.1. A Linear Damages Function

A linear instantaneous damages function can be expressed as

$$D(t) = d_1 + d_2 G(t) \tag{11}$$

The marginal damages are a constant. Given the production function in Equation (1), $-F_G(t) = d_2$. Inserting d_2 into expression (6) provides that the marginal benefits are

$$MB(t) = e^{\delta t} d_2 \int_{s=t}^{\infty} e^{-(r+\delta)s} \, ds = e^{-rt} \frac{d_2}{r+\delta} \tag{12}$$

The linear damages function, with constant marginal damages, corresponds to the case of $\alpha = 0$, discussed in the previous section. As in that case the present value of emissions reductions is declining at the social discount rate.

$$MB(0) = \frac{d_2}{r} \quad M(t) = e^{-rt} MB(0) \tag{13}$$

The value of temporary storage for linear damages is expressed as

$$VS_{0,t} = \frac{d_2}{r+\delta}\left(1 - e^{-rt}\right) \tag{14}$$

which assumes the same sign as the marginal damages. Note that in both (13) and (14) the value of reducing emissions is independent of either emissions levels or stock levels.

3.2.2. Quadratic Damages Function

The quadratic damages function can be expressed as

$$D(t) = d_1 + d_2 G(t) + d_3 G(t)^2 \tag{15}$$

Inserting the expression for atmospheric stocks from Equation (4) into Equation (15), the marginal damages follow the path defined by

$$L - F_{G(t)} = d_2 + d_3 G(t) = d_2 + d_3 e^{-\delta t}\left(G(0) + \int_0^t e^{-\delta t} E(t) dt\right) \qquad (16)$$

Combining (6) and (16), the expression for marginal benefits becomes

$$MB(t) = e^{\delta t} \int_{s=t}^{\infty} e^{-(r+\delta)s}\left(d_2 + d_3 e^{-\delta s}\left(G(0) + \int_{x=0}^s e^{\delta x} E(x) dx\right)\right) ds \qquad (17)$$

There are an infinite number of imaginable emissions paths over time. For the sake of illustration, consider the case in which it is exogenously specified that emissions are expected to follow a path such that

$$E(t) = E(0)e^{\beta t} \qquad (18)$$

where β can assume any real value. If $\beta > 0$, emissions are growing, while $\beta < 0$ implies declining emissions levels over time. Combining (18) with (17) and solving the differential, the general expression for the value of reductions in atmospheric stocks of carbon becomes

$$MB(t) = \frac{d_2}{r+\delta} e^{-rt} + \left(G(0) - \frac{E(0)}{\delta+\beta}\right)\frac{d_3}{r+2\delta}e^{-(r+\delta)t} - e^{\delta t}E(0)\frac{d_3}{(\delta+\beta)(\beta-r-\delta)}\left(e^{(\beta-r-\delta)s}\right)_{s=t}^{\infty}$$

$$(19)$$

The path of the time value of carbon depends upon several relations, including the relative values of β, and r, the relative values of $E(0)$ and $G(0)$ and the relative values of d_2 and d_3. Because of the multiple variables, there is no simple characterization of either marginal benefits or the time value of carbon. However, consider the special case where (1) the linear term $d_2 = 0$, (2) the quadratic term $d_3 > 0$, and (3) emissions are constant, i.e., $\beta = 0$. For constant levels of emissions, Equation (19) becomes

$$MB(T) = \frac{d_3\dfrac{E(0)}{\delta}}{r+\delta}e^{-rt} + \left(G(0) - \frac{E(0)}{\delta}\right)\frac{d_3}{r+2\delta}e^{-(r+\delta)t}$$

$$(20)$$

$$MB(0) = \frac{d_3\dfrac{E(0)}{\delta}}{r+\delta} + \left(G(0) - \frac{E(0)}{\delta}\right)\frac{d_3}{r+2\delta}$$

There is no simple expression for the relation between MB(t) and MB(0) as there was in the case of a linear damages function. This indicates that the value of carbon reductions in the future should not be simply discounted at the social discount rate when there is a quadratic damages function. It is clear, however, that, for any case where E(0)/$\delta \leq$ G(0), future emissions reductions are valued less than current reductions, so some form of discounting is appropriate. For the special case in which emissions are maintained at a constant level equal to the dissipation rate, i.e., E(0)/δ = G(0), so that atmospheric stocks are also constant, (20) provides that

$$MB(t) = \frac{d_3 G(0)}{r+\delta} e^{-rt} = MB(0)e^{-rt} \tag{21}$$

Like expression (13), this is a special case of constant marginal damages. For the case of E(0)/δ > G(0), where emissions are constant but greater than the atmospheric dissipation rate, so that atmospheric levels are rising, the partial of (19) shows how the rate of change of marginal benefits depends upon the level of emissions, E(0).

$$\frac{\partial MB(t)}{\partial t} = -\frac{rd_2}{r+\delta} e^{-rt} - G(0)\frac{d_3(r+\delta)}{r+2\delta} e^{-(r+\delta)t} - \frac{E(0)}{\delta} d_3 \left(\frac{r}{r+\delta} e^{-rt} - \frac{r+\delta}{r+2\delta} e^{-(r+\delta)t} \right) \tag{22}$$

The first two of the three terms on the righthand side of (22) contribute to a marginal benefit of emissions reductions that decreases as a function of time. However, at t = 0, the third term increases the partial of MB(t). For a sufficiently high E(0), the present value of the marginal benefit of later reductions can actually be higher than for earlier reductions. At the same time, the value of that term is decreasing in time, so that eventually MB(t) will be decreasing. This stands to reason, because eventually atmospheric stocks will rise to a level such that G(0) = E(0)/δ and a steady state is achieved, at which point MB(t) = MB(0)e^{-rt}.

3.3. Optimization of Emissions Path

If it is assumed that emissions will follow an optimal path, with only a general specification of the production function, F(E(t), G(t)), the analyst can derive the shadow price of emissions reduction along the optimal path (and incidentally, the optimal emissions path). Borrowing the production function of the previous two sections, the goal of the optimization problem is

$$\underset{E}{Max} \int_0^\infty e^{-rt} F(G(t), E(t)) dt \tag{24}$$

As originally assumed, $F_{EG} = 0$. Thus the production function is separable in the two variables so that $F(G,E) = B(E)-D(G)$, where $B(E)$ is the positive contribution of emissions and $D(G)$ is the damages caused by atmospheric stocks of carbon. $B(E)$ and $D(G)$ are both monotonically increasing, and $B_E > 0$, $B_{EE} < 0$, $B(0) > 0$, $D_G > 0$, $D_{GG} > 0$, and $D(0) > 0$. Note however that B_E is not necessarily a constant. $B_E' < 0$ would suggest that the marginal value of additional emissions is decreasing overtime. If, for example, it is specified that alternative energy technologies are becoming less expensive over time, then the marginal value of emissions (and the cost of controlling emissions) may also decrease with time. With the separable production function, expression (23) becomes

$$Max_E \int_0^\infty e^{-rt}\left(B(E)-D(G)\right)dt \qquad (25)$$

In an optimal control problem, $E(t)$ represents the control variable and $G(t)$ is the state (stock) variable. From Equation (2) the state equation is

$$\dot{G}(t) = E(t) - \delta G(t) \qquad (26)$$

The Hamiltonian is

$$H = e^{-rt}\left(B(E(t))-D(G(t))\right) + \lambda\left(E(t) - \delta G(t)\right) \qquad (27)$$

In this expression $\lambda(t)$ is the shadow price associated with a one-unit increase in the atmospheric stock of carbon. It represents the marginal rise in the production function associated with a marginal increase in the atmospheric stocks. Therefore, $M(t)$ is generally expected to be negative. In order to assure an autonomous problem that can be solved over the infinite horizon, the Hamiltonian can be transformed to a current value Hamiltonian,

$$H_C = B(E(t)) - D(G(t)) + m\left(E(t) - \delta G(t)\right) \qquad (28)$$

where $m(t) = e^{rt}\lambda(t)$. The necessary conditions along the optimal path include the optimality condition,

$$\frac{\partial H_C}{\partial E} = B_E + m = 0 \qquad (29)$$

and the costate equation

$$\dot{m} = rm - \frac{\partial H_C}{\partial G} = (r+\delta)m + D_G \qquad (30)$$

as well as the state Equation (25). Because the problem does not specify the details of B(E) and D(G), there is no explicit solution. Still, a phase diagram can provide insight into the path of m(t), and hence M(t), over time. From (28) ,

$$m = -B_E = Q(E) \tag{31}$$

where $Q(E) \equiv -B_E$. Since $B_{EE} < 0$ for all E, $B_E > 0$ is monotonically decreasing and $Q(E) < 0$ is monotonically increasing. Inverting $Q(E)$, leads to the expression,

$$E = q(m) = Q(m)^{-1} \tag{32}$$

Since $Q(0) < 0$, then $q(0) > 0$. Differentiating Equations (28) and (31) leads to

$$dm = -B_{EE} dE \tag{33}$$
$$q_m dm = dE \tag{34}$$

Equating dm/dE in Equations (32) and (33) leads to

$$q_m = \frac{1}{-B_{EE}} > 0 \tag{35}$$

Inserting (31) into (25) provides

$$\dot{G} = a(m) - \delta G \tag{36}$$

Equations (29) and (35) provide a pair of differential Equations in the state variable, G, and the co-state variable, m, that can be analyzed with a phase diagram to see if the problem approaches a steady state equilibrium (Figure 1). First, setting (35) equal to zero, it follows that $q(m) = \delta G$. Since $q(0) > 0$, this line must intersect the x-axis in the range $G > 0$. According to (34), the equation must have a positive slope. Now setting (29) equal to zero, it follows that $m = -D_G/(r+\delta)$. Since $D_G > 0$, m must be less than zero for all G. This makes intuitive sense since m is the current value of increasing atmospheric carbon stocks by one unit, a value we would expect to be negative. Since $D_{GG} > 0$, it follows that the line has a negative slope. In the area west of the line $q(m) = \delta G$, G is increasing, and in the area east of that line, G is decreasing. Similarly, north of the line $m = -D_G/(r+\delta)$, m is increasing, and south of that line m is decreasing. The resulting phase diagram in Figure 1 illustrates a saddlepoint in which there is a unique value of m(t) for any given starting atmospheric stock G(t), and a unique optimal path to the steady state equilibrium, at which point

$$\dot{m} = \dot{G} = 0 \qquad (37)$$

Based on the state Equation (25), we can further speculate that given that the system will reach an equilibrium level of atmospheric stocks, say G^*, there will be a constant level of emissions $E^* = G^*$. Condition (36) is consistent with the idea that carbon benefits can be discounted once equilibrium is reached, since

$$m = e^{rt}\lambda(t) = k \Rightarrow \lambda(t) = e^{-rt}k \qquad (38)$$

where k is a constant. Again, since G(t) is constant at the equilibrium point, this is a special case of constant marginal damages described in expression (13). Equation (29) can be combined with the derivative of Equation (28) to arrive at the equilibrium condition

$$B_E = \frac{D_G}{r + \delta} \qquad (39)$$

which states that the marginal costs of reduction should equal the discounted stream of marginal damages.

It is interesting to note that from the derivative of Equation (28) it follows that

$$\dot{m} = -B_{EE}\dot{E} \qquad (40)$$

so that a constant shadow price on stocks immediately implies a constant emissions level and vice versa.

This discussion leads to the conclusion, for the optimization context, that while it is appropriate to discount carbon benefits at the social discount rate once equilibrium has been reached, the current value shadow price is not constant prior to reaching long-term equilibrium. The appropriate discount rate, if any, and whether temporary storage has value both depend upon the relative values of the marginal costs of control, B(E), the marginal damages, D(G), and the first and second derivatives of these two functions.

4. DISCUSSION AND CONCLUSIONS

The purpose of this study was to improve the understanding of the appropriate treatment of the time value of carbon in the bottom-up modeling context. The analysis has demonstrated that the time value of carbon is dependent upon the policy context that is assumed. By definition, the time value of carbon will depend upon the path of marginal damages. When marginal damages are constant over

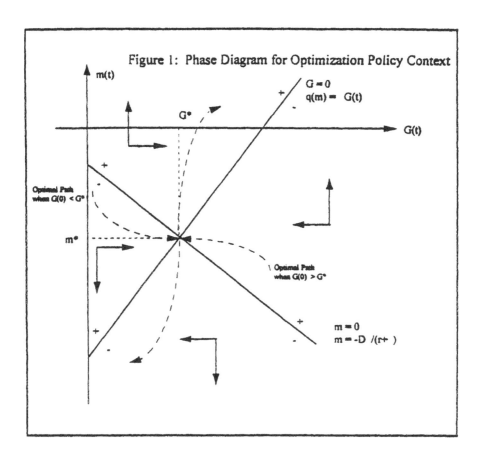

Figure 1: Phase Diagram for Optimization Policy Context

time, carbon can be discounted at the social discount rate. The more rapidly marginal damages increase, the less future carbon emissions reductions should be discounted. When marginal damages rise at a rate equal to the social discount rate, the value of future reductions does not need to be discounted.

The path of the marginal damages function is determined by the combination of an assumed instantaneous damages function and the trajectory of emissions over time. If the damages function is assumed to be linear, then marginal damages are constant, and carbon can be discounted at the social discount rate. In this case the path of emissions is irrelevant to the time value of carbon. However, if damages are nonlinear, the path of emissions is critical. Where the damages function is quadratic, for example, higher emissions levels increase the importance of future emissions reductions relative to current emissions levels.

The analysis of optimal emissions levels suggests that once the equilibrium level of emissions is reached carbon can be discounted at the social discount rate. However, until that equilibrium level of emissions is reached, the appropriate discount rate is not easily identified. It would be interesting to extend the optimization analysis by imposing policy-relevant constraints. For example, optimization subject to a constraint on emissions levels would represent the case of a politically imposed emissions cap regulation. Similarly, optimization subject to a constraint on the atmospheric stock of greenhouse gases would correspond to the commonly asserted need to limit atmospheric concentrations of carbon dioxide to less than twice the preindustrial levels.

Given the wide range of discount rates that may be appropriate, depending upon the policy context, how should bottom-up analysts approach the time value of carbon? First, whatever discount rate is adopted, analysts should be clear what assumptions they have made, and for which policy contexts their assumptions are appropriate. Second, to assure that their results are as widely applicable as possible, bottom-up cost studies should test their results for sensitivity to the time value of carbon. At a minimum, carbon discount rates should be tested for values equal to the social discount rate and zero. Presenting cost studies results in this manner alerts the policymaker who uses the study to the fact that the time value of carbon is an important issue that requires an explicit decision.

REFERENCES

Reilly, J., and K. Richards. 1993. "Climate Change Damage and te Trace Gas Index Issue." *Environmental and Resource Economics* 3:41-61, 1993.

Richards, K. and C. Stokes. 1995. "National, Regional and Global Carbon Sequestration Cost Studies: A Review and Critique." In review.

Sathaye, J., R. Norgaard, and W. Makundi. 1993. "A Conceptual Framework for the Evaluation of Cost-Effectiveness of Projects to Reduce GHG Emissions and Sequester Carbon." LBL-33859, Lawrence Berkeley Laboratory, for U.S. Department of Energy, Washington, D.C.

Critical Reviews in Environmental Science and Technology, 27(Special): S293–S307 (1997)

COERCION AND ENTERPRISE IN THE PROVISION OF ENVIRONMENTAL PUBLIC GOODS: THE CASE OF CARBON SEQUESTRATION IN THE UNITED STATES

KENNETH RICHARDS

Pacific Northwest Laboratory,* 901 D St. SW, Washington, DC, 20024, USA

ABSTRACT: Current discussions of greenhouse gas abatement strategies have focussed on the use of carbon taxes and marketable allowance schemes. Implementation of a large-scale carbon sequestration program in the United States will require broadening the scope of the instruments under consideration to include contracts, subsidies, command and control regulation and direct government production. This paper develops a taxonomy for these instruments, examines the factors that affect their relative attractiveness, and applies that analysis to carbon sequestration in the United States.

KEY WORDS: climate change, carbon sequestration, policy instruments, taxes, marketable allowances, regulation, contracts, subsidies, government production.

1. INTRODUCTION

Carbon sequestration presents both an opportunity and a challenge for the environmental policy maker. In the United States, including sequestration in a carbon emissions stabilization program provides an opportunity to reduce the costs of stabilization by as much as 80 percent relative to a system that addressed fossil fuels only (Richards, et al., 1993). The savings could amount to tens, or even hundreds, of billions dollars over three to four decades.

Given the potential of carbon sequestration as a cost-effective component of a mitigation strategy, which policy instruments are best suited to the establishment of a large-scale carbon sequestration program in the United States? Although many studies document the low cost of carbon sequestration relative to energy-related emissions abatement (e.g., DOE, 1991; IPCC, 1995), the subject of instrument choice for a carbon sequestration program has not been commonly considered. Where the subject is even mentioned, it has simply been assumed that carbon sequestration would be promoted as an appendix to a carbon dioxide source control program (e.g., Marchant, 1992; Hahn and Stavins, 1993).

*   Pacific Northwest Laboratory is operated for the U.S. Department of Energy by Battelle Memorial Institute under contract DE-AC06-76RLO 1830.

Carbon sequestration is not a typical emission abatement strategy, however. The removal of the pollutant, carbon dioxide, occurs after emission to the atmosphere, at a different site, and by someone other than the emitter. On-site field measurement is expensive, particularly for smaller operations, while the potential land area involved in a carbon sequestration program is extensive. Carbon sink models are highly uncertain. There may also be significant leakage problems. Many carbon sequestration practices require large initial investments (i.e., specialized assets) while most of the benefits are delayed for decades. Moreover, even the goal of a carbon sequestration program -removal of carbon dioxide from the atmosphere - is subject to change over time. Perceptions about the seriousness of the threat of global warming may increase or decrease as new scientific evidence becomes available.

The combination of characteristics in a carbon sequestration program presents a particular challenge for the selection of instruments to implement a large-scale carbon sequestration program in the United States. While the particular combination of all of these features in carbon sequestration are unusual, none of the characteristics is in itself unique. The purpose of this study is to develop a normative analysis of instrument choice that is broad enough to accommodate the many technological, legal and programmatic factors that are commonly ignored in the instrument choice literature, but that are fundamental to the development of a carbon sequestration program and other non-standard emissions abatement applications. The analysis will demonstrate that the common assertion that a carbon sequestration program will work as an appendix to a tax or marketable allowance system for sources is based on an over-simplified view of sinks and an overly narrow definition of candidate instruments. Instead, a cost-effective sequestration program will most likely involve some combination of command and control regulation, government production, and subsidies or contracts. This conclusion arises from an understanding of how the key characteristics of the carbon sequestration practices affect the choice of an efficient policy instrument.

The next section presents a taxonomy of instruments that recognizes two modes of operation: the coercive mode, generally analyzed in the environmental instrument choice literature, and the enterprise mode, generally examined in the transaction cost/contracts literature. The taxonomy facilitates consideration of the fundamental differences between the instruments in general, and the choice of operational mode and degree of government control in particular. Section 3 examines how the degree of government control over the production of pollution abatement affects production costs and program implementation costs. Section 4 examines those factors that determine the preferred operational mode -deadweight loss, legal constraints and political constraints. Section 5 provides a summary of the critical factors that determine the relative merits of each instrument for specific applications and suggests that a large-scale program will likely involve some combination of command and control regulation, government production and contracts or subsidies.

2. AN EXPANDED RANGE OF POLICY INSTRUMENTS

The environmental instrument choice literature has concentrated its attention on demonstrating the merits of taxes and marketable allowances. Generally, these two instruments are compared either to each other or to command and control regulation. Some studies have mentioned a few alternatives to these instruments (see, e.g., Hahn, 1989), but have done little to explain the circumstances under which a regulator might choose these alternatives. In particular, few environmental instrument studies have addressed the possibility of the government undertaking environmental protection directly, rather than through taxes, marketable allowances or regulation. For example, the current draft of the Second Assessment Report of the Intergovernmental Panel on Climate Change (IPCC, 1995) provides little more than passing mention of government direct investment in climate change mitigation.

An expanded notion of instrument choice should recognize two government modes of operation: the enterprise mode and the coercive mode (Figure 1). Government enterprise refers to that operational mode in which the government makes direct financial commitments or investments. To the extent that private firms are involved, their participation is voluntary. In contrast, the coercive mode entails mandatory participation by private firms.

Operating in its enterprise mode to abate pollution, the government can use subsidies, contracts or internal production. Here, subsidies are taken to be general offers to pay a fixed price per unit of pollution abatement to any and all takers. In contrast, a contract approach to production involves the government in setting a goal for the amount of abatement it intends to achieve, and searching out the least expensive suppliers of reductions through competitive bids or some other market mechanism. Employing the government production option, the program builds on existing and expanded organizational structures in the government to carry out abatement programs.

The coercive mode entails the more familiar environmental instruments—taxes, marketable emissions allowances, and command and control regulation. It is assumed throughout this discussion that revenues from either taxes or auctioned permits are recycled to reduce other distortionary taxes. Unlike the incentive-based instruments, command and control regulation involves the government in choosing specific technologies for regulated firms.

The horizontal dimension in Figure 1 represents the degree to which the government retains control over production decisions. On the extreme right are the instruments (government production and command and control regulation) which, in their purest form, vest all control in the government. These are hierarchical arrangements. At the other extreme, production decisions are left entirely to private parties under taxes, marketable allowances, subsidies and contracts. The government monitors only output.

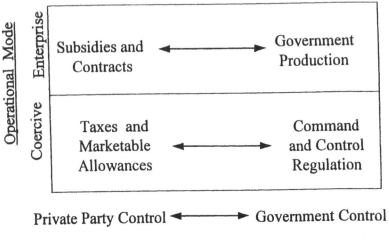

Figure I: Taxonomy of Policy Instruments

Operational Mode

Enterprise — Subsidies and Contracts ⟷ Government Production

Coercive — Taxes and Marketable Allowances ⟷ Command and Control Regulation

Private Party Control ⟷ Government Control

Extent of Control over Production Decisions

In both operational modes there are a host of intermediate instruments. Within the coercive mode, performance standards represent a modification from technology-specific regulation - one that provides firms with discretion with respect to *how* to make reductions, but not *where* the reductions are made. Marketable allowance programs that require government approval of each trade, or that restrict who may hold and trade allowances, introduce hierarchical elements into that mechanism. With respect to the enterprise mode, Stinchcombe (1985) examined a number of contractual structures that are commonly incorporated into contracts to achieve hierarchy-like outcomes.

This taxonomy helps to place instruments in their proper relation to each other. Most noticeably, subsidies, which are generally included in the instrument choice literature, are properly grouped with contracts, which are generally excluded from those studies. Subsidies and contracts are simply two different mechanisms for achieving the same goals. The instruments on the left-hand side of Figure 1 can be further differentiated according to whether instruments are price-based or quantity-based (Figure 2). Intuitively, in their pure form subsidies and contracts have the same duality relation as taxes and marketable allowances. Under perfect information either one can be used to meet a specific emissions abatement goal at the same marginal price (Weitzman, 1974).

The object of the instrument choice exercise is to minimize the social cost of implementing a given environmental policy. Consider a program to reduce emissions of a specified pollutant from their current level of E, to some target level Q. The cost of using instruments from each of the four corners of Figure 1 can be expressed as the sum of three components: (1) the cost of producing the emission

Figure II: Taxonomy of Incentive-Based Instruments

	Price-Based	Quantity-Based
Enterprise	Subsidies: 1. Payment 2. Tax credit	Contracts: 1. Payment 2. Allowance Credit
Coersive	Taxes	Marketable Allowances

(Operational Mode — vertical axis label)

reduction, (2) the cost of implementing the program, and (3) the welfare cost of raising the revenue to pay for any government expenditures under the program. The costs for each of the instruments are

Taxes/Auctioned Allowances :
$$C_T = \int_Q^E M_T(X)\,dX + \tau_T Q - DQM_T(E-Q) \tag{1}$$

Command and Control Regulation :
$$C_R = \int_Q^E M_R(X)\,dX + \tau_R(E-Q) \tag{2}$$

Government Production :
$$C_G = \int_Q^E M_G(X)\,dX + \tau_G(E-Q) + D\int_Q^E M_G(X)\,dX \tag{3}$$

Subsidies/Contracts :
$$C_s = \int_Q^E M_s(X)\,dX + \tau_s(E-Q) + D(E-Q)M_s(E-Q) \tag{4}$$

where $M_i(X)$ is the marginal cost of pollution abatement using instrument i when emissions levels are X; τ_i is the (constant) marginal transaction cost associated with instrument i; and D is the marginal cost of public funds (a measure of the deadweight loss associated with increased revenue raising).

In each of the four cost equations the first component on the right hand side expresses the cost of producing emissions abatement, the integral of the marginal cost of control. The second component expresses the implementation costs. The third component expresses the deadweight loss associated with the emissions reduction program.

There are several assumptions implicit in (1) to (4). First, the function for the marginal cost of emissions control is instrument-specific; the instruments will differ with respect to the behavioral responses they induce, the choices of pollution control practices that occur, and the incentives they provide. Thus, the production cost of reducing emissions from E to Q will depend upon the choice of instrument. Second, consistent with Kohn (1991), the implementation costs associated with command and control regulation, subsidies and contracts, and government production are proportional to the amount of emission reduction, E-Q. The implementation cost of a tax or marketable allowance program is taken to be proportional to the level of uncontrolled emissions, Q. Third, the equations reflect the deadweight loss associated with the direct payments only. They do not include deadweight loss for the government's implementation costs. Consequently, there is no deadweight loss for the command and control regulation because the government makes no direct payments with that instrument. Note that one of the consequences of these three assumptions is that the relative attractiveness of the instruments is dependent upon the level of the target emissions. In general, decreasing the level of target emissions, Q, tends to favor taxes and marketable allowances because it reduces the implementation costs.

The combination of the new taxonomy and the simple cost model illustrates that instrument selection requires choices of both the degree of government control over production decisions and the government's operational mode. These two issues are discussed in the next two sections.

3. GOVERNMENT CONTROL: PRODUCTION AND TRANSACTION COSTS

Whichever mode of operation the government chooses, it must decide whether it will retain control over the production of emissions abatement through direct government production and command and control instruments, or allow private firms discretion with respect to how the emissions reductions are made. The decision is directly analogous to a private firm's "make or buy" decision about whether to produce goods and services directly or to contract with other firms for production. In the environmental instrument choice literature the decision is generally cast as a choice between incentive-based instruments and command and control regulation, with a strong bias toward the former. The contracting/transaction cost literature, however, demonstrates that there are circumstances under which government retention of control is the cost-minimizing option. Moreover, intermediate instruments may have efficiency-enhancing attributes when compared to either pure incentive-based or pure hierarchical instruments.

3.1. Production Costs and Incentives

The first component on the righthand side of Equations (1) to (4) represents the direct cost of producing emissions reductions. The structure of the equations reflects the assumption that this cost varies among the instruments. While command and control regulation and government production rely upon the government's ability to identify low-cost technological options for pollution abatement, the incentive-based instruments shift that responsibility to private sector parties.[2] The consequence is that these private parties will search for technologies and practices that are well suited to their specific applications and will adjust the level of control of pollutants to reflect the price signals of the market place. Further, stable incentive-based instruments induce dynamic cost-effectiveness by promoting appropriate technological innovation to minimize costs over the long run.

The fact that incentive-based mechanisms tend to lower production costs is recognized by both the environmental instrument choice literature (see, e.g., IPCC, 1995; Hahn and Stavins, 1993; Hahn, 1989) and the transaction cost/contracting literature (see, e.g., Williamson, 1985; Stinchcombe, 1985). It is seldom acknowledged in the literature that the relative advantage of the incentive-based instruments is technology-specific, dependent in part upon the government's inability to identify the best technology or practice for each application. Where the range of technologies is wide, or the applications are varied, local discretion may serve to significantly reduce costs. For example, Hahn and Stavins (1993) point out that in the case of air pollution, "the cost of controlling a unit of a given pollutant may vary by a factor of 100 or more among sources." If the technology of choice is relatively easily identified and uniform across applications, however, the advantage of incentive-based instruments is less than if there is great variety in the technology and the choice of optimal applications is site-specific. Similarly, if the nature of the technology is such that there are not likely to be major cost-reducing innovations, then incentive-based instruments are less valuable.

3.2. Implemantation Costs

The second component in cost Equations (1) to (4) reflects the cost of implementing each of the instruments. A normative conclusion from the contracting literature is that allocating more discretion to private parties in the production process should be favored when innovation is important, the goal is well defined, and outputs can be measured or inputs are good proxies for output. Government production should be favored when asset specificity is high, the government is unable to credibly

---

[2]   This assumes the incentive-based instruments are based on actual output or some reasonable proxy for output.

commit not to expropriate rents, uncertainty is high, goals are ill-defined or changeable, and measurement is particularly costly.

In fact, the analysis does not necessarily lead to a choice of one or the other of the two options. The contracting/transaction cost economics literature recognizes the existence of a range of instruments between classical contracts/subsidies and pure government production. In the interior of that range the government and private firms share control over production decisions (Figure 1). The goal is to find the arrangement that minimizes the sum of production and implementation costs. These intermediate mechanisms can be designed to overcome potential difficulties related to measurement, asset specificity, and uncertainty/adaptability. These specialized governance mechanisms are costly, however, and may be less attractive than either incentive-based contracts with that are unable to accommodate investment in highly specialized assets, or government production, which sacrifices desirable incentive characteristics.

The instrument choice literature, concentrating on the coercive mode, has examined the transaction costs associated with monitoring and enforcement. What has not been as well developed in the environmental policy studies is an understanding of the tension between the goals of (1) encouraging private sector investment in the specialized assets that most cost-effectively abate pollution, (2) providing the private sector the flexibility and discretion it needs to develop and specialized, high-efficiency technologies, and (3) providing the government the flexibility to respond to changes in knowledge, technology, states of the world and preferences.

The contracting literature has provided a much fuller analysis of the factors that make it so difficult to balance the three goals, and in particular the importance of opportunistic behavior. If the regulator and regulated industries fully shared information, including information regarding uncertainty about the knowledge, technology, state of the world and regulatory preferences, they could ascertain efficient investments in emissions reducing technology, and agree to share risks in a scheme to maximize net social benefits. This arrangement would not only lead to the choice of the most cost-effective technologies given current emission targets, but also would modify investment strategies to account for potential changes in targets and other constraints.

There are two problems with this scenario. First, of course, it assumes a level of information exchange that could involve significant transaction costs. Second, the scenario ignores the role of opportunistic behavior. The fact that industry withholds information in the regulatory process is well documented. More pernicious, perhaps, is the opportunistic behavior by the government invited by private investment in specialized assets. Private firms can and should incorporate legitimate regulatory uncertainty in their investment decisions. This lowers the cost of efficiency enhancing adaptive changes in the goals of government programs. However social welfare decreases when investors alter their investments out of fear of opportunistic rent-seeking behavior by the regulator. Changes in the direction of the regulation should be based on reasonably well-defined economic and scientific criteria, not government opportunism. In EPA's emission trading program, fear of expropriation may have impeded trading, and by implication cost-effective investment (Hahn and Hester, 1989). Private firms recognized that they were revealing new cost information when they undertook interfirm trades of emissions credits. This invited confiscation by state regulators who where required to meet regional air quality standards. The ambiguity about the treatment of tradeable allowances in the regulatory process did nothing to allay investor fears.

Both the enterprise and coercive modes of operation can involve intermediate governance mechanisms. This hints at the parallelism between the two modes. Just as the government may undertake internal production involving specialized assets if it cannot make credible commitments against rent-seeking behavior, the government may rely on regulation if it can not credibly commit in the context of a tax or marketable allowance program. In both cases the basic problem is the same: how to create investment in specialized assets while retaining the adaptability to deal with uncertainty (Hahn and Hester, 1989). Identifying the role of opportunistic behavior and differentiating among types of uncertainty could be as useful in the environmental policy application as it has been in the study of contracts.

4. CHOICE OF MODE: COERCION OR ENTERPRISE

Equations (1) to (4) in the previous section generally demonstrate a lower dead-weight loss associated with the coercive mode than the enterprise mode. However, both legal and political constraints may prevent the government from employing the coercive mode. In some cases the government will use mixed modes of operation, combining instruments from both the coercive and enterprise mode.

4.1. Deadweight Loss and the Choice of Mode

Because taxes on labor, capital and consumption have distortionary effects on prices, the act of government revenue raising has a significant social cost. The extent of the distortion depends upon the precise nature of the tax and the characteristics of consumer preferences. For a given environmental tax inclusion of the revenue recycling effects will lower the apparent social cost of the program, though the tax will not generally pay for itself (Goulder, 1994).

Terkla (1984) demonstrated that the reductions in deadweight loss associated with revenue recycling from taxes gave them a substantial advantage relative to regulatory approaches. Terkla's analysis can be usefully expanded to include the instruments from the enterprise mode. The third component of Equations (1) to (4) is intended to capture the social cost of the government's revenue raising activities associated with the program. Equation (1) indicates that revenue recycling reduces the cost of using taxes or auctioned allowances by an amount equal to the product of the marginal cost of public funds and the actual amount of revenue raised by the program. In contrast, subsidies and contracts require raising additional revenue through distortionary taxes, and so impose greater deadweight losses. With respect to the deadweight loss effects, the relative advantage of taxes increases as the level of uncontrolled emissions, Q, rises. Because there no transfer payments are associated with command and control regulation, no deadweight loss impacts are assumed, as reflected in Equation (2). The deadweight loss associated with government production, Equation (3), may or may not be greater than that associated with the use of subsidies, depending upon the relative slopes of the production curves and the amount of reduction, E-Q, associated with the program.[3]

4.2. Legal Constraints

Deadweight loss considerations tend to favor the coercive mode in general, and taxes and marketable allowances in particular. There may be legal or political constraints on the use of these instruments, however. The United States Federal government employs the enterprise and coercive modes under different constitutional powers. The enterprise mode is carried out under the spending powers, while the coercive mode invokes the taxing and commerce (regulatory) powers. Despite the fact that they are both considered coercive powers, the regulatory and taxing powers face different legal constraints. The federal regulatory power is constrained by the Fifth Amendment of the U.S. Constitution from taking private property without compensation. The body of case law known as "regulatory takings" defines, albeit imperfectly, the circumstances under which a regulation acts as if to take private property, and thus require compensation for the loss.

Until recently, the "harm-benefit" rule of thumb has been a surprisingly reliable predictor of court decisions in takings cases. Under this test, if a regulation acts to prevent a "harm" it is a legitimate exercise of the regulatory power. But if a regulation requires the creation of a public "benefit", compensation is required, raising the activity to a public enterprise. This rule of thumb still leaves ample room for dispute about what constitutes a harm or benefit, but it does serve as a starting point for determining which of the two modes, coercive or enterprise, the government would use to implement a program. In the past decade, however, there has been a conservative direction to the Supreme Court's decisions on regulatory takings, broadening the circumstances under which the Court will find a taking to

---

*   This analysis is based on the assumption that there is no interaction between taxes, ignoring the complications of policy making in a second best world (for a discussion of those interactions see Goulder, 1994). Future research should incorporate the recent results of work on environmental taxes and revenue recycling. The results of that work are unlikely to change the basic relations depicted in equations (1) to (4).

have occurred. The current Congress is considering widening those circumstances further still.[4]

In contrast to the limits on the regulatory powers, there is relatively little constitutional limit to the use of the taxing power in the United States. Even where the motivation of an excise tax is clearly regulatory, and the effect of the tax is to render a business or property completely unprofitable, the Court will not find that the tax works a taking, provided that the tax raises revenue.

The differential treatment of actions under the taxing power and the regulatory power suggests that one instrument may be applied in circumstances where the other cannot. It also raises the interesting question of how auctioned marketable allowances would be treated by the Court. Since allowance programs involve direct limitations on emissions, and air pollution has always been controlled under the regulatory powers, a marketable allowance program may be subject to review as regulation. At the same time, if allowances are auctioned they clearly have revenue raising properties and may be controlled under the taxing powers. A new marketable allowance program might enjoy unfettered freedom reviewed under the taxing power, or run into the regulatory takings net, newly refined by Congress to screen out further intrusions on private property.

4.3. Political Constraints

Legal constraints are not the only limit on the use of particular powers and modes of operation. Consider the case of the tax credit, a mechanism that can be almost identical to the pure subsidy, but for two factors. First, for the tax subsidy to have any effect, the targeted parties must actually be paying taxes. And because of differential incentives among private parties (resulting from differences in marginal tax rates), tax credits represent a departure from a pure subsidy, and hence are likely less efficient instruments. In the absence of some efficiency enhancing (transaction cost reducing) justification for the departure, one must question the merit of tax credits. This leads to the second way in which tax credits differ from subsidies: tax credits and subsidies are generally initiated by different legislative committees that represent different sets of political options. It is important to understand the political constraints that direct the government toward one instrument or another. Serious political constraints may also arise as affected private parties resist the use of taxes or auctioned allowances to control emissions. The redistribution of wealth from a concentrated interest group may be difficult.

Measurement issues can also interact with political resistance to limit the use of taxes or tradeable allowances. Even if the government can develop estimates

---

4   For example, the House of Representatives recently passed the Private Property Protection Act of 1995 (H.R. 925) that, if enacted into law, would significantly expand the circumstances under which compensation would be required for reductions in private property values caused by new regulations.

that relate specific actions to changes in emissions levels, environmentalists may be reluctant to accept them. Experience with the development of the SO_2 emissions trading program, where continuous emissions monitoring was required, rather than allowing use of less expensive (albeit somewhat less precise) emissions factors, suggests that environmental groups may block adoption of any but the most precise measurement methods. Where those measurement methods are prohibitively expensive, the political resistance may be sufficient to shift the focus to one of the other instruments or a significantly modified version of taxes or tradeable allowances.

5. APPLICATION TO CARBON SEQUESTRATION

As discussed above, the choice of instruments for a carbon sequestration program includes the two extremes of surrendering complete discretion over production to private parties (auctioned marketable allowances, taxes, subsidies and contracts), or retaining all discretion for the government (command and control regulation and government production), and many intermediates between the two (e.g., performance-based standards, government approval requirements for individual trades,

TABLE 1
Summary of Factors Affecting Optimal Instrument Choice

	Instrument			
	Command and Control Regulation	Taxes and Marketable Allowances	Government Production	Subsidies and Contracts
Deadweight Loss	+	++	−	−
Level of Abatement	?	+	?	−
Implementation Cost Factors				
Asset Specificity	+	−	+	−
Program Uncertainty	+	−	+	−
Cost of Output Measurement	+	−	+	−
Production Cost Factors				
Diversity of Applications	—	+	—	+
Innovation Potential	−	+	−	+
Legal Factor				
Benefit Providing	−	+/?	+	+

contracts with hierarchical elements). These should allow tailoring the instruments to meet the specific needs of designing and implementing an effective program to enhance carbon sequestration. The summary provided by Table 1 indicates the relation between several programmatic and technology characteristics and the relative attractiveness of the individual instruments. A plus sign indicates that an increase in that factor increases the potential applicability of the instrument, while a minus sign indicates lower applicability.

Taxes and auctioned marketable allowances generally involve lower production costs and lower deadweight losses than the other instruments. The structure of these instruments in the context of a carbon sequestration program, however, is not immediately apparent. Would an allowance be required for all carbon emissions from forests? Would there be a tax on agricultural land to induce conversion to forests? Even if the considerable political obstacles to implementing these instruments could be overcome, significant technical and behavioral issues would remain. Most obvious, the measurement of carbon fluxes across the hundreds of millions of hectares of forestland in the United States would be crude at best and invite paralyzing litigation. Moreover, there would be considerable uncertainty associated with the program that would encourage cost effective investments instead. For example, agricultural landowners subject to a tax might decline to make the investment in converting to forestland, hoping to wait for the demise of the tax instead.

Rather than rely on an unworkable tax or allowance system, it may be more cost effective for the government to produce carbon sinks directly, taking advantage of its considerable holdings of federal land and the expertise embodied in the U.S. Forest Service. Internal production would allow the government to learn about this new abatement technology and develop tools for monitoring and evaluating performance of sequestration practices. Because it would be employing a hierarchical system, adjustments to the program could be made without extensive government renegotiations with private parties. As it gains experience with carbon sequestration practices and exhausts opportunities on federal land, the government may look to private landowners for expansion of carbon sinks. While the deadweight losses associated with command and control regulation are relatively low, takings law in the United States would likely bar the government from using its regulatory powers to compel landowners to create new sinks. Instead, the government could employ incentive-based instruments from the enterprise mode. Rewards could be in the form of direct payments, or if the sink is linked to a coercive mechanism applied to emissions sources, the payment could be made in the form of tax credits (in the case of a subsidy program) or emissions allowances (in the case of contracts). However, the exchange ratio between the source and the sink should not necessarily be 1:1. In addition to a number of technical issues related to risk, measurement and other implementation costs (Malik et al., 1993), the appropriate ratio should also reflect the differences in deadweight loss associated with the coercive and enterprise mode instruments.

In general, a significant economy of scale is associated with measuring carbon sequestration (Winrock, 1995). If the government does use output-based subsidies or contracts, it may have to set a minimum size restriction on eligible lands. If the program includes smaller private holdings of land, it may be most efficient to reward inputs (e.g., hectares of land planted) or another easily observable, albeit very crude, proxy for the amount of carbon sequestration.

Finally, as mentioned above, the program must account for the possibility that subsidies for new forests will induce owners of pre-existing forests to decrease the size of their holdings in anticipation of competition from the newly established timberstand (Alig et al., this issue). While constitutional restrictions may prevent the use of the regulatory powers to mandate creation of new sinks, prohibitions on conversion of forestland and regulations requiring replanting following harvest would likely be construed as harm-preventing rather than benefit-providing. Hence, the government should consider complementing government production and private contracts/subsidies with command and control regulations designed to prevent leakage.

REFERENCES

Alig, R., D. Adams, B. McCarl, J. Calloway, and S. Winnett. 1995. "Assessing Effects of Global Climate Change Strategies with an Intertemporal Model of the U.S. Forest and Agriculture Sectors." *Environmental and Resource Economics*, this issue.

Department of Energy (DOE). 1991. *Limiting Net Greenhouse Gas Emissions in the United States.* DOE/PE-0101.

Intergovernmental Panel on Climate Change. 1995. *Second Assessment Report, Working Group III.*

Goulder, L. 1994. "Environmental Taxation and the double Dividend: A Reader's Guide." Working Paper 4896, National Bureau of Economic Research, Inc. Cambridge MA.

Hahn, R.. 1989. *A Primer on Environmental Policy Design.* Harwood Academic Press,, Chur, Switzerland.

Hahn, R., and G. Hester. 1989. "Where Did All the Markets Go? An Analysis of EPA's Emissions Trading Program." *Yale Journal on Regulation* 6(1):109-153.

Hahn, R., and R. Stavins. 1993. "Trading in Greenhouse Permits: A Critical Examination of Design and Implementation Issues." An abstract paper of the John F. Kennedy School of Government, Harvard University, June, 1993.

Kohn, R. 1991. "Transactions costs and the optimal instrument and intensity of air pollution control." *Policy Sciences* 24:315-332.

Malik, A., D. Letson, and S. Crutchfield. 1993. "Point/Nonpoint Source Trading of Pollution Abatement: Choosing the Right Trading Ratio." *American Journal of Agricultural Economics* 75:959-967.

Marchant, G. "Global Warming: Freezing Carbon Dioxide Emissions: An Offset Policy for Slowing Global Warming." 22 Envtl. L., 623 (1992).

Richards, K., D. Rosenthal, J. Edmonds and M. Wise. 1993. "The Carbon Dioxide Emissions Game: Playing the Net." Presented at the 69th Western Economic Meetings, Lake Tahoe, NV, June 1993.

Stinchcombe, A. 1985. "Contracts as Hierarchical Documents." In *Organization Theory and Project Management*. Ed. Stinchcombe, A. and C. Heimer, Norwegian University Press.

Terkla, D. 1984. "The Efficiency Value of Effluent Tax Revenues." *Journal of Environmental Economics and Management* 11:107-123.

Weitzman, M. 1974. "Prices *vs*. Quantities." *Review of Economic Studies* 41(4):477-491.

Williamson, O. 1985. *The Economic Institutions of Capitalism*. The Free Press, New York.

Critical Reviews in Environmental Science and Technology, 27(Special): S309–S321 (1997)

A DYNAMIC MODEL OF FOREST CARBON STORAGE IN THE UNITED STATES DURING CLIMATIC CHANGE

BRENT SOHNGEN and ROBERT MENDELSOHN

Yale School of Forestry and Environmental Studies, 360 Prospect Street, New Haven, CT 06511, U.S.A.

ABSTRACT: As carbon dioxide increases and climate changes over the next century, will ecosystems react by storing more carbon or by releasing it? While many ecosystem models predict that natural forest systems ultimately will store more carbon, transient scenarios predict that they will release carbon during the transition. In this paper, we integrate ecosystem and management responses to a climate scenario and predict that increased amounts of carbon will be stored in both the short run and the long run. These increases will come from two areas: adaptive responses of forest managers and carbon storage within the economic system.

KEY WORDS: climatic change, dynamic optimization, carbon, forest sector models

1. INTRODUCTION

The link between forests, the carbon cycle and climatic change has been discussed throughout the literature. The role of forests in climatic change is interesting because forests are at the same time an important source of atmospheric carbon through deforestation, particularly in the tropics, and a potential sink of carbon from the atmosphere through plant growth. Understanding and quantifying ecosystem processes in order to model carbon flows into and out of forests has become a major scientific endeavor on a global scale.

Dixon *et al.* (1994), for example, show that, the net flux of carbon from forests across the globe is approximately 1.3-1.5 Pg (10^{15}) of Carbon each year, primarily as a result of deforestation. That study, plus King and Neilson (1992) and Neilson (1993), show that global net emissions of CO_2 could range from 3.4-6.0 Pg per year during climate change. Neither of these studies, however, captures how humans will adjust and adapt to climate change by harvesting, replanting, or enhancing silvicultural investment, or by storing additional carbon in forest products.

In order to capture the adjustment of economic systems to climate change, we must explicitly link ecological models with economic models, as suggested by Binkley and Van Kooten (1994) and Joyce *et al.* (1995). Our model, however, extends this earlier work in two ways. First, we incorporate both changes in productivity of ecosystems, as well as changes in ecosystem distribution. Second, we develop and utilize an explicitly dynamic economic model of timber markets in the United States. This model captures how timber markets adjust and adapt to climate change in an optimal fashion.

The objective of this paper is to compare the results of a linked, dynamic ecosystem and economic model with a "natural change" model of ecosystem response for the United States. Our paper has the following structure. First, we describe the steady-state climate and ecosystem models employed in this analysis. Then, we develop dynamic climate scenarios for the continental United States using the United Kingdom Meteorological Office (UKMO) Global Circulation Model (GCM) predictions of climate change under a doubled CO_2 scenario (Wilson and Mitchell, 1987), and transient runs from the IPCC's Scientific Assessment (1990). From this, we develop dynamic, "natural change" ecological scenarios which are used to predict carbon flows. Finally, we link the ecological models with a dynamic optimization model of U.S. timber markets to show how the timber industry in the U.S. will adapt to climatic change, thereby minimizing both the economic loss and the carbon loss.

2. STEADY-STATE RESPONSE

Both the climatological and the ecological data for this paper are derived from the Vegetation/Ecosystem Modeling and Analysis Project (VEMAP), a multiple party effort to assess the sensitivity of terrestrial ecosystems and vegetative processes to climate change (VEMAP Participants, 1995). Steady-state models of ecosystems are driven by large GCM's, which form the basis for predicting the response of ecosystems to climatic change. GCM's predict steady-state climatic variables for grid cells across the surface of the coterminous United States at a .5° × .5° resolution. In this paper, we focus on only one GCM, the UKMO model, which predicts a 6.73°C temperature change and a 15 percent increase in precipitation on average across the United States.

Two types of ecological change models are considered: biogeographic distribution models (geographic models) and biogeochemical cycle models (ecosystem production models). Although not explicitly linked, when combined, these models determine both the redistribution of ecosystems and a change in ecosystem production expected to occur under doubled CO_2 and a change in climate. We consider two combinations, the BIOME2 geographic model (Prentice et al., 1992) and TEM ecosystem production (Melillo et al., 1993) combination, and the MAPSS geographic (Neilson et al., 1992) and BIOME-BGC ecosystem production (Running and Gower, 1991) combination.

Steady-state geographic and ecosystem production changes for these are shown in Table 1. The first column provides an estimate of the relative change in geographic distribution of each ecosystem type due to climate change. This tells whether or not the ecosystem type area grows or declines during climatic change. Additional forest area for any particular type may come from other timber types or from currently non-forested areas. Column 2, then, shows how much of the original area of each ecosystem type converts to something else during climate change. This conversion can be to another forest type or to non-forest altogether. The change in total steady-state biomass per acre, for each forested ecosystem type in these two model combinations is given in Column 3 of Table 1.

TABLE 1
Steady-State Changes in Forest Area and Ecosystem Production Under Two Combinations of Ecological Models and the UKMO GCM Scenario of a Doubled CO_2 Climate. All Changes are Relative to the Initial Quantity.

	Area Change		Biomass Change
	Relative Forest Size after Climatic Change	Amount of Land to Change (proportion)	Change in Total Biomass (prop. per acre)
BIOME2 and TEM			
Northern Softwoods	0.12	0.91	0.05
Northern Hardwoods	0.58	0.97	0.11
Southern Mixed	2.30	0.07	0.11
Western Pine	0.60	0.86	0.27
PNWW Conifer	3.64	0.00	0.31
MAPSS and BIOME-BGC			
Northern Softwoods	0.26	0.94	−0.35
Northern Hardwoods	0.24	0.99	0.16
Southern Mixed	1.54	0.31	−0.16
Western Pine	1.13	0.26	−0.11
PNWW Conifer	0.55	0.56	−0.03

3. DYNAMIC RESPONSE

3.1. Dynamic Climatic Response

In order to integrate ecological models with an economic model, we must develop dynamic scenarios of climatic change. Following the IPCC's Scientific Assessment prediction that CO_2 will double by the year 2060, we assume that temperature and precipitation will increase linearly over that time period (IPCC, 1990). Our dynamic scenario assumes that climate then stabilizes.

3.2. Dynamic Ecological Scenarios

3.2.1. Dynamic Ecosystem Change

Although these models are not explicitly transient, we utilize the dynamic structure suggested by King and Neilson (1992), Neilson (1993), and Smith and Shugart (1993). We assume that all ecological change occurs proportionally to temperature and precipitation changes.

For geographic change, two separate dynamic processes must be considered. First, for forested areas that are changing from one cover type to another (either forest to forest or forest to non-forest), we assume that dieback occurs because ecological conditions change enough to make the old type unfit to continue growing in its original area. Dieback occurs in any area that converts to something else during climate change.

Second, for areas that convert into a new type (either from other forest types or from non-forest types), new species must migrate into the area. While regeneration of new ecosystem types occurs proportionally to changes in temperature and precipitation, lags, which vary regionally, are incorporated to account for limitations in the ability of species to migrate quickly.

As CO_2 in the atmosphere increases and as climate changes, plant productivity will change. This leads to changes in ecosystem production, which are translated into changes in total ecosystem biomass. Ecosystem biomass is related to carbon storage.

3.2.2. Dynamic Carbon Accounting — "Natural Change" Model

With this, a transient path of carbon storage in forest ecosystems, which ignores any human adaptation, is estimated. The pulse of carbon to the atmosphere predicted by "natural change" models is based on the difference between the rate of dieback and decline and the rate of regeneration. Because climate changes rapidly, dieback will occur relatively early, while regeneration may take longer because of tree growth functions and migration lags. In addition, ecosystem production may go up or it may decrease, leading to additional changes in carbon storage. Below, we describe how we account for forest carbon storage in this "natural change" model.

The "natural change" model includes both a soil and vegetation component in order to make the results comparable with the linked, dynamic economic model. The initial area of timberland is taken from Powell *et al.* (1993), and we assume that the proportion of land in timber remains constant. Land use in reality will depend on both economic and ecological considerations, with which we deal later.

We assume that each piece of land that converts from one type to another does so only once. Of the land that dies back, 10 percent of the carbon is lost immediately, and the remaining stock decays at a rate of 5 percent per year (King and Neilson, 1992). All of the carbon on land that converts decays before the new type begins to regrow. Before land regenerates, there is a lag, which ranges from 10 years in the south to 20 years in the north to 30 years in the west.

Biomass growth follows a logistic functional form:

$$\frac{dN_i(t)}{dt} = \lambda_i N_i(t)\left(\frac{K_i(t) - N_i(t)}{N_i(t)}\right) \qquad (1)$$

The parameter values for this growth function vary from region to region, and were determined separately. λ_i describes the rate of growth, $N_i(t)$ is the amount of carbon accumulated in the ecosystem at time t, and $K_i(t)$ is the maximum carbon storage for that ecosystem type.

$K_i(t)$ changes linearly in response to climate change. This growth function captures all aspects of ecosystem storage, except for soil storage. Soil carbon flux for land that converts from one ecosystem type to another follows:

$$soilC_i(t+1) = soilC_i(t) + \left(\frac{1}{g_i}\right)\left(soilC_i(T^*) - soilC_i(0)\right) \tag{2}$$

$SoilC_i(T^*)$ represents the new steady-state level of soil carbon expected to be present after the new type has grown on the land. Equation (2) describes the storage of carbon in soils as the land converts from one type to another. The parameter g_i describes the rate of turnover, or the number of years that it takes for forest soils to turn over. We assume that carbon in soils turns over once every 30 years (Raich and Schlesinger, 1992).

3.3. Dynamic Economic Response

3.3.1. Dynamic Economic Model

We assume that the changes predicted by the steady-state and dynamic ecosystem models described above occur exogenously to the timber market. Timber markets, thus, are left to anticipate and react to changes in an economically efficient manner. Rational behavior among independent landowners will lead them to adjust and adapt to changing ecological conditions in anticipation of future conditions.

We capture these adjustments with a dynamic model of U.S. timber stumpage markets, the objective of which is to maximize the net present value of social welfare in the market over time. Similar models have been utilized in forestry by others (Lyon and Sedjo, 1983). Social welfare is the area between the demand and the supply curves. This is a partial equilibrium model, where markets are assumed to clear in every period. We assume that the marginal cost of harvesting is the same for all land and is held constant over time, so that welfare in any period is the total area under the stumpage demand curve.

Replanting and capital costs must also be considered in the social planner's problem, because there are real costs associated with keeping land in timber. Costs for replanting vary by region because of the large differences between productivity and replanting success from area to area. Land rent describes the capital cost of maintaining land in timber. It is included explicitly so that landowners incorporate it into their harvesting decisions. If land is to remain in timber, harvesting the land must net enough to pay this amount and regenerate the land. β_i is the cost of replanting, $G_i(t)$ is the number of acres to be replanted, $R_i(t)$ is the land rent, and $X_i(t)$ is the number of acres of timber type i in the forest. The social planner's problem is

$$\underset{H_i(\cdot,t)}{Max}\int_t^\infty e^{-rt}\left\{\int_0^{Q(t)^*} D\left\{Q\left(\sum_i H_i(t)V_i(a_i,t)\right)\right\}dQ(t)\right.$$

$$\left. - \sum_i \beta_i G_i(t) - \sum_i R_i(t)X_i(t)\right\}dt \tag{3}$$

which defines a dynamic optimization problem whose objective is to determine a time varying harvest schedule that maximizes the net present value of net welfare in the timber market. $D(\cdot)$ is the stumpage demand function. $Q(t)$ is the quantity of timber harvested at moment t, $H_i(t)$ is the number of acres of timber stock i harvested, $V_i(t)$ is the yield function, and a_i is the age of the marginal piece of land harvested in each period. There are I stocks of timber species.

There are several constraints to this problem, each of which must hold for all timber types, i. A general stock constraint is placed on the system so that

$$\dot{X}_i = -H_i(t) + G_i(t) \tag{4}$$

The change in acreage each period is the difference between harvesting and replanting. In the base case, acres that are replanted can come only from acres previously harvested, or from any land that exogenously shifts into timber from other land uses. Later, when we consider climatic change, land can be regenerated from additional sources, such as other timber types or land that is unable currently to support timber but under climate change can. Regardless, land gets replanted only if certain economic criteria are met, namely the condition that the net present value of land in timber must be greater than the net present value of land in the next best alternative. This is expressed as:

$$\sum_{n_i=1}^{\infty} \left\{ P\left(t_{n_i}\right) V_i\left(t_{n_i} - t_{n_i-1}\right) e^{-r\left(t_{n_i} - t_{n_i-1}\right)} - \beta_i \right\} e^{-r\left(t_{n_i-1} - t_0\right)} \geq R_{i.ait} \tag{5}$$

where n_i is the rotation number. Additionally, $H_i(t)$, $G_i(t)$, $P(t)$ must be greater than 0, and initial and terminal conditions are given for each variable and parameter. Starting conditions include an initial age distribution over timber stocks for each timber type.

Given a national demand function for timber, plus the assumption that marginal costs are constant across every acre harvested and a transversality condition, this problem can be solved for the following first order condition:

$$\dot{P}V_i(a_i, t) + P(t)\dot{V}_i = rP(t)V_i(a_i, t) + R_i(t) \tag{6}$$

$P(t)$ is the stumpage price. Stumpage price is the log price net of the costs of delivery to the mill, which include all harvesting costs. This condition will hold only for the marginal acre of timberland harvested in each time period. The left hand side of (6) is the marginal benefit of waiting to harvest that acre, while the right hand side is the marginal cost of waiting.

In a competitive market, equation (6) defines the transition of any initial stock of timber, defined by its age distribution, to some terminal condition (Brazee and Mendelsohn, 1990). Over time, timberland owners must harvest along the path defined by this equation. It must be solved simultaneously for each timber type i in each period. With a national demand function for stumpage, prices are the same for each timber type. If the demand function is not shifting out, this equation would resolve to the more familiar, steady-state

Faustmann formula. Given population and economic growth rates, however, we assume that demand is growing continuously over time.

Because not all land in the U.S. is managed strictly for timber purposes (Powell *et al.*, 1993), we allow for two classifications of land— high intensity and low intensity managed land. The distinction between high and low intensity will be the age of the timber type. While high intensity land will be harvested according to equation (6), the rest of the land will be harvested according to the following price responsive supply equation:

$$S_i(t) = \phi_i\left(t_i > t *\right)\left(\kappa_i\right)P(t)^{\eta_{i,s}} \tag{7}$$

$\phi_i(t_i > t^*)$ is the ratio of the current period's inventory to the initial inventory above a minimum harvest age, t^*; η_i is a constant specific to each timber type; and $\eta_{i,s}$ is the price elasticity of supply. $P(t)\eta_{i,s}$ is .3, which is derived from Newman and Wear (1993). We allow low intensity land to convert to high intensity status over time, and we allow management intensity to shift over time, both according to price. As prices rise, management intensifies, as prices fall, it weakens.

The timber types that are included in this modeling effort correspond to the ecosystem types in Table 1. Softwoods are managed as both high intensity as well as low intensity timberland, while hardwoods are managed only as low intensity. A national softwood demand function was estimated from historical data on timber prices and quantities (Adams *et al.*, 1988). The price elasticity of demand in this national function was estimated as *-1.02*. We adopted this value for the hardwood price elasticity as well.

By accounting both for different ecosystem types and different management intensities, the final economic model presents a more realistic picture of timber markets than would the model presented in Equations (2) to (6). We initialized all parameters and inventories for the base year, 1990. The inventory data was obtained from a recent update of the ATLAS timber inventory model, described by Mills and Kincaid (1992).

3.3.2. Linking Ecological Changes with the Dynamic Economic Model

As in the dynamic "natural change" model, in the linked, dynamic economic model we assume that all ecosystem change occurs proportionally to temperature and precipitation. Of the land that dies back (Column 2, Table 1), some timber is available for salvage. Salvage is 50 percent of the low intensity timber that dies back, and 75 percent of the high intensity timber that dies back. Because the amount of salvage is known for each period, the model determines how much additional timber to harvest in each period according to equation's (6) and (7).

Some land that dies back will regenerate into new types of forested ecosystems, and some may remain fallow or convert to non-forest uses. Additionally, some non-forested land may convert to forest land. Regeneration lags will be incorporated on land that is managed in low intensity types. Lags vary from 10 years in the south, to 20 years in the pacific northwest, to 30 years in the north and west. Land managed under high intensity will be replanted to the new type as soon as climate has changed enough to allow the conversion.

Yield changes are proportional to changes in total biomass, as determined from the ecosystem production models. The effect of changes in biomass is further assumed to be independent of age. Because the yield of timber will depend both on the initial stock and the yearly increment, it is given by

$$V_i(a_i, t) = \hat{V}_i(\alpha_{i,t^*-1}) + \int_{t^*}^{T} \left\{ a_i(t)\hat{V}_i(a_i, t) \right\} dt \tag{8}$$

where $\alpha_i(t)$ is the climate impact, $\hat{V}_i(a_i)$ is the base yield function for a particular region, $V_i(a_i, t)$ is the climate adjusted yield function and t* is the age of the timber at the time of the shock. This adjustment accounts for the volume of trees when the climate shock begins, as well as the amount of time the trees have grown at the new, changed rates of growth.

3.3.3. Carbon Accounting

We utilize data from Birdsey (1992) to link timber inventories, timber growth, and harvests to carbon storage and flux on private timberlands in the U.S. Our timber model contains region and age specific data on timber inventories, so that we were able to link carbon storage directly to the age distribution of timber. We assume that tree, understory, and forest floor components of carbon storage are directly related to the age distribution of timber. Soil storage was assumed to be independent of the age distribution (Johnson, 1992), but dependent on the timber type. Timberland shifting out of forests was counted as a net loss in storage, while land shifting in was considered a net gain. We estimate the amount of carbon stored in timber products over time using data from Plantinga and Birdsey (1993).

4. RESULTS

Under the baseline case, linked dynamic economic model, carbon inventories in the U.S. are projected to increase throughout the simulation, with the bulk of the increases occurring the first 30 years. This results from continued conversion of forests from low to high intensity management, as well as changes in the age distribution of timber stocks. Over time, increases in carbon inventory slow significantly as more forests are harvested near their faustman rotation age. Timber removals align more closely with growth in the stock. A slight pulse of carbon from forests to the atmosphere is predicted between 2030 and 2050 as harvest are evened with timber growth.

In the BIOME2 and TEM combination, steady state carbon is substantially greater than the baseline, as the forest area expands and productivity increases. MAPSS and BIOME-BGC provide a completely different outlook, as MAPSS suggests that large quantities of forest land undergo drought induced dieback, and BIOME-BGC predicts decreases in productivity for all types except northern hardwoods.

Forest carbon fluxes for the "natural case" and the linked, dynamic economic scenario's, as well as the economic system storage are shown in Table 2. Utilizing the dynamic,

TABLE 2

Average Private Forest and Economic System (Market) Carbon Flux for Each Decade in Both the "Natural Case" and the Linked Economic Model Case. The Year Represents the Middle of the Decade. Positive Numbers Indicate that Additional Carbon is Stored, and Negative Numbers Suggest that Additional Carbon is Released

	Baseline		BIOME2 and TEM			MAPSS and BIOME-BGC		
	Dynamic			Dynamic			Dynamic	
			Natural			Natural		
	Forest	Market	Forest	Forest	Market	Forest	Forest	Market
Mid-Year	Tg Carbon per Decade							
1995	1220	797	412	1785	802	−298	72	777
2005	619	1619	−241	1584	1597	−806	−338	1508
2015	323	1620	−139	150	1614	−1005	−569	1414
2025	6	1557	40	1457	1649	−1018	−700	1309
2035	−102	1529	584	1465	1702	−794	−788	1189
2045	−9	1498	930	1087	1851	−767	−741	1119
2055	80	1408	914	760	2048	−718	−400	1090
2065	167	1381	564	159	2185	−413	496	1059
2075	198	1426	1519	586	2276	163	146	1080
2085	96	1491	1361	547	2365	376	97	1142
2095	60	1590	1192	526	2447	405	52	1198
2105	52	1598	696	451	2510	194	73	1231
2115	−35	1567	444	400	2546	154	162	1238
2125	−11	1587	443	432	2547	135	296	1216
2135	123	1525	276	276	2524	69	235	1181

economic model of forest adjustment eliminates any carbon pulse from forests in the BIOME2 and TEM combination, while the pulse is reduced significantly in the MAPSS and BIOME-BGC combination. Because the natural case does not allow either for adaptive harvest scheduling or adaptive replanting of timber, there is no way for the "natural change" model to react to climate change.

Forest carbon captures only part of the story because harvest is considered to be an immediate release of carbon to the atmosphere. Some of this carbon may be stored within the economic system, however. The decadal flux of carbon into the economic system is shown in the "market" columns of Table 2. Economic system storage occurs only for the linked models because of harvest. The BIOME2 and TEM combination involves significantly reduced prices, which leads to higher harvest levels, and to greater storage in the economic system.

We also estimated a "short-term" pulse of carbon, representing the net cumulative carbon output (input) from (to) forests and the economic system by the year 2060 (Table 3). Because we are most interested in comparing carbon storage capabilities of the "natural change" model and the dynamic model, we have not attempted to discount carbon in this model. We consider this "short-term" in light of the long adjustment periods (up to 200 years or more) suggested by ecological modelers. Long-term storage, then, is calculated for 150 years.

In each of these scenarios, additional carbon is stored in the economic system. Where much carbon in the "natural change" case is lost due to dieback, in the linked dynamic economic model scenarios, some of that gets stored in the economic system because of salvage. In fact, a pulse of carbon from forests in the MAPSS and BIOME-BGC combination is offset by storage in the economic system. Another interesting feedback appears in the BIOME2 and TEM combination. While forest storage is increasing due to ecological factors, harvests increase significantly as well, further enhancing carbon storage in the economic system.

5. CONCLUSION

Differences between the pulse estimates for the "natural change" and linked dynamic economic models indicate that timber market adaptation can ameliorate much of the harmful effects of climatic change in forests. Carbon losses are limited by two sources of adaptive behavior. First, land managers react efficiently by harvesting and replanting decisions that minimize losses. Second, carbon is stored in timber products within the timber market. What appears to be a pulse of carbon to the atmosphere, actually can be turned into a pulse of timber to markets, and increased storage in timber products. Not only

TABLE 3

Comparison of the Short-Term Pulse of Carbon and the Change in Long-Term Storage in Both the Forested Ecosystem and the Economic System. The Numbers Represent Additional Storage in Each Component Over the Base Year, 1990, by the Year 2060 or 2140

| | Carbon Pulse | | Change in Long-Term Carbon Storage | |
| | 2060 | | 2140 | |
Scenario	Natural	Dynamic	Natural	Dynamic
	Accumulated Pg Carbon			
BIOME2 and TEM				
Forest Storage	2.50	9.64	8.99	13.02
Economic Storage	0.00	11.26	0.00	30.66
Total Storage	2.50	20.90	8.99	43.68
MAPSS and B-BGC				
Forest Storage	−5.41	−3.46	−4.32	−1.91
Economic Storage	0.00	8.41	0.00	17.75
Total Storage	−5.41	4.94	−4.32	15.84

are timber inventories increased under BIOME2 and TEM, but harvest levels increase relative to the baseline as well. This further increases carbon storage.

Several interesting implications arise from this analysis. To put these numbers in perspective, yearly carbon emissions for the year 2000 in the U.S. are projected to be approximately 1.5 Pg (Clinton and Gore, 1993). In the base case, additional long-term (150 year) forest storage sequesters less than two years of carbon emissions (2.79 Pg). If, however, storage within the economic system is considered, 22.19 Pg of additional carbon may be stored. Given the climate scenarios, the possible additional carbon sequestered in forests ranges from 15.84 - 43.68 Pg (Table V).

Over the first 70 years, sequestration above the baseline in private U.S. forests range from -.018 - +.107 Pg per year. In contrast, the mitigation strategy proposed by Parks and Hardie (1995) suggests that planting additional forests may sequester an additional .015 - .049 Pg per year for 10 years. The mitigation strategies may actually just offset carbon emissions resulting from climate change, for only a short period of time, or they may be swamped by forests that are increasingly capable of sequestering additional carbon.

This result has interesting implications for carbon mitigation strategies. First, carbon mitigation strategies must be considered in light of very long-term, dynamic effects, such as climatic change. Mitigation strategies currently are proposed to help reduce net carbon emissions, but under climate change scenarios, optimal forest management may mitigate much of this impact anyway — without additional intervention.

Second, consideration must be given to where forests are planted in mitigation strategies. The ecological models discussed in this paper suggest that there will be large changes in the range of timber species. By planting in the wrong place, any additional sequestration could be lost by dieback of the species type. On the other side, mitigation strategies could take advantage of a rapidly changing climate by planting species where they are best adapted to grow. The line between mitigation strategies and adaptation strategies thus becomes clouded because it is just the marginal lands that mitigation strategies consider, which could become optimal for future timber production anyhow.

ACKNOWLEDGMENT

This research was sponsored by the Electric Power Research Institute. Three anonymous reviewers provided helpful comments, although all mistakes remaining are the fault of the authors alone. We thank Richard Haynes and John Mills for providing data on U.S. timber markets and timber inventories, and we thank Tim Kittel, Jerry Melillo, Ron Neilson, Colin Prentice, and Steve Running for their comments on previous work, and for allowing us to access the VEMAP dataset.

REFERENCES

Adams, D.M., K.C. Jackson, and R.W. Haynes (1988), 'Production, Consumption, and Prices of Softwood Products in North America: Regional Time Series data, 1950-1985', U.S. Department of Agriculture, Forest Service, Research Bulletin *PNW-RB-151*. Portland: Pacific Northwest Research Station.

Binkley, C.S. and G.C. Van Kooten (1994), 'Integrating Climatic Change and Forests: Economic and Ecologic Assessments', *Climatic Change* **28**, 91-110.

Birdsey, R. A. (1992), 'Carbon Storage and Accumulation in United States Forest Ecosystems', U.S. Department of Agriculture, Forest Service, General Technical Report *WO-59*. Washington, DC.

Brazee,R. and R. Mendelsohn (1990), 'A Dynamic Model of Timber Markets', *Forest Science* **36**, 255-264.

Clinton, W.J. and A. Gore, Jr. (1993), 'The Climate Change Action Plan', Washington, D.C.: White House Office of Environmental Policy.

Dixon, R.K., S. Brown, R.A. Houghton, A.M. Solomon, M.C. Trexler, and J. Wisniewski (1994), 'Carbon Pools and Flux of Global Forest Ecosystems', *Science* **263**, 185-190.

Intergovernmental Panel on Climate Change (1990), 'Climate Change: The IPCC Scientific Assessment', J.T. Houghton, G.J. Jenkins, and J.J. Ephraums, eds., Cambridge: Cambridge University Press.

Johnson, D. W. (1992), 'Effects of Forest Management on Soil Carbon Storage', *Water, Air, and Soil Pollution* **64**, 83-120.

Joyce, L. (1995) 'Productivity of America's Forests and Climate Change', U.S. Department of Agriculture, Forest Service. General Technical Report *GTR-RM-XXX*, Ft. Collins: Rocky Mountain Forest and Range Experiment Station.

King, G. A. and R.P. Neilson (1992), 'The Transient Response of Vegetation to Climate Change: A Potential Source of CO_2 to the Atmosphere', *Water, Air, and Soil Pollution* **64**, 365-383.

Lyon, K. S. and R.A. Sedjo (1983), 'An Optimal Control Theory Model to Estimate the Regional Long-term Supply of Timber', *Forest Science* **29**(4), 798-812.

Mills, J.R. and J.C. Kincaid (1992), 'The Aggregate Timberland Assessment System: A User's Guide', US Department of Agriculture, Forest Service, General Technical Report *PNW-GTR-281*. Portland: Pacific Northwest Research Station.

Melillo, J. M, A.D. McGuire, D.W. Kicklighter, B. Moore, III, C.J. Vorosmarty, and A.L. Schloss (1993), 'Global Climate Change and Terrestrial Net Primary Production', *Nature* **363**, 234-240.

Neilson, R. P., G.A. King, and G. Koerper (1992), 'Toward a Rule-Based Biome Model', *Landscape Ecology* **7**(1), 27-43.

Neilson, R. P. (1993), 'Vegetation Redistribution: A Possible Biosphere Source of CO_2 During Climatic Change', *Water, Air, and Soil Pollution* **70**, 659-673.

Newman, D. H. and D.N. Wear (1993), 'Production Economics of Private Forestry: A Comparison of Industrial and Nonindustrial Forest Owners', *American Journal of Agricultural Economics* **75**, 674-684.

Parks, P.J. and I.W. Hardie (1995), 'Least Cost Forest Carbon Reserves: Cost-Effective Subsidies to Convert Marginal Agricultural Land to Forests', *Land Economics* **71**(1), 122-136.

Plantinga, A.J. and R.A. Birdsey (1993), 'Carbon Fluxes Resulting From U.S. Private Timberland Management', *Climatic Change* **23**, 37-53.

Powell, D. S., J.L. Faulkner, D.R. Darr, Z. Zhu, and D.W. MacCleery (1993), 'Forest Resources of the United States, 1992', U.S. Department of Agriculture, Forest Service. General Technical Report *GTR-RM-234*, Ft. Collins: Rocky Mountain Experiment Station.

Prentice, I. C., W. Cramer, S.P. Harrison, R. Leemans, R.A. Monserud, and A.M. Solomon (1992), 'A Global Biome Model Based on Plant Physiology and Dominance, Soil Properties and Climate', *Journal of Biogeography* **19**, 117-134.

Raich, J.W. and W.H. Schlesinger (1992), 'The Global Carbon Dioxide Flux in Soil Respiration and Relationship to Vegetation and Climate', *Tellus* **44B**, 81-99.

Running, S. W. and S.T. Gower (1991), 'FOREST_BGC, A General Model of Forest Ecosystem Processes for Regional Applications. II. Dynamic Carbon Allocation and Nitrogen Budgets', *Tree Physiology* **9**, 147-160.

Smith, T.M. and H.H. Shugart (1993), 'The Potential Response of Global Terrestrial Carbon Storage to a Climate Change', *Water, Air, and Soil Pollution* **70**, 629-642.

VEMAP Participants (1995), 'Vegetation/Ecosystem Modeling and Analysis Project (VEMAP): Comparing Biogeography and Biogeochemistry Models in a Continental-Scale Study of Terrestrial Ecosystem Responses to Climate Change and CO_2 Doubling,' *Global Biogeochemical Cycles*, in press.

Wilson, C.A. and J.F.B. Mitchell (1987), 'A Doubled CO_2 Climate Sensitivity Experiment with a Global Climate Model Including a Simple Ocean', *Journal of Geophysical Research* **D11-11/20**, 13,315-13,343.

Critical Reviews in Environmental Science and Technology, 27(Special): S323–S333 (1997)

FOREST BIOMASS AS CARBON SINK — ECONOMIC VALUE AND FOREST MANAGEMENT/POLICY IMPLICATIONS

BIRGER SOLBERG

European Forest Institute, Torikatu 34, 80100 JOENSUU, Finland

ABSTRACT: All countries in Scandinavia have introduced a carbon fee on fossil fuel to reduce the emission of the climate gas CO_2. In Norway the fee is 0.82 NOK (or about 0.12 US$) per liter of gasoline, equivalent to 343 NOK (or 49 US$) per ton CO_2. The fee is an emission cost which gives a corresponding benefit for absorbing atmospheric CO_2 in forest biomass. This article shows that this benefit corresponds to a net economic value of carbon sequestration in forest biomass 2-30 times higher than the net value of timber as raw material for the forest industry in Norway, which has one of the highest timber prices in the world. If a fee high enough to stabilize the CO_2 emission in Norway were to be introduced, the value of carbon sequestration will be at least twice as high as the above estimates. It is argued that this will imply substantial changes in forest management and policy, both in rich and poor countries. Projects to restore and maintain sustainable forest ecosystems will be very profitable, and could simultaneously provide considerable other environmental benefits like increased biodiversity, soil stability, and improved watershed structures.

KEY WORDS: Carbon sequestration, climate change, forestry, economics, policy.

1. INTRODUCTION

Forests play an important role in the global carbon budget both as carbon sinks and through emission of CO_2 (Sedjo 1990, Dixon et al. 1994). One measure, among many others, that has been advocated to lower the concentration of atmospheric CO_2 is to increase the forest biomass. The main arguments for this measure are, first, that tree biomass fixes large quantities of atmospheric CO_2 over a long period — in boreal forests from 40 to 200 years depending upon species and site class. Second, forest biomass can directly (or indirectly as waste) be used for energy purposes and thereby reduce the use of fossil fuels, or can substitute for other materials such as steel or aluminium constructions whose production consume large quantities of fossil fuels. Third, increased stocks of carbon in forests provide flexibility given the uncertainty regarding the impact of global warming. Fourth, increasing the standing stock of forest biomass may in many ecosystems give several environmental benefits other than carbon sequestration, e.g. improved biodiversity, better water quality, less erosion, and improved recreation opportunities. Finally, forest production is based on a well-functioning technology and policy structure, at least in most industrialized countries, which makes increased production easy.

1064-3389/97/$.50

The third argument, among those listed above, is perhaps the most important one. If human activities are ultimately proved to cause global climate changes, increased stocks of carbon in wood biomass (as trees and/or forest industry products) will have a long-term positive marginal value regarding lowering the quantities of atmospheric CO_2. Thus one can get more time for introducing new energy technologies or other measures to decrease the emission of greenhouse gasses. If global warming is not as detrimental as some have suggested, an increased stock of timber will most likely have a positive value as timber for the production of paper, sawnwood and bioenergy, or for environmental benefits like recreation, water management, soil conservation, or biological diversity.

Several studies have examined the amount of CO_2 which could be sequestered in forest biomass (cf. Dixon et al. 1994). Only a few studies have, however, investigated how preferable the various possible measures in forestry are and how they rank compared to measures in other sectors to reduce the emission of climate gasses (EPA 1990, Lunnan et al. 1991, Hoen and Solberg 1994, Solberg and Hoen 1995). All of these studies are based on cost efficiency — i.e. they use the ratio "tons of carbon sequestered per \$", or the inverse, as criterium. These ratios give, under certain assumptions, a consistent internal ranking of the various measures considered, but they give little or no indications of how preferable the investments in these measures are for society as compared to investments in other environmental benefits, goods and services. We may, therefore, avoid a considerable nonoptimal use of resources over time if we could get realistic estimates of the economic value of fixing one unit of atmospheric CO_2 of some sort.

The first objective of this article is to estimate the economic value of a marginal increase in forest biomass as a carbon sink, and compare this value with the value of timber as raw material for forest industry production. The second objective is to discuss some important forest management and policy implications of the findings.

2. METHODOLOGY AND DATA

The main idea here is that the marginal value of fixing one unit of atmospheric CO_2 in forest biomass equals the marginal cost of reducing the emission of CO_2 in the most costly project which is implemented for that purpose. This marginal cost is referred to below as the shadow price P_t. If a carbon tax were introduced with the only purpose to reduce the consumption of fossil fuel, P_t would equal that tax.

The *gross* value (G_t) of the fixation of atmospheric CO_2 per m^3 of stemvolume produced is then calculated as:

$$G_t = P_t * S * F * M * R_t \tag{1}$$

where

G_t is the gross value of the fixation at time t of atmospheric CO_2, measured as NOK (Norwegian kroner) per m^3 stemvolume produced under bark.

P_t is the above defined shadow price at time t in NOK per ton CO_2.

S is the specific weight of the tree species measured as ton dryweight of biomass per m^3 tree volume.

F   is the fraction ton C (carbon) per ton dryweight biomass

M   is the molecular weight of CO_2 divided by the atom weight of C.

R_t is the ratio between the total tree biomass (i.e. including roots, branches, bark and stump) and the biomass in the stem at time t.

Here, F is very close to 0.50 for all tree species, and M is the constant 44/12. S and R vary with tree species. In the below calculations for Norway spruce (*Picea abies* L. Karst.) S is 0.503 ton per m^3 (Nagoda 1982) and R is 2.0 (Blingsmo 1990).

   The *net* value is calculated by considering that C cannot be stored indefinately in biomass — i.e. from the above defined gross value one has to subtract the cost of the emission of CO_2 from the decaying of the biomass at later points in time:

$$N_{t_0} = G_{t_0} - \left[\sum_{i=0}^{N} \sum_{t=t_0}^{T+a_i+d_i} P_t \cdot q_{t_i} (1+r)^{T-t} \right] (1+r)^{-(T-t_0)} \qquad (2)$$

where

N_{t_0} is the net value of the fixation of atmospheric CO_2, measured as NOK per m^3 industrial stemvolume produced under bark in the year t_0.

G_{t_0} is the gross value G in Equation (1), estimated for year t_0.

t_0 is the age of the forest stand $t_0 \, \epsilon \, [0,T]$.

q_{t_i} is the emission of CO_2 to the atmosphere at time t from assortment i of the carbon fixed at time per m^3 of industrial stem volume.

T   is the rotation length (in years) — i.e. the age of the trees when felled and used.

a_i is the anthropogenic life time (in years) — i.e. the time the wood assortment i is kept in use.

d_i is the decaying time for assortment i — i.e. the time (in years) from the end of the anthropogenic time until 99% of the wood carbon is released to the atmosphere.

P_t is the shadow price defined in Equation (1).

N   is the number of end-use assortments (including roots, branches, stump etc. as well as paper,sawnwood, fuelwood) realistic to cover the decay process, altogether 14 as defined in Hoen & Solberg (1994).

i   is end-use assortment nos. i.

r   is the calculation rate of interest.

To compare the value of forests as carbon sink with the value of forests as raw material for forest industry production, Equation (3) can be used.

$$N_T = \left[\sum_{t=0}^{T} f_t \cdot P_t (1+r)^{T-t} - \sum_{i=t}^{N} \sum_{t=0}^{T+a_i+d_i} P_t \cdot q_{t_i} (1+r)^{-t} \right] \cdot \frac{1}{V_T} \qquad (3)$$

where

N_T is the net value at time T of the total fixation of atmospheric CO_2 in a forest stand establishedat time 0 and clearfelled at time T, measured in NOK per m^3 stem volume.

T is the length of the rotation period — measured in years.

f_t is the fixation of atmospheric CO_2 in the forest stand at time t, measured in tons per ha.

V_T is the volume of the forest stand sold as industrial wood at time T, measured as m^3 under bark per ha.

P_t, T, r, a_i, d_i and q_{ti} are as defined in Equation (2).

In Norway, as well as in many other countries, the Government is committed to stabilise the anthropogenic emissions of CO_2 at the 1989 level. For this reason, a carbon tax with the purpose to reduce the consumption of fossil fuel was introduced in 1991 and it is today 0.82 NOK (about 0.12 US$) per litre gasoline, which corresponds to 343 NOK (or 49 US$) per ton CO_2, assuming 3.15 kg CO_2 per kg of gasoline and a density of 0.76 kg per litre gasoline. A similar tax is introduced in Denmark, Finland and Sweden, and both the USA and the European Union are discussing the introduction of such a tax (Dean 1993). It is also calculated, using general equilibrium growth models, that to stabilise the CO_2 emission on 1989 level in Norway, a national carbon tax corresponding to 900 NOK (or US$ 130) per ton CO_2 would be necessary in the year 2000 (Anon. 1993). The same study estimates that to stabilise the CO_2 emissions from the OECD countries in the year 2000 and the global emissions by the year 2025, a carbon tax of 650 NOK per ton CO_2 would be necessary in the year 2000 increasing to 1450 NOK per ton CO_2 in the year 2025 (all costs at 1990 price level). More recent studies indicate that taxes in the range of US$ 100-400 per ton C (i.e. US$ 27-109 per ton CO_2 or NOK 190-760 per ton CO_2) would be necessary to reduce the emission of carbon to 1990 level *globally* (Dean 1993).

The value of timber as raw material for the forest industry used for comparison in this article, is the net stumpage value — i.e. the gross price paid to the forest owner by the industry minus all the owner's variable costs for cutting and transport. This value was during 1992 in Norway on average 200 NOK per m^3 timber for the economically most important tree species in the Nordic countries, Norway spruce (*Picea abies* L. Karst.).

3. RESULTS

Table 1 gives the gross value G as a function of P_t for Norway spruce. This value should be compared to the above mentioned net stumpage value of timber of 200 NOK per m^3 as raw material for the forest industry. Table 1 shows that at the present CO_2 fee in Norway of 0.82 NOK per litre gasoline (or 343 NOK per ton CO_2), G takes the value of 478 NOK per m^3 timber — i.e. more than twice the value of the resource as production input for the forest industry. The above mentioned fees of 650 and 1450 NOK per ton CO_2 necessary to stabilise the anthropogenic emission of CO_2 in Norway correspond to a value of respectively about 900 and 2000 NOK per m^3 of timber — i.e. from 4 to 10 times the value as raw material for the forest industry.

TABLE 1

The Gross Value of Timber as CO_2 Sink[1] for *Picea abies* (K.)

Shadow price (P) NOK/ton CO_2[2]	US$/ton CO_2[3]	Gross value of timber G[1]	
		NOK/m³ stem volume	US$/m³ stem volume[3]
100 (0.23)	14	139	20
343 (0.82)*	49	478	68
500 (1.17)	71	697	99
700 (1.63)	100	975	139
900 (2.10)	129	1,254	179
1,100 (2.57)	157	1,533	219
1,300 (3.03)	186	1,811	258
1,500 (3.50)	214	2,089	299

* The present (1993) carbon fee on petrol in Norway.
[1] As defined in Equation (1).
[2] In bracket is the carbon fee in NOK per litre petrol.
[3] The exchange rate of 7.00 NOK per US$ is assumed.

Table 2 shows the net economic value of carbon sequestration N in *Picea abies* (K.) as a function of time from fixation of CO_2 to clearfelling, and interest rate, calculated by using Equation (2) and assuming P_t is NOK 100 per ton CO_2. For example it is seen that at 3% p.a. interest rate the value is increasing from NOK 50 to NOK 118 per m³ stemvolume when going from 5 years to 80 years from clearfelling respectively. The maximum net value is NOK 126 per m³, because it is 1.26 ton CO_2 per m³ stemvolume in this species (including branches, roots and top in the CO_2 content), and at high enough interest rates and time to clearfelling the value of the emission of CO_2 is close to zero.

Finally, Table 3 shows the value of N for an average stand of Norway spruce at clearfelling time calculated as defined by Equation (3). It is seen that even at a low interest rate of 2% p.a. the value of sequestering CO_2 when P is 343 NOK per ton CO_2, is 3 times the value of the timber as forest industry production input. It is also seen that the CO_2 value increases strongly with increasing r. At r equalling 7% p.a. (which is the interest rate advocated by the Ministry of Finance in Norway in public project analyses) the CO_2 value is 32 times the industry value. When comparing with the above mentioned proposed stabilising fees of 650 and 1450 NOK per ton CO_2, the CO_2 value at this interest rate is respectively 61 and 135 times higher than the value of wood for industry.

4. DISCUSSION

4.1. Estimation

The above estimated economic values for sequestration of CO_2 in forest biomass might be surprising, but there are four main reasons for the high values:(a) The quantity of CO_2

TABLE 2
The Net Economic Value[1] of Carbon Sequestration in *Picea Abies* (K.) as a Function of Time to Clearfelling and Interest Rate. (NOK per m^3 Stem-Volume Fixed).

Year to clearfelling	Interest rate (% p.a.)				
	1	3	5	7	10
5	26	50	64	73	84
10	31	61	77	89	100
20	40	77	96	107	116
30	48	90	108	117	123
40	55	99	115	122	125
50	62	106	120	124	126
60	68	111	122	125	126
70	74	115	124	126	126
80	79	118	125	126	126

[1] As defined in Equation (2) assuming P_t is 100 NOK per ton CO_2.

TABLE 3
The Net Value of Timber as Carbon Sink[1] Made Comparable to the Net Value of Industrial Roundwood for one Rotation Period of *Picea abies* (K.) at Medium Site Class[2] (NOK/m^3)

Rate of interest (% p.a.)	Shadow price P (NOK/ton CO_2)						
	100	343	400	700	1,000	1,300	1,500
1	90	316	360	630	900	1,170	1,350
2	175	609	700	1,220	1,740	2,260	2,350
3	290	1,021	1,170	2,040	2,920	3,800	4,400
4	470	1,638	1,870	3,280	4,680	6,080	7,020
5	740	2,592	3,000	5,200	7,400	9,600	11,100
6	1,170	4,101	4,700	8,200	11,700	15,200	17,600
7	1,870	6,528	7,500	13,100	18,700	24,200	28,000
8	2,990	10,480	12,000	21,000	29,900	38,900	44,900
9	4,850	16,984	19,400	34,000	48,500	63,100	72,800
10	7,940	27,784	31,800	55,600	79,400	103,200	119,100

[1] As calculated in Equation (3).
[2] Site class G17 (i.e. dominant height 17 m at 40 years breast height age), planting density 2,500 plants/ha, rotation length 80 years.

sequestered per year is high; (b) the sequestration period is long; (c) a positive r combined with the two previously mentioned reasons gives high accumulated effects over time. It should be emphasised that a positive interest rate reflects the return the society has in the best alternative investment for welfare increase — see Cline (1992) for a thorough discussion on the interest rate. Finally, (d), the estimated CO_2 values rest upon the assumption that P is positive and in the magnitude of order used in the tables, reflecting the costs (or use of resources) which we are willing to use today to lower the emission of CO_2.

The size of P depends upon our CO_2-emission target (the sooner and more we wish to lower the emission of CO_2 to the atmosphere the higher P will be), which again reflects the expected negative consequences of a possible climate change caused by the increase of greenhouse gas concentration in the atmosphere. Nearly all macro-economic analyses done till now conclude that a carbon fee in the range of 500-1000 NOK per ton of CO_2 is necessary only to stabilise the present CO_2 emission; to decrease it will demand even higher fees (Dean 1993). It is also a common result in these analyses that because of future expected growth in population and per capita income, the CO_2 fee has to increase considerably over time to stabilise the emissions (Dean, op.cit.).

The estimates of anthropogenic life and decay time assumed in this analysis, are based upon the present situation in Norway as defined in Hoen and Solberg (1994), and are burdened with many uncertain factors like future type of end-uses, climate, etc. However, sensitivity analysis show that even drastic changes of anthropogenic and decay time do not change much the estimated economic values of N in Tables 2 and 3. For example if we set anthropogenic life time to zero, the estimated economic values in table 3 will decrease only by 5%, 3%, 2%, and 1% for respectively 1, 2, 3 and 4% p.a. calculation rate of interest in Equation (3). The main reason for this is the relatively long rotation period T.

Another factor to consider regarding uncertainty of decay times is, if a carbon fee in the magnitude of 300-1000 NOK per ton of CO_2 is implemented, the value of using paper, fibreboards, second-hand sawnwood and other wood residues for energy production will increase. This will, most likely, imply substitution of fossil fuel for wood residues, which again will decrease the *net* decay rate q in Equation (3), and increase N all other factors equal. One could also imagine that the wood after clearfelling were dumped in mines or at deep sea bottom so that q approaches zero, although at the present that sounds rather farfetched.

The biomass yield estimates over time used in the calculations assume "normal" mortality and today's climate. A possible climate change in the direction of higher temperature and more rain could increase the biomass estimates, whereas the risk for calamities (high wind, fungi, and insect attacks) may increase, lowering the biomass growth. On average, however, these effects may counter balance each other.

The estimates of N in Table 3 are for T being 80 years, which is about the economic optimal rotation age for this site class of Norway spruce when industrial wood is the only production purpose (Solberg and Haight 1989). If a CO_2 fixation benefit is included, the correct T to use when estimating the relative value between N and the value of wood only for industrial purposes, is the T which maximizes the soil expectation value of the joint two benefits. This will for the values of P assumed in Table 3 give an optimal T which is considerably higher than 80 years, and that the estimates of N in Table 3 are low compared to those at optimal rotation age (Hoen and Solberg, this volume). For example, a sensitivity analysis of increasing the rotation age by 10 years to 90 years shows, all other factors equal, that the values of N in table 3 increase by 6, 26, 51, 81, and 117% for r being respectively

2, 4, 6, 8, and 10% p.a. Only for r equal to 1% p.a. decreases N in Table 3 (because the biomass growth is higher than 1% p.a.).

This question of optimal rotation is decisive for determining how the results in Table 3 depend upon site classes. Lower site class implies less biomass growth and hence less value for CO_2 fixation, whereas on the other hand it implies less timber volume and longer rotation times compared to higher site classes. Only a more detailed study can balance the simultaneous effects of these factors. One main problem is that we have little empirical evidence about how mortality changes when increasing the rotation time.

For the other main industrial tree species in the Nordic countries, Scots pine (*Pinus sylvestris*) and birch (*Betula pendula, Betula pubescens*), one get the same order of magnitudes of the value of CO_2-fixation as shown for Norway spruce in Table 3, but with some changes due to different basic densities, growth patterns, rotation ages, end uses, and industrial value of the timber. Also for tree species in other parts of the world, the same factors will decide the value of carbon sequestration.

4.2. Forest Management Implications

Growing forests for industrial wood production is and has been highly profitable in Norway. The results shown in Table 3 indicate that for P and r in the ranges defined there, the profitability of forest production increases many fold when considering the benefit of sequestration of atmospheric carbon in forest biomass. Benefits in the order of magnitude indicated in Table 3 will have several important forest management implications.

All other factors equal, it will be optimal to increase standing volume, rotation ages, and the investment intensity in silviculture. Because the environmental benefit of CO_2 fixation dominates so strongly the benefit of industrial wood, the question of timber quality will be less important relative to maximising dry weight biomass growth. This will in Scandinavia favour species with high basic density like birch compared to Norway spruce.

To keep high biomass growth it will probably be advantageous to allow more natural forest successions to take place — e.g. in Scandinavia first pioneer trees like birch followed by shadow trees like Norway spruce. To maximise biomass growth, natural regeneration could be supplemented by enrichment planting at different stages.

Silvicultural investments, like planting of abundant agriculture land and grassland, forest fertilisation, aff- and reforestation in general, will be much more profitable than without considering the carbon sequestration benefit.

The increase in rotation time is of high interest for mature or close to mature stands, and will mainly depend upon the growth and mortality of the forest stand with age, r, and its value as industrial wood. This value depends on the end-use; it is not unlikely that a high value of P will imply that one get several successions of wood end-uses: First, today's "normal" use as sawnwood, paper or boards; then recycling of some sorts for paper or construction wood; and finally as bioenergy for substitution of fossil fuels. The balance between these end-uses will depend upon P and the prices of the relevant substitution products (steel, solar energy, etc.).

An important issue here is the relation to other environmental benefits from forestry like recreation and biodiversity conservation. Prolonging the rotation time increases these two benefits in most cases. However, increased fertilisation, ditching of moor land, higher planting densities (in particular of monocultures), and afforestation decrease in most cases recreational conditions and biodiversity (Solberg and Hoen 1995). In practise, one will have to find a balance between competing benefits.

Since it is indifferent where on the globe a decrease in the concentration of atmospheric CO_2 occurs, the management implications for Nordic forestry discussed above are in principle relevant for most of the boreal and sub-boreal forests as well as for tropical forest areas. In the latter areas, stopping deforestation is one of the most severe problems. If an international market for tradable emission permits of CO_2 is created (cf. below), values of G in the order of magnitudes indicated in Table 1 could give strong incentives for building up forest biomass in many developing countries. This would involve decreased deforestation, rehabilitation of deforested areas, conservation of natural forest reserves, and tree planting for rural development. In tropical areas there could in many cases, if properly planned, be few conflicts between increased carbon sequestration in forest biomass and other environmental benefits, because increased forest biomass would in the right projects have positive environmental effects like preventing erosion, improving watershed and agriculture conditions, and increased biodiversity.

4.3. Policy Implications

With a carbon fee in the range of magnitude shown in Table 1 the economic value of forests will increase dramatically because of its potential as a carbon sink. This value is a public good, and in addition an international good in the sense that a reduction of the concentration of atmospheric CO_2 in one country benefits other countries. Today, there is no market to stimulate investments for this good. One important challenge will be to introduce policy means (including market mechanisms) that make it possible to arrive at an optimal investment level regarding forests as carbon sinks. This is not a trivial matter, and a lot of literature exists regarding environmental policy mechanisms.

For public forest lands the government can regulate the investment intensity for biomass production quite easily through direct budget allocations. The main problem will be to decide the optimal intensity on the various investments options, but this should not be too difficult, as partly discussed in the above section and by Hoen and Solberg (1994).

For private forest land (which in Scandinavia and some other parts of the world constitutes more than 70% of the total forest area) one can in principle use regulations by law or economic policy means, or combinations of these. It would, however, be very difficult in most countries to make the forest owners willing to increase the forest biomass without economic compensation of some sort — i.e. subsidies or tax incentives. These would have to be based on certain criteria for the different investments options, like quantity of carbon fixation, risk regarding implementation, ease of implementation and control. In Scandinavia it is a long tradition for regulating the investments in forestry for industrial wood production and recreation/conservation, which shows that it is possible to implement efficient forest policy means in practice (Solberg and Tikkanen 1992). One interesting issue here would be the income distribution effects, as subsidies in the order of magnitude indicated by Table 3 would imply severe income transfers to the forest land owners. This issue has to be solved politically.

Subsidies on CO_2 fixation in forest biomass have to be followed by corresponding taxes on emissions of CO_2 from the end-use of the forest biomass (including decaying in the forest of stump, roots and branches). This is discussed some further in Hoen and Solberg (this volume). In theory this may sound easy, but in practise the implementation problems might be severe.

It is not easy to predict what will happen with the forest industries if the above mentioned taxes and subsidies are implemented. In the short run prices of wood will increase because the timber supply decreases as rotation ages increases. This will press the forest industries for more use of recycled raw materials and residues, and worsen the situation for the industry. At the same time the competing products to wood will also be burdened with a fossil fee corresponding to P, and the price of these competing products will increase. The new equilibrium relative prices between wood and non-wood products will decide the future prospects of forest sector. Also, the increased build up of forest biomass will, after some time, increase the timber supply, lowering the relative price of wood and improve the situation for the forest industry. An important aspect here is that if the above mentioned subsidies and taxes are not introduced internationally, but only in one or a few countries, the forest industry in these countries will be severely hampered, and most likely competed out. In addition, the reduction in net emission of carbon will be negligible, as the production will just shift to other countries.

It is very costly to lower the concentration of greenhouse gases in the atmosphere, and it is important to do this as cost efficiently as possible (Dean 1993). Introduction of tradable emission rights represents a possible strategy. Although beyond the scope of this paper, it should be mentioned that such a system can be introduced on domestic level as well as internationally between two countries or more; needless to say it is more cost efficient the more countries are involved (Torvanger et al. 1994).

Regarding forestry a system of tradable permits for emission of CO_2 could, properly managed, be of considerable value for many developing countries. One could e.g. imagine that some of the emission fees collected from fossil fuel use in industrialised countries, were invested in building up forest biomass in developing countries (e.g. stopping deforestation, rehabilitation of degraded land, establishment of fuelwood or industrial wood plantations, etc.), thus improving agriculture, water catchment, and biodiversity, and providing valuable wood. This would require international agreements, control, and monitoring systems.

Most wealthy countries give technical assistance to poor countries. With a value on carbon sequestration as discussed, many projects on restoring and building up sustainable forest ecosystems in poor countries will be highly profitable investments. Properly planned, both the donor and the receiving countries could gain considerably from such projects.

REFERENCES

Anon. (1991), 'The greenhouse effect, impacts and policy means' [Drivhuseffekten, virkninger og tiltak], Report from Ministry of Environment, Norway. (In Norwegian).

Blingsmo, K.R. (1990), 'Transformation from production figures to biomass figures' [Omregning fra produksjonstabeller til biomassetabeller], Report from Norwegian Forest Research Institute, Ås, Norway. (In Norwegian).

Cline, W.R. (1992), *The economics of global warming,* Institute for International Economics, Washington D.C.

Dean, A. (1993), 'Macroeconomic model results', in Y. Kaya, N. Nakincenovic, W.D. Nordhaus, F.L. Toth eds., Costs, Impacts and Benefits of CO_2 Mitigation, Report CP-93-2, IIASA, Laxenburg, 213-233.

Dixon, R.K., S. Brown, R.A. Houghton, A.M. Solomon, M.C. Tiexler, and J. Wisniewski (1994), 'Carbon pools and flux of global forest ecosystems', *Science* **263**, 185-190.

EPA (1990), *Policy options for stabilizing global climate change, Vol.I & II,* Environmental Protection Agency, Office of Policy, Planning and Evaluation, Washington D.C.

Hoen, H.F. and B. Solberg (1994), 'Potential and economic efficiency of carbon sequestration in forest biomass through silvicultural management', *Forest Science* **40**, 429-451.

Hoen, H.K. and B. Solberg (1995), 'CO_2-taxing,timber rotations, and market implications', *Environmental and Resource Economics,* this volume.

Lunnan, A., S. Navrud, P.K. Rørstad, K. Simensen, and B. Solberg (1991), 'Forestry and forest production in Norway as a measure against CO_2-accumulation in the atmosphere' [Skog og skogproduksjon i Norge som virkemiddel mot CO_2-opphopning i atmosfÆren], Aktuelt fra Skogforsk 6/1991, Norwegian Forest Research Institute, Ås, Norway. (In Norwegian).

Nagoda, L. (1982), 'Chemical structure and properties of wood' [Kjemisk struktur og egenskaper til trevirke], lecture notes, Department of Wood Technology, Agricultural University of Norway. (In Norwegian).

Sedjo, R. (1990), 'The global carbon cycle', *Journal of Forestry* **88**, 33-34.

Solberg, B. and R.G. Haight (1991), 'Analysis of optimal economic management regimes for Picea abies stands using a stage-structured optimal-control model', *Scandinavian Journal of Forest Research* **6**, 559-572.

Solberg, B. and H.F. Hoen (1995), 'Economic aspects of carbon sequestration — some findings from Norway', in press.

Solberg, B. and I.Tikkanen (1992), 'Evolution of forest policy science in Finland and Norway', paper for IUFRO Centennial Meeting, Berlin, September 1992.

Torvanger, A., J.S. Fuglestvedt, C. Hagem, L. Ringius, R. Selrod, and H.A. Aaheim (1994), 'Joint implementation under the climate convention: phases, options and incentives', Report 6/1994, CICERO, Oslo.

Critical Reviews in Environmental Science and Technology, 27(Special): S335–S350 (1997)

INCREMENTAL COSTS OF CARBON STORAGE IN FORESTRY, BIOENERGY AND LAND-USE

JOEL N. SWISHER

UNEP Collaborating Centre, Risø National Laboratory, Box 49, DK-4000 Roskilde, Denmark

ABSTRACT: This paper presents a comprehensive and consistent methodology to account for the incremental costs and net changes in carbon stocks for different categories of forestry and biomass energy projects. The methodology allows consistent comparisons of the costs and quantities of carbon stored in different types of projects and/or national programs, facilitating the inclusion of forestry and biomass energy projects in a possible CO_2 emission reduction regime. The framework presented includes guidelines for defining a reference case against which carbon-storage projects are compared, a carbon-storage accounting method based on simple analytic techniques for one-time terrestrial sinks, an endowment approach to incremental costing, and a discussion of local benefits and risk factors.

KEY WORDS: Terrestrial carbon sinks, carbon storage costs, global climate change, carbon stock models, forest conservation, reforestation, biomass energy.

1. INTRODUCTION

Deforestation is presently a significant source of global carbon dioxide emissions. The possibility of reducing or reversing this process, through protection of existing forested lands, reforestation and afforestation, is therefore a potentially important component of a global climate-change mitigation strategy, and possibly a relatively cost-effective component. Moreover, the prospect of using harvested biomass as an energy source to replace fossil fuels makes the potential for using forest lands to prevent carbon emissions even larger over the long term.

The potential importance and cost-effectiveness of forestry, bioenergy and other land-use measures in climate change mitigation indicates the need for detailed evaluation of cost and performance in increasing terrestrial carbon storage. These measures involve the choice of land-use options that store more carbon than the land-use expected without the measures in place. Thus, the cost and carbon-storage benefit must be measured and evaluated as net values compared to a reference case. The need for this type of evaluation has led to the development of incremental costing frameworks for evaluating climate-change mitigation strategies and projects. Incremental cost analysis of such projects is a key evaluation tool for the Global Environment Facility (Ahuja 1994).

The objective of this paper is to demonstrate an analytic framework for the consistent evaluation of the magnitude and cost of carbon storage by a "bottom-up" approach, beginning at the project level. Rather than hypothesize about reforesting areas of continental dimensions, the approach here is to consider carbon storage in a context in which it might succeed: that of a local or national sustainable forest development strategy, with an emphasis on tropical developing countries.

2. TERRESTRIAL CARBON STORAGE: A METHODOLOGICAL FRAMEWORK

The most direct use of carbon-storage costs is to compare such costs to other measures to mitigate global climate change, such as in the energy sector. Carbon storage by maintaining and enhancing carbon sinks is, by its nature, a different process from reducing an annual flow of emissions from an energy conversion system. Energy-sector emission-reduction measures prevent the emission of a quantity of irretrievable carbon emissions. The only valid analogue to such emission prevention is permanent carbon storage in terrestrial biomass; temporary short-term storage is not comparable.

Thus, the goal of a project to enhance carbon sinks is long-term sustainable carbon storage. However, terrestrial carbon sinks do not accumulate carbon indefinitely, but approach a limiting value. The carbon-storage benefit of a carbon sink is a one-time increment in the carbon stock on land. A project that simply removes some carbon from the atmosphere by growing biomass today, if the biomass will be decomposed or burned after a short time, is not sufficient as a long-term increment in carbon storage and does not sustainably reduce the carbon stock in the atmosphere (see Section 6).

For the purpose of this analysis, the dynamics of the carbon stocks over time are not considered in detail. Rather, only the long-term average carbon storage is analyzed (Swisher 1991, Winjum et al., 1992, Sathaye et al., 1995). Although the year-to-year dynamics of carbon stocks and flows may be important for analyses such as emission inventories, they are not highly relevant to a carbon-storage methodology, and they introduce considerable complexity and uncertainty. To contribute to slowing global climate change, forestry and land-use projects must keep carbon out of the atmosphere indefinitely (bioenergy projects can be an exception). Thus, the long-term net carbon storage is of concern here, rather than questions of the precise timing of carbon uptake and release.

The development of a comprehensive incremental costing methodology requires a consistent framework (Swisher 1991). This framework should provide for the evaluation of:

1. Reference case from which all changes and costs are measured. This step is inherently ambiguous, but project costs and impacts can only be evaluated relative to what they replace (Section 3).
2. Carbon accounting for all forestry, bioenergy and land-use project types, even if not all project types are relevant to a particular analysis, as a consistent basis for comparison (Section 4).
3. Cost accounting for the full requirements to endow a project over its lifetime, net of any benefits beyond those needed to cover costs and incentives to keep the project operating (Section 5).

4. Local environmental and social benefits, not just to assess the positive local impact of a forestry project, but because the provision of tangible local benefits is essential for success and thus directly related to the realistic potential for long-term carbon storage (Section 7).

The net carbon-storage credit is a one-time value that is assumed to be sustainable. The corresponding cost value should therefore also be a one-time (rather than an annualized) value. This approach is comparable to the application of annualized cost values to the regular and roughly uniform annual emission reductions from energy-sector measures. In both cases, the result can be expressed in terms of $/tC. The cost value should include the project's opportunity cost and an endowment sufficient to cover the project's development and the expenses and incentives for on-going operation, sufficient to assure the sustainability of the project and its carbon storage (see Section 5).

$$\text{Incremental unit cost} = C'_{net}/R'_{net} \tag{1}$$
$$R_{net} = A \ R'_{net} \tag{2}$$

where:

C'_{net} = Net endowment cost of the carbon-saving project, instead of the reference ($/ha)
R'_{net} = Net sustainable carbon storage in the project, compared to the reference (tC/ha)
R_{net} = Total net sustainable carbon storage in the project, compared to the reference (tC)
A = Land area of the project (ha)

3. REFERENCE CASES AND OPPORTUNITY COSTS

The carbon-accounting and costing procedure should explicitly account for the costs and carbon storage of the land-use in the absence of the project, a difficult but necessary step in estimating the impact of both forest and energy projects (UNEP 1992). One aspect of the reference case is the land value for the various types of land-uses identified. This is necessary to determine the opportunity cost of establishing other land-uses such as forestry projects. If possible, rural land-value data should be taken from national statistical data. The data should be statistically corrected to provide estimates of land-values for rural land areas with minimum services, differentiated by bio-climatic region.

The reference land-use type may not be not static, especially for land areas that are subject to deforestation pressure and therefore of interest with respect to forest protection projects. Although changes in land cover and use can be complex, the long-term average carbon storage is analyzed here. This requires conversion of year-by-year estimates of changes in land cover and land-use into more stable values. One can base the reference condition on a simple estimate of the share of land area with a given land-use type. This is reasonable for plantations, where the existing land-use represents the reference condition, but it is difficult if deforestation is expected in the reference case.

Whether the reference land use resembles the existing conditions, or represents a counter-factual case meant to be prevented, depends on the type of project. In the case of forest reserves, the reference is the land-use that would result without the reserve, and net carbon accumulation will be positive if the reference case includes biomass reductions, i.e., deforestation. If carbon-storage credit were not given for protecting natural forests, there would be counter-productive incentives to remove the existing forest and then replant. Although this strategy might increase the rate of carbon removal by trees, it would likely result in a net increase of the atmospheric carbon stock (Harmon et al., 1990).

For the reference, one must estimate the long-term average shares of land area occupied by different land-uses such as primary forest, secondary forest, logged forest, agriculture, pasture, and fallow or abandoned land. One quantitative approach is to use a Markov matrix to analyze the trajectory of existing and future land-use types (Fearnside 1992). Based on estimated rates of land-use change from one type to another, one can construct a Markov matrix that shows that probability, in any given year, of one land-use category being converted to another. The matrix can be solved to yield an estimate of the steady-state shares of land area occupied by each land-use type.

4. METHODOLOGY FOR CARBON-STORAGE ACCOUNTING

This section classifies the different types of forestry and land-use projects and provides a framework for carbon accounting in Section 4.1. Specific details regarding each type of carbon-stock calculation are given in Sections 4.2–4.5, where simple but credible techniques are suggested for consistently analyzing a comprehensive range of projects and related carbon stocks. In each case, more complex models can be used, but the problem with such models is that often they were designed to answer a specific research question about a certain type of forestry or land-use issue. As a result, they have far more detail than needed here in some areas, while they completely omit other relevant areas, making consistent comparisons difficult. Care must be taken when using such models to ensure that a consistent evaluation is made, based on the net increment in carbon stock for a unit of land area (tC/ha), sequestered from the atmosphere, compared to a reference condition for that land area.

4.1. Categories of Projects and Carbon Stocks

Reversing the trend toward deforestation involves several different forestry project classifications, including plantations, agroforestry, forest reserves, etc., each with different costs and different net impacts on carbon stocks. Some accumulate carbon in new biomass grown in the project, some store carbon in standing natural forest, some accumulate carbon in harvested products that enter long-term storage, and biomass energy plantations store net carbon in unburned fossil fuel by preventing carbon emissions from fossil fuel use. Each different project classification can be explicitly distinguished by types of carbon stock (standing biomass, new biomass, harvested biomass, soil carbon, fossil fuel) that are relevant, as shown in Table 1.

The basic formula for the net carbon-storage credit must estimate and sum these carbon stocks for a particular project, and compare them to the carbon stocks for a reference land-use:

$$R'_{net} = CV_p + CS_p + CF_p - CV_r - CS_r \qquad (3)$$
$$CV_p = CV_{nat(p)} + CV_{av(p)} + CV_{h(p)} \qquad (4)$$
$$CV_r = CV_{nat(r)} + CV_{av(r)} + CV_{h(r)} \qquad (5)$$

where:

CV_p = Carbon stored in vegetation and products by the project (tC/ha)
CS_p = Carbon stored in soil by the project (tC/ha)
CF_p = Carbon content of fossil fuel replaced by biomass fuel from the project (tC/ha)
CV_r = Carbon stored in vegetation and products for the reference state (tC/ha)
CS_r = Carbon stored in soil for the reference state without the project (tC/ha)
CV_{nat} = Carbon storage in natural forest (tC/ha)
CV_{av} = Average biomass carbon during the rotation of plantation (tC/ha)
CV_h = Average carbon value of the harvested biomass (tC/ha)

For each project classification, we can use Table 1 to relate the terms in equations 3-5 to the different changes in carbon stocks for a project. Each type of carbon stock (standing biomass, new biomass, harvested biomass, soil carbon, fossil fuel) corresponds to one term (CV_{nat}, CV_{av}, CV_h, CS, CF).

4.2. Carbon in Standing Biomass and Soils

A simple estimate of the carbon storage in the biomass and soil of natural forests in tropical bio-climatic zones, as a function of climate, can be given by (Brown and Lugo 1982):

$$CV_{nat} = 298 + 239 \log Z - 112 \qquad \text{(tC/ha)} \qquad (6)$$
$$CS_{nat} = 154 \exp(-0.45 Z) \qquad \text{(tC/ha)} \qquad (7)$$

where

Z = Average annual (temperature/rainfall) ratio in °C-year/dl

Note that the forest stock estimates derived from these relationships have been revised by their authors. The revised (lower) average stock estimates include secondary forests and their (lower) carbon content (Brown et al., 1989). The relationships in equations 6 and 7 are thus appropriate only for primary forests, which is the intended application here. For other existing reference conditions to which a proposed project might be compared, such as farm or pasture land, similar types of carbon storage estimates are available.

Alternatively, more sophisticated techniques can be used to incorporate the effects of soil, slope, drainage, etc. A reasonable compromise might be to use the algorithms developed for the Biome model, which combines climate and soil data to predict potential land cover and can be used to estimate carbon storage (Prentice et al., 1992, Klein-Goldewijk et al., 1994). One can also use more detailed maps of soil types and land-use capacity that are available in some countries such as Costa Rica (see Tosi et al., 1985).

TABLE I
Parameters for Calculation of Carbon Storage by Project Classification

Type of Project: Variable:	Standing Biomass CV_{nat}	New Biomass CV_{av}	Harvested Biomass CV_h	Soil Carbon CS	Saved Fossil Energy CF
Forest Reserves/Protection	+	0	0	+	0
Natural Forest Management	+	0	+	+	0
Timber Plantations	0	+	+	+	0
Forest/Ecosystem Restoration	0	+	0	+	0
Fuelwood Farms	+	+	0	+	0
Agroforestry	+	+	+	+	0
Biomass Energy Plantations	0	+	0	+	+

Note: "+" means the carbon stock applies to the project classification, "0" means it does not.

Source: Swisher 1991

The same relationships used to estimate carbon storage in forestry projects can be applied to the reference case if the reference includes forest land. However, the reference condition may also include some amount of secondary forest and fallow or abandoned land, which does not fit one of the above categories well. The relevant carbon storage parameters for possible reference land-use categories are shown in Table 2. The biomass and soil carbon densities for these land-use conditions can be simply estimated as fractions of CV_{nat} or CS_{nat} in the same climate (Houghton et al., 1987, Brown et al., 1989).

$$CS = N_s \, CS_{nat} \qquad (8)$$
$$CV_r = N_v \, CV_{nat} \qquad (9)$$

where:

N_s = 0.50 for steep and highly erodible areas
N_s = 0.75 for pasture, cropland and fallow woodland
N_s = 0.90 for logged or secondary forest and timber plantations
N_s = 1.00 for natural forest management, forest restoration and agroforestry
N_v = 0.40 for mature fallow woodland
N_v = 0.75 for land converted to logged or secondary forest

4.3. Carbon Accumulated by Afforestation

The carbon accumulated in plantations by new biomass, CV_{av}, is the long-term average biomass carbon over the period of rotation. CV_{av} depends on CV_m, the carbon stored in vegetation planted by the project, upon maturity. The value for CV_m depends on the mean annual biomass increment (MABI), the ratio of total-to-stemwood biomass, and the carbon density of wood (Brown et al., 1986). These values, especially the MABI that measures growth potential, vary with species, soil and climate and can be adjusted according to bio-climatic zone or land-use capacity (Tosi et al., 1985, Prentice et al., 1992).

TABLE 2
Parameters for Calculation of Carbon Storage by Reference Land-Use Category

Reference Land Uses:	*Carbon Stock*: *Variable*:	Standing Biomass CV_{nat}	New Biomass CV_{av}	Harvested Biomass CV_h	Soil Carbon CS
Pasture/Agriculture		0	0	0	+
Fallow/Shifting Cultivation		0	+	0	+
Secondary Forest		0	+	+	+
(Selective) Logged Forest		+	0	+	+
(Selective) Fuelwood Removal		+	0	0	+

Note: "+" means the carbon stock applies to the land-use category, "0" means it does not.

A simple estimate of CV_m, based on growth in a young forest, at which time annual growth is 6 percent of CV_m, is (Brown et al., 1986, Cooper 1983):

$$CV_m = p_c \, MABI_{max} \, SWM/g \qquad (10)$$

where:

p_c = Carbon density of wood (typically 0.23-0.29 tC/m^3)
$MABI_{max}$ = Maximum MABI for a young forest (m^3/yr)
SWM = Stemwood multiplier to convert to total biomass (ranges from 1.6 to 2.5)
g = Maximum growth rate (typically 0.06/year from Brown et al., 1986)

The value for CV_{av}, the average biomass carbon storage over the length of a rotation (t_h), is proportional to the average ratio of standing biomass to biomass at maturity (B_{av}/B_m). This ratio depends on t_h and on the fraction of biomass remaining after harvest (B_o/B_m).

$$CV_{av} = CV_m \, (B_{av}/B_m) \qquad (11)$$

where:

$$(B_{av}/B_m) = \sum_{t=0}^{t=t_h} (B_t/B_m)/t_h \qquad (12)$$

and

(B_t/B_m) = Ratio of standing biomass at time t to biomass at maturity
t_h = Rotation time between harvests

The ratio (B_t/B_m) is a function of the growth rate, the time since harvest, and the amount of biomass remaining after harvest. Although sometimes represented as a simple linear function, this relationship can be estimated with more reasonable accuracy from simple non-linear expressions, such as given in Cooper (1983):

$$(B_t/B_m) = \{ \, 1 - \exp(-g \, t) \, [\, 1 - (B_o/B_m)^{-n} \,] \, \}^{-1/n} \qquad (13)$$

where:

(B_o/B_m) = Ratio of biomass remaining after harvest to biomass at maturity
n = -0.17 for fast-growing species (from Cooper 1983)
g = Maximum growth rate (typically 0.06/year from Brown et al., 1986)

Alternatively, more complex forest-growth models can be used, such as the CO2fix model from the Dutch Institute for Forestry and Nature Research (Nabuurs and Mohren 1993). This model simulates in greater detail the year-to-year dynamics of forest growth, management and decay, and it can be used to calculate the carbon storage parameters given in the above methods. Although the CO2fix model deals with growth and harvest of timber plantations in detail, it is not well suited to analyzing forest systems other than plantation projects. Critical parameters such as the existing natural biomass and soil carbon must come from outside the model. For plantation and restoration projects, however, this type of model gives more detailed and precise results than the simpler relationships given above.

4.4. Carbon in Forest Products

For projects in which a share of the biomass harvested from each rotation is used as timber, rather than paper or fuelwood, this biomass enters long-term storage. CV_h is the average value of stored biomass carbon net of decay. If CV_h is assumed to accumulate until balanced by decay, which occurs at a constant annual rate depending on the type of wood and where it is used (Nabuurs and Mohren 1993), the average accumulation approaches the steady state value. This value is proportional to the difference between (B_o/B_m) and (B_h/B_m), the biomass fraction before and after harvest and N_{lts}, the fraction that enters long-term storage (Cooper 1983).

$$CV_h = CV_m \, N_{lts} \, [(B_h/B_m) - (B_o/B_m)] \, [\, \exp(-d \, t_h)/(1 - \exp\{-d \, t_h\} \,) \,] \qquad (14)$$

where:

$$(B_h/B_m) = (B_t/B_m) \text{ for } t = t_h \qquad (15)$$

and

N_{lts} = Fraction of the harvest that enters long-term storage with decay rate d
d = Annual decay rate of harvested timber
t_h = Rotation time between harvests

where

d = 0.01–0.03 for timber
d = 0.05–0.1 for particle board and related uses
d = 0.3–0.5 for paper and packaging (can generally be considered insignificant)

This approach does not capture potential feedbacks driven by the price of wood. For example, increased tree planting and timber production could decrease the price of timber and lead to reduced planting and carbon storage elsewhere. On the other hand, in a future that includes significant climate-change mitigation, the development of biomass energy technology could lead to greatly increased demand for plantation wood as fuel, thus increasing prices and tree planting rates.

4.5. Carbon in Saved Fossil Fuel

The carbon content of the fossil fuel replaced by a biomass energy project, CF, is that of the fossil fuel that would be consumed to produce the equivalent amount of commercial energy, over the life of the project, such as a power plant. This value depends on the type of fossil fuel replaced, the heating value of the biomass fuel, the relative efficiencies of biomass and fossil fuel combustion (including losses in transport) and the carbon embodied in fossil fuel used to grow, harvest and transport the biomass fuel (Swisher 1993, Turhollow and Perlack 1991).

$$CF = CV_m \left[(B_h/B_m) - (B_o/B_m) \right] \left[(t_p/t_h) (E_{nrg}/E_{fos})/(X_{nrg}/X_{fos}) - N_h \right] \quad (16)$$

where:

(t_p/t_h)	=	Ratio of energy project life to harvest cycle
(E_{nrg}/E_{fos})	=	Ratio of bioenergy system efficiency to that of fossil fuel system
(X_{nrg}/X_{fos})	=	Ratio of biomass carbon content to that of fossil fuel
N_h	=	Ratio of carbon emissions from biomass production to biomass carbon content

Assuming equal efficiencies for biomass and fossil fuel production and combustion, using a heating value of 19 GJ/ton-biomass, or 38 GJ/tC (26 kgC/GJ), and taking the carbon content of the fuel replaced as 24 kgC/GJ for coal, 20 kgC/GJ for oil, and 14 kgC/GJ for natural gas, the carbon saved is 0.91 tC for coal, 0.76 tC for oil, and 0.53 tC for natural gas, per tC in the wood burned (Swisher 1993). If the value of N_h is 0.06, these results are reduced to 0.85 for coal, 0.70 for oil, and 0.47 for natural gas (Marland and Marland 1992, Turhollow and Perlack 1991). Biomass energy plantations also can accumulate carbon as saved fossil fuel, beyond the one-time carbon-storage values for forest biomass, indefinitely into the future, provided that the project can be replicated and the biomass fuel can continue to be harvested sustainably (Hall et al., 1991, Marland and Marland 1992, Swisher 1993).

4.6. Comparison of Carbon Stock Results

As an example of the comparability of the simple methods presented above with more complex methods, results of carbon storage estimates for several projects are compared to simulations with CO2fix. To the extent possible, the inputs were harmonized for each run, although the detail required for CO2fix input files is beyond that of the data available for some projects. Table 3 shows the results of the comparisons, all of which are based on plantations of Caribbean pine (*pinus caribaea*).

The values given for CV_{av} in Table 3 agree well for most cases, although the values from CO2fix are significantly lower in the cases based on Brazil data from the CO2fix document (Nabuurs and Mohren 1993) and from projects in Ecuador (McCormick 1987). These cases include heavy thinning of the plantations at early stages of the rotation, and the resulting decrease in CV_{av} is not fully captured in the method presented here. The values for CS_{nat} agree reasonably well, although the net soil carbon storage cannot be compared because CO2fix does not calculate these values. The values given for CV_h in Table 3 are generally in good agreement between the two models.

TABLE 3

Comparison of Carbon Stock Results: C=CO2fix (Nabuurs and Mohren 1993), S = Stock Method Presented Here

Project	CV_{av} tC/ha		CV_h tC/ha		Total of $CV_{av}+CV_h$		CS_{nat} tC/ha		Increase in CS_{nat}	
	C	S	C	S	C	S	C	S	C	S
1. Brazil pine plantation	62	78	34	36	96	114	98	112	1	17
2. Brazil pine plantation — poor site	39	55	24	25	63	80	76	74	1	11
3. Costa Rica pine plantation — good site	72	68	73	88	145	154	107	112	1	17
4. Costa Rica pine plantation	53	47	53	62	106	109	76	74	1	11
5. Ecuador pine plantation — good site	52	65	80	80	132	145	112	112	1	17
6. Ecuador pine plantation	45	51	56	61	101	112	82	82	1	12
7. Venezuela pine plantation	41	42	33	31	74	73	83	85	1	13
8. Venezuela pine plantation — poor site	28	29	22	21	50	50	53	42	1	6

[1] CO2fix does not calculate changes in soil carbon compared to a reference case.

Data sources: 1,2: Nabuurs and Mohren 1993; 3,4: Swisher 1991; 5,6: McCormick 1987; 7,8: CVG 1993. Total carbon storage values given by Swisher (1991) are smaller because they include reduction factors for project risks.

5. COSTS: THE ENDOWMENT METHOD

The cost should include the net opportunity cost of the land used for the project, rather than the land-use that the project replaces, and an endowment sufficient to cover the project's establishment and the expenses and incentives for its on-going operation, including maintenance, management and monitoring, sufficient to assure the sustainability of the project and its carbon storage:

$$C_{net} = A\ C'_{net} = C_{opp} + C_{estab} + C_{maint} + C_{mngmt} + C_{monit} + C_{energy} \qquad (17)$$

where:

C_{net} = Total net endowment cost of the project, instead of the reference
C_{opp} = Net opportunity cost of land
C_{estab} = Cost of establishing the project
C_{maint} = Cost of management and extension services
C_{mngmt} = Cost of endowment for project maintenance
C_{monit} = Cost of endowment for performance monitoring
C_{energy} = Net cost of biomass energy production

A comprehensive assessment of project costs is emphasized because it can be a difficult objective to fulfil in developing countries, where both the cost information and the markets they describe may be incomplete or nonexistent. In existing assessments of carbon storage costs in forestry, it is noted that some cost values are incomplete or inconsistent (Sedjo et al., 1995). Such estimates may therefore be underestimates of carbon storage costs according to the methods presented here, which are explicitly designed to be comprehensive and consistent. A difficulty with a comprehensive approach, compared to simply using establishment costs, is that needed data are not available for a range of project types in many developing countries. This difficulty may help explain the incomplete studies to date.

The net opportunity cost is the difference between the value of the land without the project and the present value of project benefits. In the simplest case (protected reserve only), the opportunity cost is the land value. If the project produces marketable products together with the carbon storage, the benefits from these products can be considered to reduce the opportunity cost. Some projects with potentially large benefits might show negative net opportunity costs, suggesting that non-market barriers are preventing these profitable projects from being implemented.

Capturing the importance of overcoming such barriers and maintaining on-going incentives over the life of a project is a reason to use an endowment approach to costing (Swisher 1991, IPCC 1995). The main difference in this approach, compared to simple present-value analysis, is that net opportunity cost is not allowed to be negative. This means that the recurring direct costs, such as maintenance and monitoring costs, can always be covered without borrowing against future revenues, which is analogous to accounting for working capital in financial analysis.

$$Copp = Maximum\ (0,\ P_{land} - B_{prod}) \qquad (18)$$

and

$$B_{prod} = \sum_{t=t_b}^{t=t_h} (GR_t - PC_t)/(1+r)^{t-1} \tag{19}$$

where:

P_{land} = Land value
r   = Discount rate
GR_t = Gross revenues from the land's output in year t
PC_t = Production costs of project inputs in year t
t_b = First year that revenues are sufficient to cover costs

If the net opportunity cost is zero, it indicates that the project represents the highest value use of the land. Additional revenues are not deducted from the endowment cost, under the assumption that such local benefits are retained, rather than repaid into the project, as necessary incentives for continued participation in the project. In the endowment approach, benefit values should be calculated for each year beginning with the first year that gross revenues are sufficient to cover costs. Costs (and benefits) incurred before that time are included in the cost of establishing and maintaining the project.

The costs of establishing and maintaining plantations and other forestry projects are covered well in the literature (Durst 1987, Sedjo 1983). In addition, management and monitoring costs must be treated as necessary costs to ensure project success and verify carbon storage (Swisher 1991). The costs for biomass energy projects are the capital and operating costs of bioenergy production in excess of present value of energy savings (Swisher 1993, Hall et al., 1991).

6. TIMING AND DISCOUNTING

The endowment costs are one-time values, which are compatible with the carbon-storage values, and they give incremental cost values for carbon storage in $/tC. It should be noted that in general these cost values are undiscounted costs. The costs are generally incurred when a project is initiated, and the net carbon storage can occur almost immediately, gradually over the project life, or decades later, depending on the type of baseline or reference land use. The endowment cost can be discounted forward to the time when carbon-storage credit is verified, increasing the cost value in $/tC.

This approach suggests that a one-time increase in carbon storage would be less expensive if one waited until sometime in the future rather than implement the project today. The carbon storage value is the same, but future money is worth less than today's money. Thus, reforesting a given area would appear cheaper the longer one waits to establish the project and incur the costs.

This does not mean that forestry projects in general have less urgency, because many of the least expensive options, and ones with other important benefits such as biodiversity, involve preventing deforestation. Protection and prevention can only be accomplished before the deforestation has occurred, while the forest remains. The protection option is thus a one-time opportunity. Such measures cannot be delayed until later when they might be less expensive. Rather they would be missed entirely, and more expensive measures (e.g., restoration) would be needed.

An alternative to using strictly one-time carbon-storage values is to assign a time-value to carbon emissions. However, it is problematic to attempt to discount carbon, which may have a different time-value than the monetary time-value indicated by the discount rate. As the atmospheric concentration of greenhouse gases increases in the future, the potential damage value of additional emissions is likely to increase, suggesting that preventing future emissions may be worth more than preventing present emissions. If we discount carbon, however, the time when the discounted damage value reaches zero is about when greenhouse gases are expected to begin affecting the global climate.

There are also risks of discounting carbon at the project level. For example, assume that a forest plantation stores 2 tC/ha-year uniformly for 50 years, and that the forest is then completely cleared for pulpwood and returned to its initial case without replanting, thus releasing 100 tC/ha. The long-term carbon-storage benefit in this is zero. However, if the carbon flows are discounted at 4% per year, the "present value" of the 50 annual positive flows exceeds that of the final negative flow by 33 tC/ha, suggesting a significant positive benefit in terms of carbon storage.

7. LOCAL BENEFITS AND RISK FACTORS

A carbon forestry strategy that tends to exclude human inhabitants should be suspect, as the demand for land and income make it difficult and inappropriate to put global benefits before local subsistence. Instead, the strategy should foster the development of sustainable land uses from which local inhabitants derive tangible benefits. The local socio-economic effect is the key to the success of tropical forestry projects and their prospect of providing long-term carbon storage, especially in developing countries (In industrialized countries, these effects can more reliably be assumed to be expressed by market forces). Indeed, it is difficult to imagine tropical forest development achieving sufficient longevity for long-term carbon storage without providing local benefits. This requirement is what makes it difficult to distinguish explicitly between a planned project and its reference case which, it might be argued, should include the project if its benefits are substantial.

At the same time, a project's carbon storage estimate should account for the possibility that some part of the project either will not succeed, or would have been done regardless of the project (Swisher 1991). In either case, the effective carbon stored would decrease, and the cost per tC would increase. These risks depend in part on the local barriers to implementation and the provision of local benefits. The barriers include land tenure uncertainty and the lack of long-term incentives, political support, technical experience, management plans, facilities for sustainable uses such as eco-tourism, and qualified personnel to carry out projects and provide extension services needed to successfully involve local people. The project risk resulting from the presence of these barriers depends on the adequacy of project management, maintenance and monitoring, all of which are essential to overcome such barriers, and the adequacy of related project expenditures.

REFERENCES

Ahuja, D., (1994), *The Incremental Cost of Climate Change Mitigation Projects*, Global Environmental Facility Working Paper, Washington.

Brown, S., A.E. Lugo, (1982), "The Storage and Production of Organic Matter in Tropical Forests and their Role in the Global Carbon Cycle," *Biotropica*, vol. 14, pp. 161-178.

Brown, S., A.E. Lugo, J. Chapman, (1986), "Biomass of Tropical Tree Plantations and its Limitations for the Global Carbon Budget," *Canadian Journal of Forest Research*, vol. 16, pp. 390-394.

Brown, S., A.J.R. Gillespie, A.E. Lugo, (1989), "Biomass Estimations for Tropical Forests with Applications to Forest Inventory Data," *Forest Science*, vol. 35, no. 4, pp. 881-902.

Cooper, C.F., (1983), "Carbon Storage in Managed Forests," *Canadian Journal of Forest Research*, vol. 13, pp. 155-166.

Corporación Venezolana de Guayana (CVG), (1993), "Programa de Desarrollo Forestal del Oriente de Venezuela," CVG, Caracas.

Durst, P.B., (1987), "Financial Aspects of Contract Reforestation in the Philippines," Southeastern Center for Forest Economics Research, FPEI Working Paper no. 10, Raleigh, NC.

Fearnside, P.M., (1992), "Carbon Emissions and Sequestration in Forests: Brazil," Lawrence Berkeley Lab., LBL-32758.

Hall, D.O., H.E. Mynick, R.H. Williams, (1991), "Cooling the Greenhouse with Bioenergy," *Nature*, vol. 353, pp. 11-12.

Harmon, M.E., W.K. Ferrel, J.F Franklin, (1990), "Effects on Carbon Storage of Old-Growth Forests Converted to Young Forests," *Science*, vol. 247, pp. 699-702.

Houghton, R.A., R.D. Boone, J.R. Fruci, J.E. Hobbie, J.M. Melillo, C.A. Palm, B.J. Peterson, G.R. Shaver, G.M. Woodwell, (1987), "The Flux of Carbon from Terrestrial Ecosystems to the Atmosphere in 1980 Due to Changes in Land Use," *Tellus*, vol. 39B, pp. 122-139.

Intergovernmental Panel on Climate Change (IPCC), (1995), *Methods for Assessment of Mitigation Options*, IPCC Working Group II, UNEP, Geneva.

Klein-Goldewijk, K., J.G. van Minnen, G.J.J. Kreileman, M. Vloedbeld, R. Leemans, (1994), "Simulating the Carbon Flux between the Terrestrial Environment and the Atmosphere," *Water, Air and Soil Pollution*, vol. 76, pp. 199-230.

Marland, G., and S. Marland, (1992), "Should We Store Carbon in Trees?" *Water, Air and Soil Pollution*, vol. 64, pp. 181-195.

McCormick, I., ed., (1987), *Analisís Económico de Inversiones en Plantaciones Forestales en Ecuador*, Dirección Nacional Forestal, Quito, Ecuador.

Nabuurs, G., and G. Mohren, (1993), *Carbon Fixation through Forestation Activities*, Institute for Forestry and Nature Research, IBN Report 93/4, Wageningen, Netherlands.

Prentice, I.C., W.P. Cramer, S.P. Harrison, R. Leemans, R.A. Monserud, A.M. Solomon, (1992), "A Global Biome Model Based on Plant Physiology and Dominance, Soil Properties and Climate," *Journal of Biogeography*, vol. 19, pp. 117-134.

Sathaye, J., W. Makundi, K. Andrasko, (1995), "A Comprehensive Mitigation Assessment Process for the Evaluation of Forestry Mitigation Options," *Biomass and Bioenergy*, in press.

Sedjo, R.A., (1983), *The Comparative Economics of Plantation Forestry*, Resources for the Future, Washington.

Sedjo, R.A., J. Wisniewski, A. Sample, J.D. Kinsman, (1995), "The Economics of Managing Carbon via Forestry: Assessment of Existing Studies," *Environmental and Resource Economics*, vol. 5, pp. 1-27.

Swisher, J.N., (1991), "Cost and Performance of CO2 Storage in Forestry Projects," *Biomass and Bioenergy*, vol. 1, no. 6, pp. 317-328.

Swisher, J.N., (1993), "Bottom-Up Comparisons of CO_2 Storage and Costs in Forestry and Biomass Energy Projects," *Proceedings Biomass Conference of the Americas*, Burlington Vt.

Tosi, J. A., et al., (1985), *Sistema para la Determinación de la Capacidad de Uso de las Tierras de Costa Rica*, Centro Científica Tropical, San José.

Turhollow, A., R. Perlack (1991), "Emissions of CO_2 from Energy Crop Production," *Biomass and Bioenergy*, vol. 1, no. 2, pp. 129-135.

UNEP, (1992), *UNEP Greenhouse Gas Abatement Costing Studies. Phase One Report.* UNEP Collaborating Centre on Energy and Environment, Risø, Denmark.

Winjum, J.K., R.K. Dixon, P.E. Schroeder, (1992), "Estimating the Global Potential of Forest and Agroforest Practices to Sequester Carbon," *Water, Air and Soil Pollution*, vol. 64, pp. 213-227.

Critical Reviews in Environmental Science and Technology, 27(Special): S351–S364 (1997)

ASSESSING TIMBER AND NON-TIMBER VALUES IN FORESTRY USING A GENERAL EQUILIBRIUM FRAMEWORK

WILLIAM A. THOMPSON, G. CORNELIS van KOOTEN and ILAN VERTINSKY

Forest Economics and Policy Analysis Research Unit, University of British Columbia, Vancouver, B.C. V6T 1Z3

ABSTRACT: Net benefits of alternative forest policies are examined using a simulation model that links a model of the biology and economics of the forest with a GE model of the B.C. economy, thereby recognizing the effects of changes in timber supply on prices. Results indicate that: (1) price effects from the GE model reinforce policies oriented towards managing for non-timber values; (2) carbon uptake values can generally be ignored if other nonmarket values are included; and (3) untargeted incremental silviculture is often uneconomic, but it increases C sequestration by 3,000-13,000 tonnes per year.

KEY WORDS: Cost-benefit analysis; forest management; general equilibrium model; climate change; non-timber values.

1. INTRODUCTION

More than 95 percent of the productive forest lands in British Columbia (B.C.), or 45.6 million hectares, are owned and administered by the Provincial government. While the objective of managing forests for multiple use and existence values has been included in the B.C. Forest Act since 1978, forest management has emphasized timber values. Indeed, the models used by the Province to evaluate alternative forest management policies focus on timber production, while ignoring other forest values or representing them as constraints imposed on forest practices. These models also ignore the market power of the B.C. forest sector, treating timber prices as exogenous variables, although B.C. accounts for one-third of global softwood lumber exports.

This paper describes a new simulation model of the forest system that links the biology and economics of the forest, explicitly recognizing the effects of changes in timber supply on prices and revenues. This is done by linking a comprehensive forest system model, that describes the dynamics of the biological system and the direct economic effects of alternative harvesting policies, to a general equilibrium model of the Province as an open economy. The simulation model shows the costs and benefits of competing uses of the forest (logging and recreation), as well as carbon uptake and existence values. The model is used to examine the consequences of alternative policy options, where nonmarket values are included in the cost-benefit evaluation criterion.

2. THE FOREST MODEL

The simulation model used in this study is an updated version of the timber supply model FOREST (version 6), originally developed by researchers at FEPA and a private consulting firm for the B.C. Ministry of Forests (hereafter MoF) (Thompson et al., 1992). FOREST projects (a) the forest inventory and timber harvest for B.C., given specification of forest protection, forest growth and yield, silviculture, economics, and harvest targets, priorities and constraints; and (b) the outcomes of management practices on inventory and timber yield. The current version of FOREST incorporates a new feature that projects non-timber benefits associated with the current state of, and changes to, the forest (including additions or deletions of land to the commercial forest land base). Further, the model has been linked with a general equilibrium (GE) model of the B.C. economy to simulate timber price responses to changes in regulated timber harvest rates or other exogenous factors.

Only the basic elements of FOREST are described here, as details are available elsewhere (Thompson et al., 1992; Phelps et al., 1990). The model uses forest inventory for the forest lands of B.C. regulated and administered by the MoF as Tree Farm Licenses (TFLs) and Timber Supply Areas (TSAs)—two forms of public forest land tenure that rely on private and public management, respectively. The inventory covers 25.7 million hectares judged suitable for commercial forestry, and MoF classifies it by region, predominant tree species (or "growth type"), site quality, past silvicultural treatment, delivered wood cost, and age class.

In the FOREST model, provincial forest lands are divided into three forest regions (aggregations of the MoF administrative regions): *Coast*, which combines the Vancouver region and the small coastal part of the Prince Rupert region; *Northern Interior*, which includes the Prince George region, the interior part of the Prince Rupert region, and the small northern part of the Cariboo region; and *Southern Interior*, corresponding to the Kamloops, Nelson and the remainder of the Cariboo regions. This aggregation best accommodated available data on tree species and growing conditions, and the regional structure of the forest industry.

Within each region, forest types are aggregated into five coniferous and one deciduous *growth types*, designated by the dominant species. Growth types are further subdivided into good, medium and poor *site qualities*. On better sites, volume growth is faster, and the timber sells for a higher price per cubic metre (m^3) because of the larger dimensions of the trees. The forest inventory is divided into 10-year *age classes*. The model projects the inventory by decades, calculating the changes due to timber harvests, natural losses, regeneration and stand improvement, and advancing age classes. Forest growth is accounted for by volume-age relationships that are specific to each combination of region, growth type, site class and silvicultural treatment category. Average growth rates range from 1.2 m^3 ha^{-1} for Douglas-fir on poor sites in the Southern Interior to 12.5 m^3 ha $^{-1}$ for Sitka spruce on good Coastal sites. Volume-age relationships for planted and silviculturally-enhanced stands are hypothetical, based upon the judgment of professionals, a stand growth and yield model (Mitchell et al., 1992), and limited data. All harvested lands are assumed to receive "basic" silviculture— appropriate treatment to ensure successful regeneration of commercial species on the site within one decade— and that the scheduled treatment of not-sufficiently restocked (NSR) "backlog" lands—

clearing, site preparing and planting of backlog NSR lands—is completed by the year 2000 (Thompson et al., 1992). FOREST also calculates losses of forest area caused by fire and insect pests.

The harvest module determines the classes of forest to cut based on the current state of the forest land base, harvest targets set by the government through the allowable annual cut (AAC), timber prices, and a user-specified set of cutting priorities and constraints. Both clear cuts and selective cuts are available options, but selective cuts are applied only to Douglas-fir in the Southern Interior. The harvest priorities and constraints used in this study are: maximum fraction of cut to be taken from salvage (25%); minimum volume per hectare (150 m^3/ha for Coast and N. Interior; 100 m^3/ha for S. Interior); minimum age and cutting cycle for selective cut (age 100 every 20 years); and priority of cut within growth type (lowest cost first). The AAC target annual harvest is varied according to the scenario that is chosen (see below). Financial costs and returns to simulated management activities are also calculated.

3. THE GENERAL EQUILIBRIUM MODEL

To account explicitly for the effects of market forces on timber production, we use a general equilibrium model of B.C. as an open economy (Binkley et al., 1994). The model is based on Percy's (1986) model with newer data and a modified production function for the forest industry where capital or labour cannot substitute for land in the short-term. In this study, it is used to forecast timber price changes for each decade.

The GE model includes nine sectors, five of which are used to represent the forest industry, as forestry is a dominant activity in the economy, accounting for almost 10 percent of provincial GDP and 29.5 percent of goods (non-services) GDP (B.C. Ministry of Finance and Corporate Relations, 1991). The forest sector was disaggregated to account for regional differences between products, production technology, and factor and output prices. The first two sectors are Coastal and Interior timber harvesting, the third and fourth sectors are Coastal and Interior solid wood production. It is assumed that the input of logs to the Interior wood products sector is exclusively from Interior logging and that Coastal wood products use logs only from Coastal logging. The lumber and panel output from the two sectors goes to domestic and export markets, while the chips go to the B.C. pulp and paper sector (the fifth). Production from this sector goes primarily to the export market.

The sixth sector is composed of other primary products, that is, the renewable and nonrenewable resource industries (including agriculture, fishing, mining and energy). This sector relies heavily on world markets and faces highly elastic demand for its output. The seventh sector consists of the non-forest manufacturing industries. Imports create significant competition for this sector, but domestic manufactured goods are treated as imperfect substitutes for imports. Thus, prices for domestic manufactured goods are endogenous to the model. The eighth sector is composed of industries not engaged in trade (e.g., retail and wholesale trade), housing and personal services. The ninth sector is government, modeled separately from other service industries because the two differ in production technology. Output from the government sector is treated as if sold on a fee-for-use basis. Finally, output demand consists of final demand and intermediate-use demand by other industries.

As output is consumed in the region in which it is produced, prices are determined within the provincial market.

Wages are assumed to be fixed in the short-run, in which case the labour market adjusts through changes in unemployment or labour force participation rates. Both capital and land are treated in the model as sector specific and fixed in the short run. Forest land receives special treatment. Forest land is a primary input to the Coastal and Interior timber harvesting sectors, but its supply is completely inelastic, being set by government regulation through the AAC. In this short-run model, it is assumed that there are no substitutes for forest land. Thus, changes *within one decade* in forest production technology (e.g., silvicultural investments) or land use (e.g., expansion of the commercial forest land base to include currently uneconomic lands) are excluded from the model. Changes *between decades* are incorporated in the forest supply model.

Parameter values for the model are based on data provided by the B.C. Ministry of Economic Development (now Ministry of Finance) from their 1984 input-output table (Binkley et al., 1994). For timber production, sectoral data for 1988 and 1989 were used to reflect changes made in 1987 to the stumpage system, including reforestation requirements. In cases where published data are not available for the Coast and Interior separately, regional estimates are based on the size of the labour force or industry capacity. Over the short term, the international market in which B.C. forest products are sold (mainly the U.S.) is modeled as having excess demand elasticities of -2.0 for Canadian lumber and -0.5 for pulp and paper (Phelps, 1993; Binkley et al., 1994).

4. VALUING ALTERNATIVE FOREST USES FOR MULTIPLE USE

Provincial-level management strategies for the Coast, S. Interior and N. Interior regions are evaluated using cost-benefit analysis. The value ascribed to a unit of forest land is determined by its contribution to (1) timber production, (2) carbon uptake, (3) preservation (nonuse), and (4) recreation.

4.1. Timber Benefits

Benefits of timber production are reported here as net (private plus public) financial returns to timber production. Initial average log prices for each timber growth type, age and region are estimated from current Vancouver log market prices and data on net logging profits, stumpage rates and operating costs (Nawitka, 1987; Sterling Wood, 1989; H.A. Simons and Cortex, 1993). Changes in log prices are modeled as a composite of two processes: *short-term* response of log prices to changes in B.C. timber supply and *long-term* response of log prices to changes in world demand for forest products. In the *short term*, supply responses to changes in prices are estimated using the GE model, while *long-term* changes in log prices are modeled using an assumed fixed trend for world timber prices.

Operating costs of harvesting and delivering wood to the mill are split into two components: delivered wood cost (*dwc*), which varies directly with harvest volume, and road (development and maintenance) cost (*rc*), which varies directly with the area cut.

Average *rc* for each of the three simulated regions is estimated from MoF Annual Reports: $135, $116 and $181 per ha for the Coast, S. Interior and N. Interior, respectively. Average *dwc* is estimated as $47.54 per m³ for the Coast and $39.86 for the Interior (Williams and Gasson, 1986; Williams, 1987; but see also Sterling Wood, 1989). To reflect the variation in operating cost, the land base is divided into two distance categories, "near" and "far." Average *dwc* and *rc* are taken as 10% lower (higher) than the regional average for the near (far) land classes. The number 10% is determined by finding the best two-point fit to the cumulative operating cost distribution for coastal B.C. (Williams and Gasson, 1986) and expressing the result as variation around the mean to the nearest 1 percent. This limited representation of variation in operating cost may mask the impact of small changes in price or cost on the extensive margin.

4.2. Carbon Uptake Benefits

Forests absorb CO_2 from the atmosphere and store it as carbon. Hence, forest policies have an important role in mitigating climate change, and management should take account of these benefits (Moulton and Richards, 1990; Richards and Stokes, 1995). Our estimates of carbon uptake benefits consider only the merchantable wood in stems, which we assume begins accumulating carbon in significant amounts 25 years after planting, with different tree species sequestering different amounts of carbon (see van Kooten et al., 1993 for details). This is the part of the trees that gets converted into products that retain carbon for relatively long periods; it is also the component of trees that is most amenable to measurement and monitoring for carbon tax/subsidy schemes (Binkley and van Kooten, 1994). In essense, we treat foliage, branches, bark and roots as being equal to the biomass that would accumulate in vegetation other than commercial tree species that would grow on sites if there were no trees. Thus, the merchantable wood represents the incremental gain from forestation. The carbon in stems is assumed to remain sequestered for all time. Since wood products from B.C. are mainly exported, subsequent carbon release is attributed to consumers.

A review of the literature suggests that carbon can be valued anywhere from about $5 to $300 per tonne. This divergence arises because the marginal costs of CO_2 absorption through tree plantations are initially low, but then rise rapidly thereafter (Richards and Stokes, 1995). Estimates of the marginal damages caused by an aggravated greenhouse effect are not well known and the carbon value one chooses depends on the point on the marginal cost curve that is chosen. Nordhaus (1991) estimates both a marginal benefit and marginal cost of abating CO_2 emissions, concluding that carbon is valued at about $20/ tonne; van Kooten et al. (1993) considered carbon values of $20, $50 and $300 per tonne (see also Maddison this volume). In this study, we employ a value of carbon of $50/tonne, which appears a reasonable estimate. No attempt was made to conduct sensitivity analysis on this figure, although evidence from van Kooten et al. suggests that carbon values would have to be over $200/tonne to make a large impact on silvicultural investment and harvesting decisions, mainly due to the discrete nature of such decisions. In the analysis, the value of carbon remains constant in real terms and carbon sequestration benefits appear annually.

4.3. Preservation Benefits

To determine the value that residents place on wilderness protection, a survey was mailed to 1,230 randomly-selected households throughout B.C. in January, 1993. A total of 275 surveys were returned for a response rate of 22.4%; there was no follow-up to the first mailout (Watson, 1994). Using respondents' stated willingness to pay (WTP) for various levels of wilderness protection, the total WTP for wilderness protection was calculated as a function of the amount of wilderness protected (Table 1). Based on the amounts indicated in column 2 for protection of the amount of wilderness indicated in column 1, total *extra* WTP was obtained by assuming: (1) individuals who revealed a higher WTP would also be willing to pay the lower amounts; (2) individuals who revealed a small WTP would pay this amount if an area larger than that indicated were protected; (3) there are 3 million residents in B.C.; (4) average household size is 2.568 (as indicated by 250 respondents who answered this question); and (5) old-growth timber area accounts for 1/12 of total wilderness in B.C. Marginal WTP for wilderness (= marginal WTP for old growth multiplied by 12) was regressed on area:

$$\text{Marginal WTP} = 81.0 \; e^{-0.2746 \; \text{area}} \quad R^2 = 0.9844$$

Total estimated WTP for wilderness preservation was found by integrating marginal WTP (Table 1).

The marginal WTPs for wilderness protection are provided in the last column of Table 1. To distinguish between preservation on the Coast versus the Interior, the marginal WTPs are allocated according to the present amount of wilderness protected in the region. That is, for 2 million ha protected on the Coast and 5 million ha in the Interior, the Coast protection value (marginal WTP) would be based on 2 million ha (i.e., $46.70/ha) and not on 7 million ha (province total that is protected) (i.e., $11.80/ha). This assumes separability of the regions and respondents. There is some anecdotal evidence from the survey that people thought this way; however, it is possible that this assumption results in overestimation of the value of preservation.

Finally, it was necessary to determine existing levels of mature and over-mature forest lands. Data about the current levels of forest land protection in British Columbia are difficult to obtain. Part of the problem is that the government is currently reviewing which areas are to be protected, with the objective being to attain a level of protection amounting to at least 12% of the Province's land base, compared to the approximately 6% in place prior to B.C.'s 1992 Protected Areas Strategy. Table 2 provides some information regarding levels of protection. This information is used to determine the allocation of marginal values for preservation.

4.4. Recreation Benefits

Data on recreation benefits (both use and option values) are also provided in Table 2. These are by forest district and are aggregated into the three forest regions considered in this analysis—Coast and the two Interior regions. We employ the sum of the values in the last two columns of Table 2 for areas with trees 120 years or greater, and half that value for areas with trees of age 30 or less. A linear interpolation is used for stand ages between 30 and 120 years. The marginal value is assumed to be constant in real terms; that is, the incremental value on a per capita basis of land in recreation is assumed to be independent of how much is available. These assumptions may bias the analysis in favour of preservation since recreation and logging may have some complementarity.

5. MANAGING FOREST LANDS FOR MULTIPLE USE: SCENARIOS

Each simulation scenario is characterized by one of two *long-term value trends* and one of eighteen *policy options*. The two *long-term value trends* differ with respect to projections of timber and non-timber values. The first assumes that long-term timber and non-timber values will remain constant in real dollar terms. The second assumes that long-term timber values will increase at 1% per annum in real terms and that recreation and preservation values will increase in proportion to population growth (2.5, 1.9 and 1.2% per annum for Coast, Southern Interior and Northern Interior, respectively; Statistics Canada 1994). Equilibration of supply and demand in the short term are determined by the link between the GE model and FOREST.

For each *long-term value trend*, eighteen *policy options* are examined. These consist of all combinations of one of three forest land bases, three allowable annual cuts relative to the land base (i.e., m³/ha/year), and two levels of investment in incremental silviculture (see Table 3). The three forest land bases are: (1) the current commercially operable forest, technically known as the "net presently productive, available and suitable land base" (B.C. Ministry of Forests 1984), which comprises approximately 25.7 million hectares; (2) a 10% reduction in the current operable forest; and (3) a 20% reduction in the current operable forest. In the latter two options, the simulated operable forest is reduced equally across all regions and forest types. The current AAC is taken as the average AAC over the last five years, or 78.0 million m³. Thus, the AAC relative to the land base is 3.0 m³/ha/year. The three AAC options examined are the current practice, 10% higher than the current level, and 10% lower than the current level; that is, 3.0, 3.3 and 2.7 m³/ha/year. Two levels of investment in incremental silviculture (spacing, pruning and fertilization) are considered: (1) zero (somewhat less than current practice), and (2) all technically appropriate silvicultural treatments, regardless of cost. We recognize that this latter investment policy is economically unrealistic, but it is a necessary simplification to make the present study feasible and still gain insights about the benefits attributable to intensive silvicultural efforts. As noted above, all the scenarios assume that silvicultural treatments achieve legally-mandated reforestation following harvest and that the current reforestation targets for backlog NSR forest land will be met.

For each scenario, the policy options are compared using net present value (NPV) of all resources (a cost-benefit or economic efficiency criterion). Simulated results for 200 years are obtained for the entire Province taking into account the following values: the timber harvest (combined net private and public returns); recreation benefits; carbon uptake benefits; and the value of old-growth preservation. For each resource, both financial and nonmarket (non-financial) measures are simulated. Net present values are calculated in 1992 dollars using real discount rates of 1%, 4% and 7%, and summed for each discount rate to obtain the NPV for the entire forest resource. Summaries of the monetary and *non-monetary* results are provided in Tables 3 and 4, respectively.

6. COST-BENEFIT ANALYSIS: RESULTS

For the constant *value scenario*, the net present value of all forest benefits over 200 years discounted at a rate of 1% is $95 billion or more under the three policies in which the land base is reduced by 20% and no incremental silviculture is done (Table 3). At the intermediate (4%) and high (7%) discount rates, policies that reduce the land base and the AAC, and where no incremental silviculture is performed, yield higher benefits than the other policies. Regardless of the discount rate, when world timber prices remain constant and non-timber values do not increase, maximization of societal benefits suggests that policy makers in B.C. ought to reduce timber harvests below current levels. Further, given constant timber prices and non-timber values, policies that attempt to compensate for a reduced land base by massive investments in incremental silviculture have lower NPV than their counterparts with no incremental silviculture, despite the carbon benefits that such investment provides.

TABLE 1
Total, Average and Marginal Willingness to Pay per Year for Additional Wilderness Preservation in British Columbia, $1992

Area Preserved (mil ha)	Extra Cost ($/yr)	Number of Responses	Old-Growth Forest Extra WTP ($ mil)	Old-Growth Forest Marginal WTP ($/ha)	Wilderness Area Total WTP ($ mil)	Wilderness Area Marginal WTP ($/ha)
0	—	—	—	—	0	81.0
1	—	—	—	—	70.8	61.5
2	—	—	—	—	124.6	46.7
3	—	—	—	—	165.5	35.5
4	—	—	—	—	196.5	27.0
5	—	—	—	—	220.1	20.5
6	0	43	0	—	238.0	15.6
7	6.25	231	6.2	6.2	251.7	11.8
8	20.00	179	17.3	11.1	262.0	9.0
9	33.33	143	24.9	7.6	269.9	6.8
10	50.00	110	31.8	6.8	275.9	5.2
11	60.50	66	35.3	3.5	280.4	3.9
12	75.00	55	38.7	3.4	283.9	3.0
13	87.50	39	40.8	2.1	286.5	2.3
14	100.00	37	42.7	1.9	288.5	1.7

Source: Calculation based on Watson (1994).

TABLE 2
Undeveloped Watersheds by Forest Region and Protected Area[a]

Region	Area of Mature Timber ('000s ha)[1]	Total Area of Undeveloped Watersheds ('000s ha)[2]	Total Area of Protected Undeveloped Watersheds ('000s ha)[2]	Area of Undeveloped Watersheds in Parks & Wilderness Study Areas ('000s ha)[2]	% of B.C.'s Adult Population in Region[3]	Recreation Use Value[1] ($ mil per yr)	Recreation Option Value[1] ($ mil per yr)
Vancouver	3,402	1,529	438	280	74.5%	4.54	111.13
Prince Rupert	6,367	13,437	465	948	2.5	4.97	4.49
Kamloops	2,373	890	422	264	11.3	10.03	11.23
Prince George	9,596	6,174	435	955	5.0	6.83	8.22
Nelson	1,390	1,107	168	320	4.9	8.15	9.40
Cariboo	3,565	527	31	308	1.8	5.11	2.87
Total	26,69	23,664	1,959	3,075	1000.0	39.62	147.34

[1] Source: B.C. Ministry of Forests (1991). Option value for preserving areas for future recreation and wildlife viewing.

[2] Source: B.C. Ministry of Forests (1992).

[3] Total 1991 B.C. population was 2,883,365, with adult population equal to 2,125,818.

The *scenario* with increasing timber, recreation and preservation values gives somewhat different results. The NPV at a 1% discount rate is highest for the policy that keeps the land base and AAC at current levels, and where full-scale incremental silviculture is implemented (Table 3). Nearly as large NPVs are simulated for the current land base, full-scale incremental silviculture and either a 10% decrease or increase to the current AAC (i.e., 70.2 or 85.8 million m³). At 4% and 7% discount rates, simulated NPV varies little among policy options, although the simulations do indicate a reduction in net benefits from practising incremental silviculture.

In general, the 20% land base reduction yields the highest NPVs when either timber and non-timber values remain constant or the discount rate is high (or both). In contrast, the current land base yields the highest NPVs when values are rising and the discount rate

TABLE 3
Net Present Value of All Forest Resources ($ Billions 1992)

Land base mil ha	AAC mil m³	Inc. silv. Yes /No	Values constant[1]			Values increase[2]		
			Discount rate					
			1%	4%	7%	1%	4%	7%
25.7	85.8	N	76	28	18	341	54	26
25.7	78.0	N	81	32	21	387	57	28
25.7	70.2	N	85	34	23	380	58	30
25.7	85.8	Y	75	26	16	394	53	25
25.7	78.0	Y	76	30	19	406	56	27
25.7	70.2	Y	77	32	21	396	57	28
23.1	78.0	N	86	32	21	322	55	28
23.1	70.2	N	89	34	23	371	58	30
23.1	62.4	N	92	35	24	371	58	30
23.1	78.0	Y	86	30	20	388	55	27
23.1	70.2	Y	85	32	22	388	56	29
23.1	62.4	Y	84	34	23	374	56	29
20.6	70.2	N	97	34	23	321	56	30
20.6	62.4	N	95	34	24	353	57	30
20.6	54.6	N	97	36	25	349	56	31
20.6	70.2	Y	93	32	22	366	55	29
20.6	62.4	Y	91	34	23	368	55	29
20.6	54.6	Y	91	34	24	356	55	30

[1] Timber and non-timber values.
[2] Timber by 1% per annum; non-timber proportionate to population increase (1.5, 1.0 and 0.5% per annum for Coast, S. Interior and N. Interior, respectively).

is low. Generally, investment in incremental silviculture is found to be uneconomic except in the case when long-term prices and non-timber values increase and discount rates are relatively low. This conclusion must be qualified since we assume all technically appropriate silvi-cultural treatments are to be applied, regardless of their efficiency. Clearly, the performance of incremental silviculture would be improved by implementing only those treatments for which benefits exceeded costs.

The contribution of non-timber values to economic efficiency is provided in Table 5 (see also Table 4). The percentage contribution of non-timber values to the NPV varies much more with discount rate and *value scenario* than with land base, harvest rate or silvicultural investment. For the constant *value scenario*, non-timber values contribute 57 to 68% of the NPV at a 1 percent discount rate, 31 to 40% at a 4 percent rate, and 26 to 33% at a 7 percent rate. In contrast, for the rising timber *value scenario*, non-timber values constitute about one quarter of the NPV regardless of the discount rate or policy option. Incremental silviculture enhances non-timber values. This is clear from Table 5 where, with the exception of increasing timber and non-timber values, incremental silviculture increases the proportion of social welfare contributed by non-timber values.

TABLE 4
Non-Monetary Measures of Forest Resource Values (200-Year Time Horizon and Initial Land Base of 25.7 Million ha)

Land Base mil ha	AAC mil m³	Inc. silv. Yes /No	Values constant[1]				Values increase[2]			
			Cut[3]	OG[4]	CO2[5]	Rec[6]	Cut[3]	OG[4]	CO2[5]	Rec[6]
25.7	85.8	N	57.3	763	112	168	58.9	760	111	168
25.7	78.0	N	56.4	803	113	168	58.2	762	110	168
25.7	70.2	N	55.0	915	115	168	58.0	762	112	168
25.7	85.8	Y	60.8	938	125	168	62.4	762	121	168
25.7	78.0	Y	59.7	964	122	168	62.2	865	122	168
25.7	70.2	Y	56.6	1118	121	168	58.9	1044	123	168
23.1	78.0	N	52.1	951	105	218	53.0	948	100	218
23.1	70.2	N	50.7	987	102	218	52.4	950	99	218
23.1	62.4	N	49.3	1095	104	218	52.3	950	101	218
23.1	78.0	Y	55.7	1104	112	218	56.1	950	109	218
23.1	70.2	Y	53.7	1136	109	218	56.0	1044	110	218
23.1	62.4	Y	51.6	1420	107	218	52.7	1211	111	218
20.6	70.2	N	46.4	1158	92	267	47.2	1136	89	267
20.6	62.4	N	45.1	1171	90	267	46.5	1137	88	267
20.6	54.6	N	43.9	1404	90	267	46.3	1158	90	267
20.6	70.2	Y	48.2	1214	99	267	49.8	1137	97	267
20.6	62.4	Y	47.7	1302	97	267	48.8	1220	98	267
20.6	54.6	Y	46.1	1648	94	267	46.4	1442	98	267

[1] See note 1, Table 3.
[2] See note 2, Table 3.
[3] Mean annual timber harvest ('000,000 m³ yr⁻¹).
[4] Old growth forest area ('000 ha) in the year 2100.
[5] Mean annual carbon dioxide uptake ('000 tonnes yr⁻¹).
[6] Mean annual recreation use of the forest ('000,000 user-days yr⁻¹).

As expected, the amount of carbon that is sequestered by B.C. forests increases as more is invested in silviculture (Table 4). Investments in silviculture increase carbon uptake by between 3,000 and 13,000 tonnes per year, although the amount per hectare is small. However, when timber prices rise by 1% and recreation and preservation values rise, the amount sequestered declines. This is because trees are harvested somewhat sooner than would otherwise be the case, primarily because timber values become more important in some regions.

Finally, it is important to consider the contribution of timber values, which can be determined from Tables 3 and 5. Surprisingly, a ranking of policies by NPV based solely on timber values is similar to that for multiple use management, with only one exception. Assuming an objective of maximizing financial returns from timber and constant future prices, the simulations indicate that massive investment in incremental siviculture should not be undertaken unless investments are carefully targetted. The same is true for the case where timber values rise by 1% per annum, with the exception of the case where a 1% discount rate is used. With increasing real timber prices, maintenance of the current AAC and land base plus heavy investment in incremental silviculture yields the greatest timber-harvest-only NPV at the 1% discount rate. For increasing prices and a 4% discount rate, the current land base with 10% AAC reduction and no incremental silviculture produces the highest NPV, while AAC and land base reductions perform best at the higher discount rate. These simulated results reflect the price feedback from the GE model. A moderate reduction in harvests can increase timber prices sufficiently to yield a net increase in timber revenues.

7. CONCLUSIONS

The objectives of this study were (1) to examine the changes that may occur in the evaluation of different forest management options when forest values other than timber production are considered explicitly, (2) to recognize consequences of market power on the

TABLE 5
Proportion of Simulated Net Present Value Attributed to Non-Timber Forest Resources (%)

Land base million ha	AAC million m³	Inc. silv. Yes /No	Values constant[1]			Values increase[2]		
			Discount rate					
			1%	4%	7%	1%	4%	7%
25.7	85.8	N	63	37	33	30	26	27
25.7	78.0	N	60	33	29	26	25	25
25.7	70.2	N	57	31	26	27	24	24
25.7	85.8	Y	65	40	36	26	27	29
25.7	78.0	Y	64	36	31	25	26	26
25.7	70.2	Y	63	33	28	26	25	25
23.1	78.0	N	65	36	31	34	27	27
23.1	70.2	N	62	34	28	30	26	25
23.1	62.4	N	61	32	27	30	26	25
23.1	78.0	Y	65	39	33	28	28	28
23.1	70.2	Y	66	36	29	29	27	26
23.1	62.4	Y	66	34	28	30	27	26
20.6	70.2	N	64	36	30	37	29	27
20.6	62.4	N	65	35	28	33	28	26
20.6	54.6	N	64	34	27	34	28	26
20.6	70.2	Y	67	38	31	32	29	27
20.6	62.4	Y	68	37	29	32	29	27
20.6	54.6	Y	67	35	28	33	29	27

[1] See note 1, Table 3.
[2] See note 1, Table 3.

price and supply of timber, and (3) to consider the role of carbon sequestration in these circumstances. When long-term average timber and non-timber values are kept constant, under all discount values considered, inclusion of non-timber values does not significantly alter the set of dominant policy options. This results because reductions in the land base and harvest levels result in higher timber prices, an increase in the area available for recreation (and associated values), higher nonuse values (greater protection of old growth), and enhanced CO_2-uptake benefits. This complementarity in timber and non-timber benefits is an unanticipated result obtained from the linkage of a GE model with a timber supply model.

Nonetheless, the effect of carbon sequestration on optimal decisions about forest use is marginal. That is, when managers take into account the recreation and preservation values of forests, which tend to lengthen rotation ages, the contribution of carbon uptake values is relatively minor. Only when carbon sequestration values are considered as the sole non-timber value to balance against commercial timber values is the importance of this aspect magnified.

REFERENCES

Binkley, C.S. and G.C. van Kooten (1994). 'Integrating Climatic Change and Forests: Economic and Ecological Assessments', *Climatic Change* 28(October): 91-110.

Binkley, C.S., M. Percy, W.A. Thompson and I.B. Vertinsky (1994). 'A General Equilibrium Analysis of the Economic Impact of a Reduction in Harvest Levels in British Columbia', *Forestry Chronicle* 70(Jul/Aug): 449-454.

B.C. Ministry of Finance and Corporate Affairs (1991). *The Structure of the British Columbia Economy*. Report prepared for the B.C. Round Table on Environment and the Economy. Victoria. March. pp.51.

B.C. Ministry of Forests (1991). *Outdoor Recreation Survey 1989/90. How British Columbians Use and Value their Public Forest Lands for Recreation*. Recreation Branch Tech. Rep. 1991-1. Victoria.

B.C. Ministry of Forests (1992). *An Inventory of Undeveloped Watersheds in British Columbia*. Recreation Branch Technical Report 1992:2. Victoria.

H.A. Simons Strategic Services Division and Cortex Consultants (1993). *Historical and Future Log, Lumber and Chip Prices in British Columbia*. FRDA Report 207. B.C. Ministry of Forests, Victoria. 70 pp.

Mitchell, K.J., S.E. Grout, R.N. MacDonald and C.A. Watmough (1992). *User's Guide for TIPSY: A Table Interpolation Program for Stand Yields*. Victoria: B.C. Ministry of Forests.

Moulton, R. and K. Richards (1990). *Costs of Sequestering Carbon Through Tree Planting and Forest Management in the United States*. Gen. Tech. Rep. WO-58. Washington: U.S.D.A. December.

Nawitka Resource Consultants (1987). *Impact of Intensive Forestry Practices on Net Stand Values in British Columbia*. FRDA Report 014. Victoria: B.C. Ministry of Forests. 79 pp.

Nordhaus, W.D. (1991). 'The Cost of Slowing Climate Change: A Survey', *The Energy Journal* 12(1): 37-66.

Percy, M. (1986). *Forest Management and Economic Growth in British Columbia*. Report to the Economic Council of Canada. Ottawa: Minister of Supply and Services Canada. 88 pp.

Phelps, S.E., 1993. *A Summary of Elasticities of Demand and Supply for North American Softwood Lumber*. Ottawa: Policy and Economics Directorate, Forestry Canada. 13 pp.

Phelps, S.E., W.A. Thompson, T.M. Webb, T.M., P.J. McNamee, D. Tait and C.J. Walters (1990). *British Columbia Silviculture Planning Model: Structure and Design*. Unpublished Report. Victoria: B.C. Ministry of Forests.

Richards, K.R. and C. Stokes (1995). 'Regional Studies of Carbon Sequestration: A Review and Critique', *Forest Science* In Press.

Statistics Canada (1994). *Annual Demographic Statistics, 1993*. Cat. #91-213. Ottawa.

Sterling Wood Group Inc. (1989). *Expected Delivered Log Costs for Areas Treated under the Canada-British Columbia Forest Resource Development Agreement*. FRDA Report 079. Victoria. 89 pp.

Thompson, W.A., P.H. Pearse, G.C. van Kooten and I. Vertinsky (1992). 'Rehabilitating the Backlog of Unstocked Forest Lands in British Columbia: A Preliminary Simulation Analysis of Alternative Strategies'. In *Emerging Issues in Forest Policy* by P.N. Nemetz. Vancouver: UBC Press.

van Kooten, G.C., W.A. Thompson and I. Vertinsky (1993). 'Economics of Reforestation in British Columbia When Benefits of CO_2 Reduction are Taken into Account'. In *Forestry and the Environment: Economic Perspectives* by W.L. Adamowicz, W. White and W.E. Phillips. Wallingford, UK: C.A.B. International.

Watson, V. (1994). *Economics of Biodiversity and Forest Preservation*. Unpublished M.Sc. Thesis. Vancouver: Dept of Agricultural Economics, Univ. of B.C.

Williams, D.H. (ed.) (1987). *The Economic Stock of Timber in the Coastal Region of British Columbia: Technical Appendices*. Report 86-11 Vol. II. Vancouver: FEPA Research Unit, Univ. of B.C.

Williams, D.H. and R. Gasson (1986). *The Economic Stock of Timber in the Castal Region of British Columbia*. Report 86-11 Vol. I. Vancouver: FEPA Research Unit, Univ. of B.C.

Printed and bound by CPI Group (UK) Ltd, Croydon, CR0 4YY

23/10/2024

01778246-0015